3D Printing for Energy Applications

# 3D Printing for Energy Applications

*Edited by*

*Albert Tarancón*
*Catalonia Institute for Energy Research and ICREA*
*Barcelona, Spain*

*and*

*Vincenzo Esposito*
*Department of Energy Conversion and Storage*
*Technical University of Denmark*
*Fysikvej*
*Lyngby, Denmark*

The American Ceramic Society

WILEY

Published by John Wiley & Sons, Inc., Hoboken, New Jersey.
Published simultaneously in Canada.



*Library of Congress Cataloging-in-Publication Data:*

Names: Taranco´n, Albert, editor. | Esposito, Vincenzo, editor.
Title: 3D printing for energy applications / edited by Albert Tarancón and Vincenzo Esposito.
Description: First edition. | Hoboken, New Jersey : Wiley-American Ceramic Society, [2021] | Includes bibliographical references and index.
Identifiers: LCCN 2020031685 (print) | LCCN 2020031686 (ebook) | ISBN 9781119560753 (cloth) | ISBN 9781119560760 (adobe pdf) | ISBN 9781119560784 (epub)
Subjects: LCSH: Three-dimensional printing. | Energy industries–Technological innovations.
Classification: LCC TS171.95 .A165 2021 (print) | LCC TS171.95 (ebook) | DDC 621.9/88–dc23
LC record available at https://lccn.loc.gov/2020031685
LC ebook record available at https://lccn.loc.gov/2020031686

Cover Design: Wiley
Cover Image: © Used with permission with IREC

MIX
Paper from responsible sources
FSC® C013604

# Contents

**Contributor** *xiii*
**Introduction to 3D Printing Technologies** *xvii*

**Part I 3D printing of functional materials** *1*

**1 Additive Manufacturing of Functional Metals** *3*
*Venkata Karthik Nadimpalli and David Bue Pedersen*
1.1 Introduction *3*
1.1.1 Industrial Application of Metal AM in the Energy Sector *5*
1.1.2 Geometrical Gradients in AM *6*
1.1.3 Material Gradients in AM *6*
1.2 Powder Bed Fusion AM *7*
1.2.1 Geometric Gradients in PBF *8*
1.2.2 Material Gradients in PBF *9*
1.3 Direct Material Deposition *12*
1.3.1 Powder and Wire Feedstock for Near-Net-Shape AM *12*
1.3.2 Functional Material Gradients in DED *13*
1.4 Solid-State Additive Manufacturing *16*
1.5 Hybrid AM Through Green Body Sintering *19*
1.5.1 Common AM Technologies for Green Body Manufacturing *19*
1.5.2 CAD Design and Shrinkage Compensation *20*
1.5.3 Additive Manufacture *20*
1.5.4 Debinding and Sintering *21*
1.5.5 Functionally Graded Components in Sintered Components *22*
1.6 Conclusions *22*
Acknowledgment *24*
References *24*

2 **Additive Manufacturing of Functional Ceramics** *33*
*José Fernando Valera-Jiménez, Juan Ramón Marín-Rueda, Juan Carlos Pérez-Flores, Miguel Castro-García, and Jesús Canales-Vázquez*
2.1 Introduction *33*
2.1.1 Why 3D Printing of Technical Ceramics? *35*
2.1.2 Materials and Applications *35*
2.2 Ceramics 3D Printing Technologies *36*
2.2.1 Lamination Object Modeling (LOM) *37*
2.2.2 Ceramics Extrusion *38*
2.2.2.1 Robocasting/Direct Ink Writing *39*
2.2.2.2 Fused Deposition of Ceramics *42*
2.2.3 Photopolymerization *44*
2.2.4 Laser-Based Technologies *47*
2.2.5 Jetting *49*
References *52*

3 **3D Printing of Functional Composites with Strain Sensing and Self-Heating Capabilities** *69*
*Xin Wang and Jihua Gou*
3.1 Introduction *69*
3.2 Carbon Nanotube Reinforced Functional Polymer Nanocomposites *70*
3.2.1 Strain Sensing of CNT Reinforced Polymer Nanocomposites *70*
3.2.2 Resistive Heating of CNT Reinforced Polymer Nanocomposites *71*
3.3 Printing Strategies *72*
3.3.1 Spray Deposition Modeling and Fused Deposition Modeling *72*
3.3.2 Printing of Highly Flexible Carbon Nanotube/Polydimethylsilicone Strain Sensor *73*
3.3.3 Printing of Carbon Nanotube/Shape Memory Polymer Nanocomposites *73*
3.4 Strain Sensing of Printed Nanocomposites *73*
3.5 Electric Heating Performance Analysis *79*
3.6 Electrical Actuation of the CNT/SMP Nanocomposites *82*
3.7 Conclusions *85*
References *87*

**Part II 3D printing challenges for production of complex objects** *91*

4 **Computational Design of Complex 3D Printed Objects** *93*
*Emiel van de Ven, Can Ayas, and Matthijs Langelaar*
4.1 Introduction *93*
4.2 Dedicated Computational Design for 3D Printing *95*
4.2.1 Overhang Angle Control Approaches *96*
4.2.1.1 Local Angle Control *96*

4.2.1.2 Physics-Based Constraints *97*
4.2.1.3 Simplified Printing Process *97*
4.2.2 Design Scenarios *98*
4.3 Case Study: Computational Design of a 3D-Printed Flow Manifold *99*
4.3.1 Fluid Flow TO *100*
4.3.2 Front Propagation-Based 3D Printing Constraint *102*
4.3.3 Fluid TO with 3D Printing Constraint *103*
4.4 Current State and Future Challenges *104*
References *105*

**5 Multicomponent and Multimaterials Printing: A Case Study of Embedded Ceramic Sensors in Metallic Pipes** *109*
*Cesar A. Terrazas, Mohammad S. Hossain, Yirong Lin, and Ryan B. Wicker*
5.1 Multicomponent Printing: A Short Review *109*
5.2 Multicomponent Printing: A Case Study on Piezoceramic Sensors in Smart Pipes *111*
5.2.1 Brief Introduction to AM of Embedded Sensors for Smart Metering *111*
5.2.2 Fabrication of the Embedded Piezoceramic Sensor in Metallic Pipes *114*
5.2.2.1 Smart Coupling Fabrication Process Using EPBF Technology *114*
5.2.2.2 Materials *116*
5.2.2.3 Sensor Housing *117*
5.2.2.4 Re-poling of PZT *118*
5.2.2.5 Impact in Sensing Properties Due to Heat-Treatment Induced By AM Process *119*
5.2.2.6 Smart Coupling Component *119*
5.2.2.7 Compressive Force Sensing *119*
5.2.2.8 Temperature Sensing *120*
5.2.3 Impact of the AM and Performance of the Multicomponent Printed Device *122*
5.2.3.1 Compressive Force Sensing *122*
5.2.3.2 Temperature Sensing *124*
5.2.3.3 Crystalline Structure Analysis *126*
5.3 Summary and Outlook *128*
Acknowledgments *129*
References *130*

**6 Tailoring of AM Component Properties via Laser Powder Bed Fusion** *135*
*Simon Ewald, Maximilian Voshage, Steffen Hermsen, Max Schaukellis, Patrick Köhnen, Christian Haase, and Johannes Henrich Schleifenbaum*
6.1 Introduction *135*
6.2 Machines, Materials, and Sample Preparation *138*

6.3 Sample Preparation and Characterization Techniques *139*
6.4 Material Qualification and Process Development *140*
6.5 Tailoring Grain Size via Adaptive Processing Strategies *143*
6.6 Tailoring Material Properties By Using Powder Blends *146*
6.7 Tailoring Properties By Using Special Geometries Such As Lattice Structures *148*
Funding *150*
Conflicts of Interest *150*
References *150*

**7 3D Printing Challenges and New Concepts for Production of Complex Objects** *153*
*Hayden Taylor, Hossein Heidari, Chi Chung Li, Joseph Toombs, and Sui Man Luk*
7.1 Introduction *153*
7.2 Geometrical Complexity *154*
7.3 Material Complexity *155*
7.4 Energy Requirements *156*
7.5 Promising Metal Deposition Approaches *157*
7.6 Multimaterial and Multi-property SLA *159*
7.7 Temporal Multiplexing *159*
7.8 Resin Formulations with Multiple End-States *160*
7.9 Associated Processing Considerations *160*
7.10 Bioprinting of Realistic and Vascularized Tissue *162*
7.11 Emerging Volumetric Additive Processes *163*
7.12 Computation for CAL *166*
7.13 Material–Process Interactions in CAL *167*
7.14 Current Challenges in CAL *169*
7.15 Expanding the Capabilities of CAL *170*
7.16 Concluding Remarks and Outlook *171*
Acknowledgments *172*
References *172*

**Part III 3D printing of energy devices** *181*

**8 Current State of 3D Printing Technologies and Materials** *183*
*Poul Norby*
8.1 3D Printing of Energy Devices *183*
8.1.1 Batteries *183*
8.1.1.1 3D Printing Structured Electrodes *186*
8.1.1.2 3D Printing Solid Electrolytes *195*
8.1.1.3 3D Printed Full Batteries *197*

8.1.1.4 Conclusion and Outlook 200
References 200

**9 Capacitors** 205
*Lukas Fieber and Patrick S. Grant*
9.1 Introduction 205
9.2 Capacitors and Their Current Manufacture 206
9.2.1 Capacitor Classifications, Operating Principles, Applications, and Current Manufacture 206
9.2.1.1 Electrostatic Capacitors 206
9.2.1.2 Electrolytic Capacitors 209
9.2.1.3 Electrochemical Capacitors 210
9.2.2 Capacitor Components: Function and Requirements 211
9.2.3 Performance 213
9.2.4 The Challenge of Manufacturing Capacitors 214
9.3 The Promise of Additive Manufacturing 215
9.4 Additive Manufacturing Technologies: Considerations for Capacitor Fabrication 217
9.4.1 AM Process Categories 217
9.4.1.1 Material Extrusion – Fused Filament Fabrication 217
9.4.1.2 Material Extrusion – Direct Ink Writing 221
9.4.1.3 Vat Polymerization 223
9.4.1.4 Powder Bed Fusion 225
9.4.1.5 Material Jetting 227
9.4.1.6 Binder Jetting 228
9.4.2 Multi-technology or Hybrid Printing 229
9.4.3 Complete Capacitor Devices Fabricated by Additive Manufacturing 230
9.5 Summary and Outlook 232
Acronyms 233
References 235

**10 3D-Printing for Solar Cells** 249
*Marcel Di Vece, Lourens van Dijk, and Ruud E.I. Schropp*
10.1 Introduction 249
10.2 Examples of 3D-Printing for PV 250
10.3 Geometric Light Management 255
10.3.1 Background 255
10.3.2 Optical Model for External Light Trapping 257
10.3.3 Design and 3D-Printing of the External Light Trap 260
10.3.4 Characterization 261
10.4 Conclusions 266
References 267

**11 3D Printing of Fuel Cells and Electrolyzers** *273*
*A. Hornés, A. Pesce, L. Hernández-Afonso, A. Morata, M. Torrell, and Albert Tarancón*
11.1 Introduction *273*
11.2 3D Printing of Solid Oxide Cells Technology *274*
11.2.1 Solid Oxide Fuel Cells *275*
11.2.1.1 SOFC Electrolyte *276*
11.2.1.2 SOFC Electrodes *278*
11.2.2 Solid Oxide Electrolysis Cells *283*
11.2.3 SOC Stacks and Components *284*
11.3 3D Printing of Polymer Exchange Membranes Cells Technology *286*
11.3.1 Polymeric Exchange Membrane Fuel Cells *287*
11.3.1.1 PEMFC Electrolyte *288*
11.3.1.2 PEMFC Catalysts Layer *288*
11.3.1.3 PEMFC Gas Diffusion Layer *289*
11.3.1.4 PEMFC Bipolar Plates and Flow Fields *290*
11.3.2 Polymer Exchange Membrane Electrolysis Cells *293*
11.3.2.1 PEMEC Liquid Gas Diffusion Layer *293*
11.3.2.2 PEMEC Bipolar Plates and Flow Fields *293*
11.3.2.3 Fully Printed PEMEC *294*
11.4 3D Printing of Bio-Fuel Cells Technology *294*
11.5 Conclusions and Outlook *297*
References *297*

**12 DED for Repair and Manufacture of Turbomachinery Components** *307*
*S. Linnenbrink, M. Alkhayat, N. Pirch, A. Gasser, and H. Schleifenbaum*
12.1 Introduction *307*
12.2 DED Based Repair of Turbomachinery Components *309*
12.2.1 DED Process *310*
12.2.2 Work Environment *310*
12.2.3 Process Chain for the Repair of Turbine Blades *310*
12.2.3.1 Step 1: "Machining & Preparation" *310*
12.2.3.2 Step 2: "Reverse Engineering" *311*
12.2.3.3 Step 3: "Generation of Tool Paths" *313*
12.2.3.4 Step 4: "DED Process" *313*
12.2.3.5 Step 5: "Adaptive Machining" *314*
12.3 DED Based Hybrid Manufacturing of New Components *314*
12.3.1 Hybrid Additive Manufacturing *315*
12.3.2 Turbocharger Nozzle Ring *317*
12.3.3 Hybrid Production Cell *319*
12.3.4 Process Chain for Hybrid Additive Manufacturing of Nozzle Rings *320*

12.3.4.1 Step 1: "Choice of DED Strategy" *320*
12.3.4.2 Step 2: "DED Process" *321*
12.3.4.3 Step 3: "Optical Metrology" *322*
12.3.4.4 Step 4: "Adaptive Milling" *322*
12.3.4.5 Step 5: "Joining of Top Ring" *322*
12.4 Summary *323*
Acknowledgments *324*
References *324*

**13 Thermoelectrics** *327*
*Fredrick Kim, Seungjun Choo, and Jae Sung Son*
13.1 Introduction *327*
13.2 Additive Manufacturing Techniques of Thermoelectric Materials *328*
13.2.1 Extrusion-Based Additive Manufacturing Process *328*
13.2.2 Fused Deposition Modeling (FDM) Technique *336*
13.2.3 Stereolithography Apparatus (SLA) Process *337*
13.2.4 Selective Laser Sintering (SLS) Process *339*
13.2.5 Summary and Outlook *345*
Acknowledgements *345*
References *345*

**14 Carbon Capture, Usage, and Storage** *351*
*Jason E. Bara*
14.1 Introduction *351*
14.2 Can 3D Printing Be Used to Fabricate a $CO_2$ Capture Process at Scale? *354*
14.3 A Brief Note on 3D Printing and $CO_2$ at Smaller Scales & Research Efforts *356*
14.4 Conclusions *358*
References *358*

**Index** *361*

# Contributors

***Venkata Karthik Nadimpalli***
Department of Mechanical Engineering,
Technical University of Denmark, Kgs. Lyngby, Denmark

***David Bue Pedersen***
Department of Mechanical Engineering,
Technical University of Denmark, Kgs. Lyngby, Denmark

***José Fernando Valera-Jiménez***
3D-ENERMAT, Materials for Energy & 3D Printing Lab, Renewable Energy Research Institute, Universidad de Castilla-La Mancha, Albacete, Spain

***Juan Ramón Marín-Rueda***
3D-ENERMAT, Materials for Energy & 3D Printing Lab, Renewable Energy Research Institute, Universidad de Castilla-La Mancha, Albacete, Spain
And
Print3D Solutions, Albacete, Spain

***Juan Carlos Pérez-Flores***
3D-ENERMAT, Materials for Energy & 3D Printing Lab, Renewable Energy Research Institute, Universidad de Castilla-La Mancha, Albacete, Spain

***Miguel Castro-García***
3D-ENERMAT, Materials for Energy & 3D Printing Lab,
Renewable Energy Research Institute, Universidad de
Castilla-La Mancha, Albacete, Spain

***Jesús Canales-Vázquez***
3D-ENERMAT, Materials for Energy & 3D Printing Lab,
Renewable Energy Research Institute, Universidad de
Castilla-La Mancha, Albacete, Spain

***Xin Wang***
Composite Materials and Structures Laboratory, Department of Mechanical and Aerospace Engineering,
University of Central Florida,
Orlando, FL, USA

*Jihua Gou*
Composite Materials and Structures Laboratory, Department of Mechanical and Aerospace Engineering, University of Central Florida, Orlando, FL, USA

*Emiel van de Ven*
II Optomechatronics Group, Netherlands Organisation for Applied Scientific Research TNO, Delft, The Netherlands

*Can Ayas*
I Structural Optimization and Mechanics Group, Delft University of Technology, Delft, The Netherlands

*Matthijs Langelaar*
I Structural Optimization and Mechanics Group, Delft University of Technology, Delft, The Netherlands

*Cesar A. Terrazas*
W.M. Keck Center for 3D Innovation, The University of Texas at El Paso, El Paso, TX, USA

*Yirong Lin*
W.M. Keck Center for 3D Innovation, The University of Texas at El Paso, El Paso, TX, USA

*Ryan B. Wicker*
W.M. Keck Center for 3D Innovation, The University of Texas at El Paso, El Paso, TX, USA

*Mohammad S. Hossain*
W.M. Keck Center for 3D Innovation, The University of Texas at El Paso, El Paso, TX, USA

*Simon Ewald*
Digital Additive Production, RWTH Aachen University, Aachen, Germany

*Maximilian Voshage*
Digital Additive Production, RWTH Aachen University, Aachen, Germany

*Steffen Hermsen*
Digital Additive Production, RWTH Aachen University, Aachen, Germany

*Max Schaukellis*
Digital Additive Production, RWTH Aachen University, Aachen, Germany

*Patrick Köhnen*
Steel Institute, RWTH Aachen University, Aachen, Germany

*Christian Haase*
Steel Institute, RWTH Aachen University, Aachen, Germany

*Johannes Henrich Schleifenbaum*
Digital Additive Production, RWTH Aachen University, Aachen, Germany
And
Fraunhofer Institute for Laser Technology ILT, Aachen, Germany

*Hayden Taylor*
Department of Mechanical Engineering, University of California, Berkeley, CA, USA

*Hossein Heidari*
Department of Mechanical Engineering, University of California, Berkeley, CA, USA

***Chi Chung Li***
Department of Mechanical Engineering, University of California, Berkeley, CA, USA

***Joseph Toombs***
Department of Mechanical Engineering, University of California, Berkeley, CA, USA

***Sui Man Luk***
Department of Mechanical Engineering, University of California, Berkeley, CA, USA

***Poul Norby***
Department of Energy Conversion and Storage, Technical University of Denmark, Lyngby, Denmark

***Lukas Fieber***
Department of Materials, University of Oxford, Oxford, UK

***Patrick S. Grant***
Department of Materials, University of Oxford, Oxford, UK

***Marcel Di Vece***
Interdisciplinary Centre for Nanostructured Materials and Interfaces (CIMaINa) and Physics Department "Aldo Pontremoli", University of Milan, Milan, Italy

***Lourens van Dijk***
Soluxa B.V., Nijmegen, The Netherlands

***Ruud E.I. Schropp***
Department of Physics and Astronomy, University of the Western Cape, Belville, South Africa

***A. Hornés***
Department of Advanced Materials for Energy, Catalonia Institute for Energy Research (IREC), Barcelona, Spain

***A. Pesce***
Department of Advanced Materials for Energy, Catalonia Institute for Energy Research (IREC), Barcelona, Spain

***L. Hernández-Afonso***
Department of Chemistry, University of La Laguna, Tenerife, Spain

***A. Morata,***
Department of Advanced Materials for Energy, Catalonia Institute for Energy Research (IREC), Barcelona, Spain

***M. Torrell***
Department of Advanced Materials for Energy, Catalonia Institute for Energy Research (IREC), Barcelona, Spain

***Albert Tarancón***
Department of Advanced Materials for Energy, Catalonia Institute for Energy Research (IREC), Barcelona, Spain
and
Catalan Institution for Research and Advanced Studies (ICREA), Barcelona, Spain

***S. Linnenbrink***
Fraunhofer Institute for Laser Technology (ILT), Aachen, Germany

***M. Alkhayat***
Fraunhofer Institute for Laser Technology (ILT), Aachen, Germany

***N. Pirch***
Fraunhofer Institute for Laser Technology (ILT), Aachen, Germany

***A. Gasser***
Fraunhofer Institute for Laser Technology (ILT), Aachen, Germany
and
Chair for Laser Technology (LLT), Aachen, Germany

***H. Schleifenbaum***
Fraunhofer Institute for Laser Technology (ILT), Aachen, Germany
and
Digital Additive Production (DAP), Aachen, Germany

***Fredrick Kim***
School of Materials Science and Engineering, Ulsan National Institute of Science and Technology (UNIST),
Ulsan, Republic of Korea

***Seungjun Choo***
School of Materials Science and Engineering, Ulsan National Institute of Science and Technology (UNIST),
Ulsan, Republic of Korea

***Jae Sung Son***
School of Materials Science and Engineering, Ulsan National Institute of Science and Technology (UNIST),
Ulsan, Republic of Korea

***Jason E. Bara***
University of Alabama,
Tuscaloosa, AL, USA

# Introduction to 3D Printing Technologies

*Albert Tarancón*[1,2], *Kyriakos Didilis*[3], *and Vincenzo Esposito*[3]

[1] *Catalonia Institute for Energy Research (IREC), Barcelona, Spain*
[2] *Catalan Institution for Research and Advanced Studies (ICREA), Barcelona, Spain*
[3] *Technical University of Denmark, Department of Energy Conversion and Storage, Fysikvej, Lyngby, Denmark*

3D printing is considered one of the technologies that will change the world in the next future. The capability of producing series of free-shape customized objects using almost all relevant families of materials (plastics, metals, and ceramics) is considered a revolution in the field of manufacturing, especially because it can be done even by individuals. The low investment required for simple 3D printers and the open availability of design files make this technology an entire change in the way of understanding product fabrication and prototyping.

At a first stage, 3D printing was mainly developed for structural parts made of plastics but, nowadays, the technology evolved into a complete additive manufacturing chain, covering design, simulation, optimization, fabrication and rapid prototyping of functional objects as well as complete devices made of metals and ceramics. This evolution represents an enormous competitive advantage since the fabrication of high value-added products such as devices and functional parts will open the use of this technology to the vast majority of application scenarios and industrial sectors, including the energy field.

Moreover, a real deployment of 3D printing of functional materials will benefit in promoting a circular economy by preventing the loss of valuable materials and reducing the energy consumption in the manufacturing process. In general, the uses of additive manufacturing techniques can represent a reduction of up to 80% of waste material and 70% of energy consumption. Besides, 3D printing promotes the simplification of the manufacturing processes as well as reducing the environmental impact of distribution. This decentralized manufacturing approach combined with the open distribution of digital models will represent a technological revolution that might bring marginal costs to near zero, if raw materials are widely available.

## I.1 3D Printing Technologies

3D printing allows the fabrication of three-dimensional objects by deposition of successive layers of material using a digital model. Intensive research on additive manufacturing has been carried out during the last decades to allow the fabrication of three dimensional objects by assembling material without the use of tooling or molds. 3D printing started in 1981 at Nagoya Municipal Industrial Research Institute publishes, where Hideo Kodama reported the first photopolymer system. Based on the Kodama's concept, Charles developed in 1984 the stereolithography (SLA) printer based on photopolymers. The first FDM (fused deposition modeling) machine was developed in the 1990s. FDM is based on the extrusion of melting plastic filaments that is deposit as a thread layers on a print bed. In 1992, SLA evolves into SLS (selective laser sintering) machine, where powder and lasers replace the photopolymers and the UV light, respectively. Further development took place, in 1997, into the first laser additive manufacturing. In 2000, the already consolidated inkjet printing converges toward 3D printing methods, evolving into the first 3D inkjet printer for drop-by-drop deposition of complex object and vertical extension. Since those years, the number of technical solution of these methodologies have been multiplied, covering a very large range of technical solutions, including multi-materials printing by FMD, desktop 3D printers, open source inexpensive solutions, hybridization of deposition methods and development of novel starting materials for the fabrication of ceramics, metals, and composite materials.

The major 3D printing processes commercially available are briefly introduced in the next paragraphs and described in Figure I.1. They can be grouped into different categories according to the dimensional order of the material deposition, namely, point, line, or plane (Figure I.2).

**Stereolithography (SLA)** is an additive manufacturing process based on the photo-polymerization of resin material upon exposure to a laser or UV light source. As the name suggests, -graphy is a Greek work that means -writing, something that is reflected on the way the photocurable material solidifies. The light source passes through the necessary focus lenses for controlling and adjusting the output and is reflected from a movable mirror system onto the photosensitive material surface to cure and solidify the area between the build platform and the liquid surface. The laser beam movement controlled by the mirrors will write the designed pattern on the build platform. When the first layer is formed by the solidified material, the build platform will move down at a distance equal to the layer height, to create the area and form the next layer. Adjusting the laser beam focus and spot size will determine the resulting curing depth and pitch of the solidified material line both in the

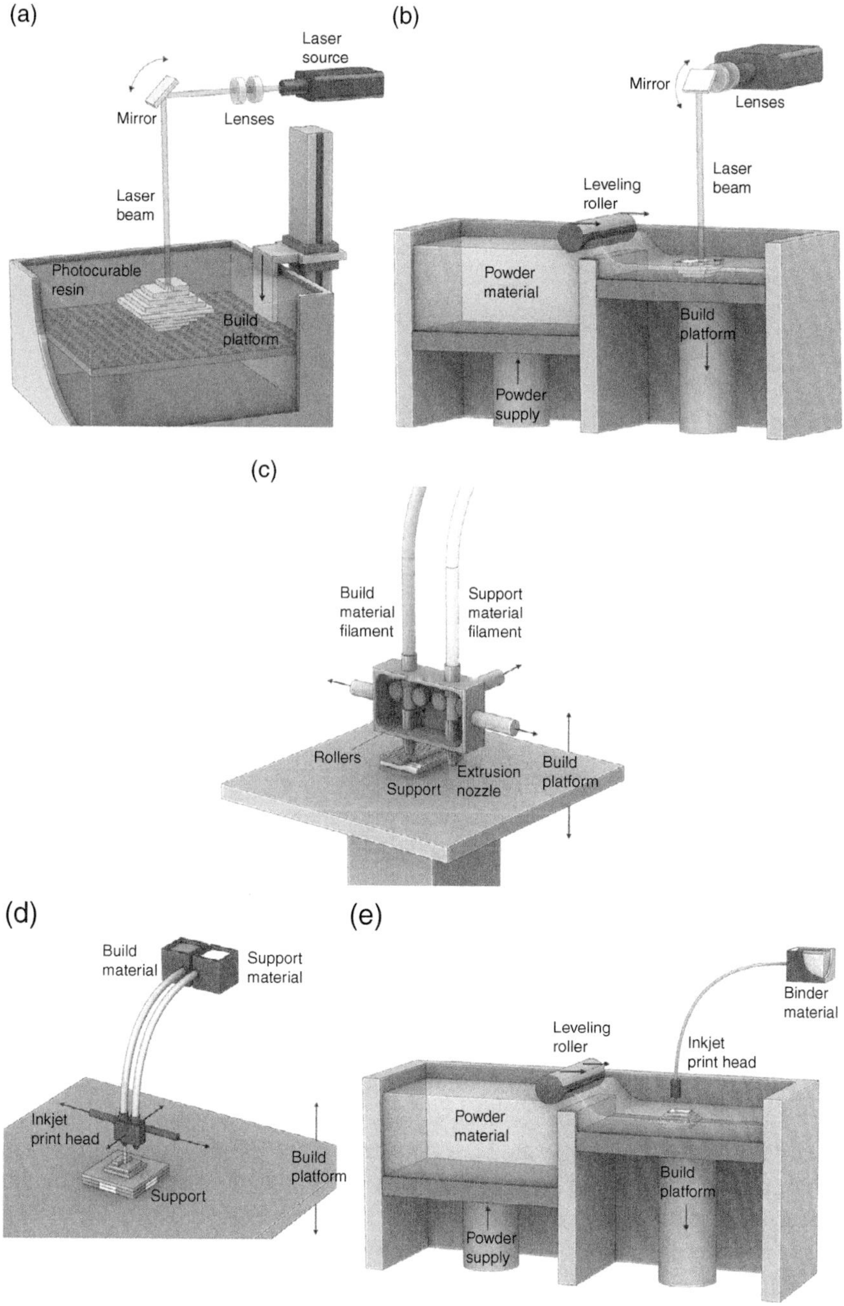


**Figure I.1** An overview of the basic components and materials used on the most popular Additive Manufacturing processes: (a) SLA, (b) SLS, (c) FDM, (d) DIP, and (e) IIP.

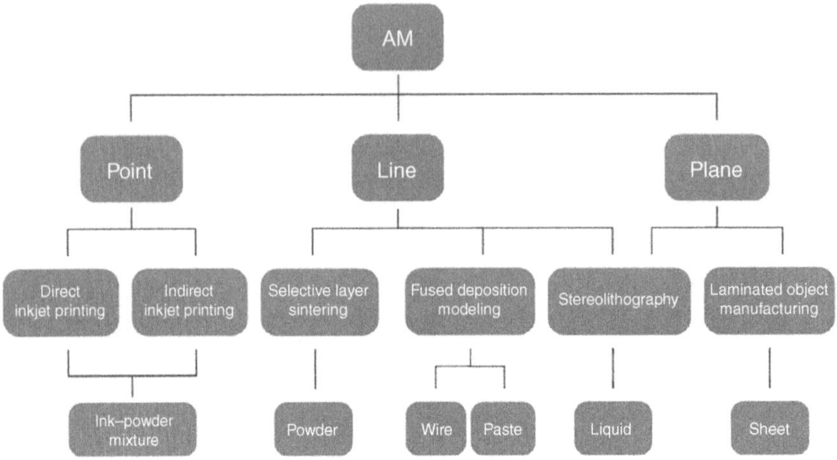


**Figure I.2** Classification of commercially available additive manufacturing methods according to dimensional order, process, and material.

$X/Y$ plane and the $Z$ axis, offering high vertical and lateral resolution in the order of 10–25 μm [1, 2]. As a result, high surface finish can be achieved with excellent mechanical properties, of porous and dense structures. In terms of material, when using ceramics, photocurable pastes with or without solid loading can be used, normally >50 vol% and grain sizes that range from 0.5 to 5 μm. The high viscosity of the ceramic paste is beneficial for providing support on the structure, however, the nature and the configuration of the process can make it difficult to print multi-material structures. The chemistry of the formulations consisting of several additives, such as UV absorbers/blockers, stabilizers, and pore formers [3], making recycling and reusing of uncured material challenging. Moreover, post-processing such as cleaning is required after printing, which adds to the expected de-binding and sintering that is common in manufacturing of ceramics.

**Selective Laser Sintering (SLS)** is a process based on the same principle, that a laser beam will be reflected and directed by a mirror system, to fuse together powder particles that are placed inside a powder chamber. The fused particles will form a layer at the end of the writing path and the build platform will be lowered at a distance equal to the selected layer thickness of the 3D model. At this point, a leveling roller will transfer powder from an adjacent chamber, to the main build chamber and provides the next powder layer to be fused. In this case, the resolution of the process is controlled by the synergic action of the build chamber and the roller that levels the new powder layer, normally a resolution between 80 and 100 μm [1, 4] can be achieved. The powder that has not been exposed to the laser path will be removed at the end of the process, providing this way support for the proceeding layers with overhang features. The high energy provided by the laser is responsible for the solid state diffusion of the powder particles and the resulting densification, with final parts characterized with good mechanical properties. However, when higher temperatures are required, depending on the material requirements, thermal stresses can be introduced and practices such as preheating of the powder bed can be applied [5]. Surface finish is dependent on the powder grain size that ranges from 0.3 to 10 μm [1] and normally no post processing is required. Overall, the process utilizes several components for defining and handling the material layer therefore the equipment is expensive with increased manufacturing time.

**Fused Deposition Modeling (FDM)** uses thermal energy to melt the filament material that is feed through a heating element with the aid of a roller system and is extruded directly onto the build platform or a substrate. The extruded line of material will adhere to the adjacent and underlying material upon cooling and will form the layered structure. For the deposition of the proceeding layers the build platform will move down as specified by the layer thickness settings which vary between 50 and 200 μm [4], while for the $X$ and $Y$ plane parameters the extruder diameter and the molten material pitch can be adjusted by the process parameters. It is important to mention that parameter settings will affect the resulting quality and performance of the final part. The extrusion nature of the process gives rise to several print defects and makes small features challenging to print. Cooling related issues can be solved with a control temperature build platform, while the extrusion parameters are responsible for structural and geometrical issues [6]. The filament materials used are thermoplastics with solid particle loading above 40% and grain sizes ranging around 1–5 μm [1]. Post processing to remove organic components and sintering is required for densification to occur, however it is a cost effective solution with fairly simple equipment.

**Inkjet printing** can be divided to two process groups, however, both processes are based on the same principal that is widely known from conventional 2D ink printers, a jetting nozzle controls the amount of material or binder deposited on the substrate/material that is build layer by layer. Direct Inkjet Printing (DIP) utilizes a ceramic suspension to be used as ink, where droplets are deposited onto a substrate to form the material layer, along with the support material that is deposited when overhangs or cavities are printed. Depending on the ink formulation an appropriate drying step has to be introduced before commencing with the next layer, such as cooling or evaporation [5], while several suspensions contain additives to counteract clogging of the nozzle that is a common issue [1]. In this case the jetting parameters, travel speed, and the layer thickness will affect the resulting printing resolution that is higher compared to the previously mentioned processes, ranging from 1 to 10 μm [1]. Several systems have been developed for controlling both material deposition and the build platform movement in order to achieve such low resolutions, in the expense of printing time when 3D structures with high aspect ratios are required. Highly diluted and stable inks containing nanoparticles are used, with a solid concentration less than 5 vol% and grain sizes range from 10–50 nm [1].

**Indirect Inkjet Printing (IIP)** as the name suggests is similar to DIP, with the difference that liquid binder the jetting material deposited on ceramic powder to form the material layer. It is a powder bed process that can offer several advantages such as structural support during printing, reuse of powder material, increasing shape complexity, and reducing printing time compared to DIP, however print resolution is in the order of 100 μm [1], similar to SLS. In this case the mechanical performance of the printed parts is a significant disadvantage and post-process hardening is a way to counteract it.

## I.2 3D Printing Hierarchical, Material and Functional Complexity

All types of industry will benefit from the introduction of complex geometries generated by additive manufacturing but it will be even more interesting to take advantage of other unique capabilities of additive manufacturing, such as hierarchical, material of functional complexity [7].

The hierarchical complexity involves the fabrication of features with shape complexity across multiple size scales. This multiscale approach can cover up to five orders of magnitude for some of the available 3D printing techniques. For instance, SLA printing scale ranges from tens of micrometres to almost one meter, which allows including small details in high aspect ratio printed parts. This feature is extremely relevant for many applications like energy where the performance of devices is typically proportional to the active area such as in chemical reactors or electrochemical cells.

A high level of material complexity can also be introduced in printed pieces by processing in a different way at different points of the part, for example, introducing different levels of porosity all along the object by changing the laser parameters in SLA or the gray level in inkjet printing. This material complexity can also refer to a more advanced multi-material printing. This has been traditionally employed in inkjet implementing a matrix of printheads for colorful printing but it is still under development for other types of technologies. This approach allows the deposition of different materials or, eventually, materials of different nature in a single process. The multi-material capability enables the fabrication of graded compositions or, ultimately, complete devices. Independently on the printing deposition technique, one of the unsolved critical issues in processing of multi-material parts is the quality of the interfaces between dissimilar materials [8, 9]. This represents one of the most interesting fields of current research on 3D printing.

Finally, 3D printing is able to lend a high level of functional complexity to the parts by design or by direct implementation of functional materials. By design, one can easily imagine the fabrication of specific shapes with certain functionality such as plasmonic patterns or cantilevers for energy harvesting as well as the deposition of thin layers acting as antireflective or water-proof coatings for photovoltaics. By using functional materials, the conceptual design of 3D printed parts with high level of functional complexity is even more straightforward since the printed material has specific functionalities by itself. Despite the enormous potential of this approach, its implementation is very much limited by the short list of available advanced materials for printing uses [8]. Just to give an example of the low availability of printable functional materials, Table I.1 shows a comprehensive compilation of most of the advanced oxides reported so far (beyond structural materials or bioceramics).

This migration from structural to functional materials will result in the fabrication of advanced devices with high value added, which will extend the markets where 3D printing is applied from prototyping to manufacturing. Up to now, most of the 3D printing techniques have been developed and commercialized for the fabrication of polymeric and metallic structural parts but recently the focus has moved to the production of functional-quality components made of advanced materials including, for instance, composites, ceramics, and nanomaterials.

In the case of composites, enhanced properties are expected from the fabrication of complex shapes using inorganic-polymer matrix based materials [8]. The most evident application is probably the printing of fibre-reinforced polymeric composites for improving mechanical properties of structural parts [17] but also other applications in which the inorganic loading has functional properties are envisaged, for example, 3D-printed dielectric/plastic composites [18]. Beyond polymer-based materials, it is of great interest the printing of metal–ceramic or ceramic–ceramic composites since they can have a strong impact in strategic fields such as electronics or energy [19].

The relevance of the recent progress on 3D printing of ceramics lies in the broadest spectrum of functional properties of this type of materials compared to all other classes, such as metals or polymers. The unique functional properties of advanced ceramics (electrical, optical, or magnetic) make them of critical importance to face upcoming technological challenges especially in the fields of electronics, information and communication technologies (ICT), and energy and environment. In this regard, recent advances in printing piezoelectric materials [20], dielectrics [21] or ionic conductors [22, 23] represent the beginning of a big revolution.

**Table I.1** Functional ceramics processed in the past with additive manufacturing technologies

| AM technique | Inkjet | SLS | SLA | FDM | LOM |
|---|---|---|---|---|---|
| Functional ceramic | $BaTiO_3$<br>PZT<br>$TiO_2$<br>LSMO/YBCO<br>YSZ/CGO<br>SDC/SSC<br>$BaTiO_3$ [10]<br>$BaTiO_3$/P(VDF-TrFE) [11]<br>ZnO ink [12] | PZT<br>$BaTiO_3$<br>SiCN<br>YSZ | PZT<br>$Fe_2O_3/Fe(C_2O_4)$<br>SiCN $BaTiO_3$/photo-resin [2]<br>PZT@Ag/photo-resin [13]<br>ZnO/photo-resin [14]<br>YSZ [15, 16] | $BaTiO_3$<br>PZT, PMN<br>$LiFePO_4/Li_4Ti_5O_{12}$<br>$BaZrO_3$, $SrTiO_3$<br>$BaMn_2Al_{10}O_{19-x}$<br>ITO, ZnO<br>$La(Mg_{0.5}, Ti_{0.5})O_3$<br>$Zr_{0.8}Sn_{0.2}TiO_4$<br>$TiO_2$<br>$BaTiO_3$/PVA paste [15] | $SiO_2$-$Al_2O_3$-RO-glass<br>LZSA-Glass<br>PZT |

*Source:* Based on Mueller et al. [8]. © 2017 John Wiley & Sons.

The use of nanomaterials for 3D printing will also bring interesting advantages, especially if they are functional nanomaterials, because they will enable the increase of complexity in shape (fine surface finishing, high vertical resolution, or improved layer assembly), hierarchy (multi-length-scale structures or graded porosity), materials (low sintering temperature in ceramics and metals or multi-material deposition in suspension), and functionality (use of nanocomposites or nanomaterials with high surface area or core-shell structures).

## I.3 3D Printing for Energy

The possibility of printing functional materials with applications in energy has been attracting a growing attention since 3D printing technologies represent a new paradigm for the manufacture of energy conversion and storage technologies [13]. Among other advantages, additive manufacturing offers unique capabilities for increasing the specific performance per unit mass and volume of energy devices by implementing high levels of hierarchical and shape complexity. While the implementation of multiscale features can be of great interest for chemical reactors or batteries, a high level of shape flexibility becomes crucial in harvesting applications where the efficiency of the generators strongly depends on their capability to properly couple to variable scenarios, for example, for a good adaptability of thermoelectric generators to the source of waste heat. Complementary, the opportunity to implement fully controlled graded compositions or tuneable porosity, that is, a certain level of materials complexity, also represents a big advantage for those energy devices with multiple interfaces or porous electrodes, such are fuel cells, gas separation membranes, or electrochemical capacitors. Despite this potential, the fabrication of highly complex devices for the energy sector by using 3D printing is just an emerging field [17]. This is probably due to the complexity of the devices usually employed. However, since the complexity adds cost to traditional processes, the more complex the final part, the more likely that AM will be of benefit for the sector [16].

This emerging application of 3D printing already end up with remarkable examples of the fabrication of components for solid oxide fuel cells [14, 24, 25], batteries [15, 26], or photovoltaic systems [27, 28]. Despite most of the existing examples in energy applications correspond to low-aspect ratio deposition of functional layers (mainly by inkjet), increasing activity is devoted to proper 3D printing of complex shape multi-material parts and devices. In this direction, the most inspiring examples were reported by Kim et al. [21] and Pesce et al. [29] who were able to fabricate high-aspect ratio interdigitated Li-ion microbatteries by inkjet printing of concentrated viscoelastic inks of complex lithium oxides and solid oxide cells based on corrugated oxide-ion YSZ electrolytes by SLA, respectively. More activity in complex shapes has been recorded in the field of heterogeneous catalysis or solar systems. Different authors have reported 3D fabrication of catalyst supports based on alumina [30–32] and plastic-made light concentrators [23]. Although in these cases the 3D printing techniques were employed to fabricate the structural but functional parts of the systems, they proved the interest of 3D printing for resolving problems related to limitations in the classical fabrication of relevant parts. Regarding chemical engineering, internally structured reactors available by 3D printing can play a very important role allowing solutions that were not reachable previously [33]. The main goal is a balanced integration of mass, heat, and momentum transfer in the 3D printed reactor [34] by generating an internal structure by design rather than chance, like in packed bed, monolithic- or foam-type reactors. In the case of solar concentration, 3D-printed traps made of smoothened silver-coated thermoplastic resulted in a relevant improvement of the external quantum efficiency of crystalline silicon, thin film nanocrystalline silicon and organic solar cells [23].

With the expected increase of such successful experiences in printing relevant parts for the energy industry, it will progressively start adopting 3D printed devices and systems. The energy manufacturing sector is expected to be more reluctant to adopt the printing technology than end-users to integrate printed objects (if their performance is kept similar or is improved). On the other hand, more inconveniences are foreseen for end-users regarding required standards and certifications of their products. These standardization efforts should go together with quantitative Life Cycle Assessment studies on the reduction of waste material and energy consumption as well as the evaluation of the recyclability and ecological impacts of these alternative products. Moreover, a full deployment of the 3D printing technologies for a huge market as the energy requires an industry laying on a robust value chain. Currently, there is an absence of redundant big players in almost all the steps of this value chain and a lack of skilled workforce (from designers to operators), more noticeably for functional materials such as ceramics.

Another important point to cover to reach the energy market, is the development of big size and high speed 3D printing processes able to cover a high and increasing demand. This evolution from a prototyping to a manufacturing approach will be driven by increasing the complexity of the printed parts (extending capabilities to fabricate high value-added parts such are devices) and ultimately pointing the upcoming mass customization of energy products. Mass customization refers the fabrication of custom-made products at a competitive price. In this regard, developing "tabletop factories" such as 3D printers will generate a competitive manufacturing process in terms of product flexibility and short time-to-market for energy devices boosting the idea of an industry 4.0 based on mass customization by 3D printing.

## I.4 Scope of the Book

Large scientific and technical production in the field of 3D printing has risen in the last years. The fame of 3D printing comes from the actual power of the technique combined with its coverage in a broad range of applications. Such a potential is also boosted by the current viability of affordable 3D printers for consumers and prosumers, where the market often escorts the actual possibilities with the promise of revolutionizing the way we consume and create. This combination of factors has indeed created great expectations. However, for the advanced uses, 3D printing firmly refers to a complex chain of additive manufacturing procedures and, besides some exceptions, it relies on boundary interdisciplinary research. Form the scientific and technical point of views, the variety of the topics and the possibilities appear endless and often it results difficult to identify the real possibility.

In view of such preliminary considerations, in this book, we try to shade light on the real possibilities of 3D printing in one of the most charming opportunities opened by additive manufacturing methods: the energy field. To construct a critical discussion around this fascinating topic, we define in Part I the link between most recent advances in the field of additive manufacturing with functional materials to be used in the energy systems. Under such a frame, selected contributions with the typical material science approach, we introduce the three classes of materials used in 3D printing: polymers, metals, and ceramics. We also highlight the need of use combinations of materials; multi-materials manufacturing is the key in energy systems and the actual potential and limits of manufacturing multifunctional materials. Such an approach is reviewed in Part II, with key contributions about consolidated techniques, hybrid printing technologies, and new opportunities in the field. We especially focus on 3D printing challenges for production of complex objects, suiting a wide range of energy systems devices, not only for functional parts but also for accessory components. Finally, elements of interdisciplinary and consideration on consolidated trends in the energy research field are reported in Part III, with key-examples of 3D printing of energy devices. In this last section, we have selected some important cases covering both consolidated energy technologies, such as turbines, batteries, capacitors, solar and nuclear as well as emerging technologies (piezoelectric energy harvesting, electrochemical fuel cells, and thermoelectric energy generation), and environmental solutions (chemical conversion and $CO_2$ capture) that bring the promise of clean and sustainable solutions for the environment.

We edited the book to address readers with different backgrounds and aims, including graduated students in the materials science and engineering, chemists, mechanical engineers interested in manufacturing methods but also for a wider readership, seeking in 3D new opportunities of research and business.

## References


1 Ruiz-Morales, J. C., Tarancón, A., Canales-Vázquez, J., Méndez-Ramos, J., Hernández-Afonso, L., Acosta-Mora, P., … Fernández-González, R. (2017). Three dimensional printing of components and functional devices for energy and environmental applications. *Energy & Environmental Science*, 10, 846–859.

2 Yang, Y., Chen, Z., Song, X., Zhu, B., Hsiai, T., Wu, P.-I., … Shung, K. (2016). 3D printing of piezoelectric element for energy focusing and ultrasonic sensing. *Nano Energy*, 27, 78.

3 Eckel, Z. C., Zhou, C., Martin, J. H., Jacobsen, A. J., Carter, W. B., & Schaedler, T. A. (2015). Additive manufacturing of polymer-derived ceramics. *Science*, 351, 58.

4 Wang, X., Jiang, M., Zhou, Z., Gou, J., & Hui, D. (2017). 3D printing of polymer matrix composites: A review and prospective. *Composites Part B: Engineering*, 110, 442–458.

5 Travitzky, N., Bonet, A., Dermeik, B., Fey, T., Filbert-Demut, I., Schlier, L., … Greil, P. (2014). Additive manufacturing of ceramic-based materials. *Advanced Engineering Materials*, 16, 729.

6 Nadgorny, M., & Ameli, A. (2018). Functional polymers and nanocomposites for 3D printing of smart structures and devices. *ACS Applied Materials and Interfaces*, 10, 17489.

7 Gibson, I., Rosen, D. W., & Stucker, B. (2010). *Additive manufacturing technologies*. New York: Springer Science + Business Media LLC.

8 Mueller, J., Courty, D., Spielhofer, M., Spolenak, R., & Shea, K. (2017). Mechanical properties of interfaces in inkjet 3D printed single- and multi-material parts. *Additive Manufacturing*, 4, 193–199.

9 Lopes, L. R., Silva, A. F., & Carneiro, O. S. (2018). Multi-material 3D printing: The relevance of materials affinity on the boundary interface performance. *Additive Manufacturing*, 23, 45.

10 Gaytan, S. M., Cadena, M. A., Karim, H., Delfin, D., Lin, Y., Espalin, D., … Wicker, R. B. (2015). Fabrication of barium titanate by binder jetting additive manufacturing technology. *Ceramics International*, 41, 6610.

11 Zhou, X., Parida, K., Halevi, O., Liu, Y., Xiong, J., Magdassi, S., & Lee, P. S. (2020). All 3D-printed stretchable piezoelectric nanogenerator with non-protruding kirigami structure. *Nano Energy*, 72, 104676.

12 Tubío, C. R., Guitián, F., & Gil, A. (2016). Fabrication of ZnO periodic structures by 3D printing. *Journal of the European Ceramic Society*, 36, 3409.

13 Yang, Y., Chen, Z., Song, X., Zhu, B., Hsiai, T., Wu, P.-I., … Shung, K. K. (2016). Three dimensional printing of high dielectric capacitor using projection based stereolithography method. *Nano Energy*, 22, 414.

14 Yee, D. W., Lifson, M. L., Edwards, B. W., & Greer, J. R. (2019). Additive manufacturing of 3D-architected multifunctional metal oxides. *Advanced Materials*, 31, 1.

15 Chavez, L. A., Wilburn, B. R., Ibave, P., Delfin, L. C., Vargas, S., Diaz, H., … Lin, Y. (2019). Optimization of 3D printing parameters for BaTiO3 piezoelectric ceramics through design of experiments'. *Materials Research Express*, 6, 8.

16 Hoag, C., Spradling, D., & Shulman, H. (2012). *Introduction to additive manufacturing of ceramics* (pp. 15). Ceramic Industry.

17 Compton, B. G., & Lewis, J. A. (2014). 3D-printing of lightweight cellular composites. *Advanced Materials*, 26, 5930–5935.

18 Castles, F., Isakov, D., Lui, A., Lei, Q., Dancer, C. E. J., Wang, Y., … Grant, P. S. (2016). Microwave dielectric characterisation of 3D-printed BaTiO3/ABS polymer composites. *Scientific Reports*, 6, 22714.

19 Scheithauer, U., Weingarten, S., Johne, R., Schwarzer, E., Abel, J., Richter, H.-J., … Michaelis, A. (2017). Ceramic-based 4D components: Additive manufacturing (AM) of ceramic-based functionally graded materials (FGM) by thermoplastic 3D printing (T3DP). *Materials*, 10, 1368.

20 Cui, H., Hensleigh, R., Yao, D., Maurya, D., Kumar, P., Kang, M. G., … Zheng, X. R. (2019). Three-dimensional printing of piezoelectric materials with designed anisotropy and directional response. *Nature Materials*, 18, 234–241.

**21** Kim, H., Renteria-Marquez, A., Islam, M. D., Chavez, L. A., Rosales, C. A. G., Ahsan, M. A., … Lin, Y. (2019). Fabrication of bulk piezoelectric and dielectric BaTiO3 ceramics using paste extrusión 3D printing technique. *Journal of the American Ceramic Society*, 102, 3685–3694.

**22** Masciandaro, S., Torrell, M., Leone, P., & Tarancón, A. (2019). Three-dimensional printed yttria-stabilized zirconia self-supported electrolytes for solid oxide fuel cell applications. *Journal of the European Ceramic Society*, 39, 9–16.

**23** McOwen, D. W., Xu, S., Gong, Y., Wen, Y., Godbey, G. L., Gritton, J. E., … Wachsman, E. D. (2018). 3D-printing electrolytes for solid-state batteries. *Advanced Materials*, 30, 1707132.

**24** Esposito, V., Gadea, C., Hjelm, J., Marani, D., Hu, Q., Agersted, K., … Jensen, S. H. (2015). Fabrication of thin yttria-stabilized-zirconia dense electrolyte layers by inkjet printing for high performing solid oxide fuel cells. *Journal of Power Sources*, 273, 89.

**25** Sukeshini, A. M., Cummins, R., Reitz, T. L., & Miller, R. M. (2009). Inkjet printing of anode supported SOFC: Comparison of slurry pasted cathode and printed cathode. *Electrochemical and Solid-State Letters*, 12, B176.

**26** Sun, K., Wei, T. S., Ahn, B. Y., Seo, J. Y., Dillon, S. J., & Lewis, J. A. (2013). 3D printing of interdigitated Li-Ion microbattery architectures. *Advanced Materials*, 25, 4539.

**27** Ahn, B. Y., Duoss, E. B., Motala, M. J., Guo, X., Park, S., Xiong, Y., … Lewis, J. A. (2009). Omnidirectional printing of flexible, stretchable, and spanning silver microelectrodes. *Science*, 323, 1590.

**28** van Dijk, L., Paetzold, U. W., Blab, G. A., Schropp, R. E. I., & Di Vece, M. (2016). 3D-printed external light trap for solar cells. *Progress in Photovoltaics*, 24, 623.

**29** Pesce, A., Hornés, A., Núñez, M., Morata, A., Torrell, M., & Tarancón, A. (2020). 3D printing the next generation of enhanced solid oxide fuel and electrolysis cells. *Journal of Materials Chemistry A*, 8, 16926–16932.

**30** de Hazan, Y. (2012). Robotic deposition of 3D nanocomposite and ceramic fiber architectures via UV curable colloidal inks. *Journal of the European Ceramic Society*, 32, 1187.

**31** Stuecker, J. N. (2004). Advanced support structures for enhanced catalytic activity. *Industrial and Engineering Chemistry Research*, 43, 51.

**32** Van Noyen, J. (2011). Catalyst design with porous functional structures. *WIT Transactions on Ecology and the Environment*, 154, 93.

**33** Scheffler, M., & Colombo, P. (Eds.) (2005). Heterogeneously catalyzed processes with porous cellular ceramic monoliths (Chapter 5.4). In *Cellular ceramics: Structure, manufacturing, properties and applications* (pp. 454). Weinheim: Wiley-VCH GmbH.

**34** Dautzenberg, F. M., & Mukherjee, M. (2001). Process intensification using multifunctional reactors. *Chemical Engineering Science*, 56, 251.

# Part I

# 3D printing of functional materials

# 1

# Additive Manufacturing of Functional Metals

*Venkata Karthik Nadimpalli and David Bue Pedersen*

*Department of Mechanical Engineering, Technical University of Denmark, Kgs. Lyngby, Denmark*

## 1.1 Introduction

Additive manufacturing (AM) technologies [1, 2] comprise a family of manufacturing methods that colloquially are known by the common appellation of "3D Printing." AM has created a strong linkage between digital and physical manufacturing, thus nourishing by its nature, a wider trend, digitization, and the automation of the manufacturing industry. For this reason, the increasing adoption of AM within the manufacturing industry is pushing companies to research new ways of adapting their manufacturing models and optimize their manufacturing strategies by integrating these manufacturing technologies of tomorrow into existing production and bolster their strategies toward a digital to physical conversion [3]. To illustrate the digital to the physical linkage of AM, Figure 1.1 serves as an overview of the gross elements for a generic AM process.

The digital nature of AM processes lends itself to the possibility of adding functionality to the components across the process chain, herein functionality relates to the form or geometry, as the geometry of the workpiece is built up from digital data, so does the functionality as relates to the material placement and material composition. Hofmann et al. [4] and Sobczak and Drenchev [5] explored the various classes of functionally graded metal components. It is useful to classify the functional gradients in AM components according to different kinds of material and geometric gradation. Figure 1.2 shows a schematic of four types of material gradients in AM technologies. Type-I is with a single material, and the functionality comes from the design and geometry of the structure. Type-II deals with at least two materials used during the AM process, forming a discrete

*3D Printing for Energy Applications*, First Edition. Edited by Albert Tarancón and Vincenzo Esposito.

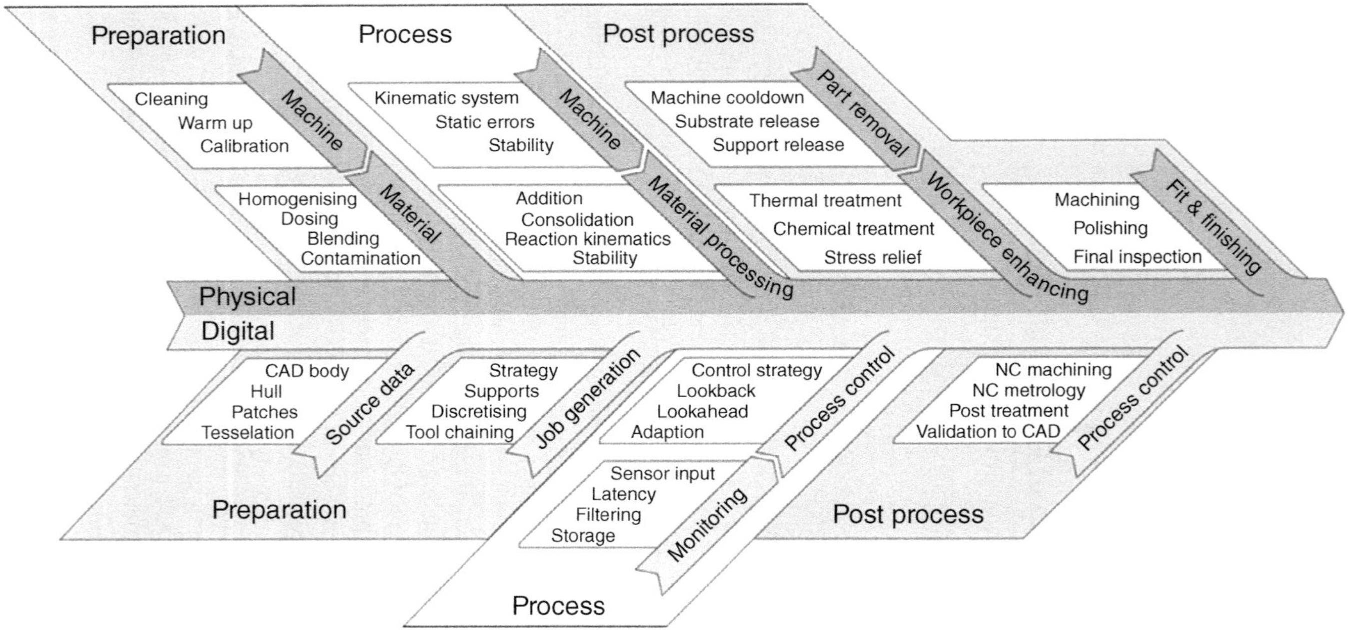


**Figure 1.1** Overview of physical and digital links of an AM process chain.

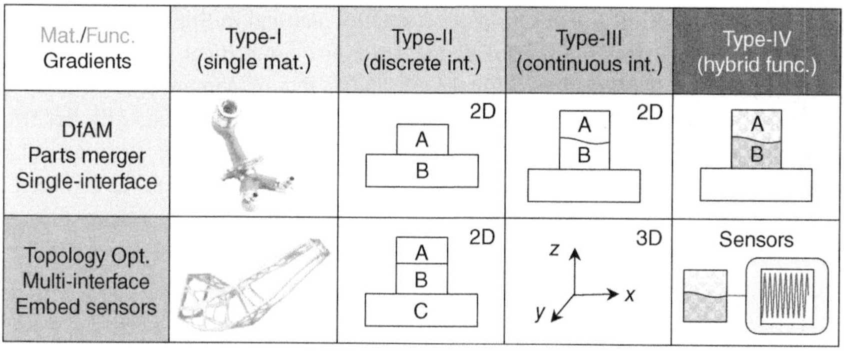


**Figure 1.2** Schematic overview of functional AM components.

interface with an abrupt transition between the two materials. Type-III involves at least two materials with a gradient interface between them. The material gradient can be introduced by process parameter change (microstructural control) or by in situ physical addition of multiple materials. Type-IV refers to any hybrid functionality that is introduced by a combination of Types I–III or by the addition of sensors/other functional mechanisms. Functionality can also be achieved by integrating several AM processes into a hybrid process chain [6]. While the geometric gradients apply to most AM components (single/multi-material), the material gradients are process dependent. Specifically, AM machine tools can change the material either at the voxel, layer, or part level. The capabilities of the process thus determine the kinds of functional gradients in AM components, which can be classified into geometric and material gradation.

### 1.1.1 Industrial Application of Metal AM in the Energy Sector

Industrial applications of functional metal AM components can be found across various sectors including nuclear power, oil & gas, turbine components, wind & tidal energy, fuel cell components, and electromagnetic energy to name a few. Some such applications are discussed in this subsection. The GE Leap fuel nozzle was one of the first certified metal AM components to undergo high-critical testing and deployment into production [7]. Siemens has been at the forefront of metal AM applications with replacement parts for nuclear plants, sealing rings for steam turbine blades and high-efficiency gas turbine burners [8]. Oerlikon has showcased topology optimized turbine blades and drill bits for oil & gas applications with integrated sensors [9]. Biome renewables has designed and retrofitted an AM part to increase efficiency of existing wind and tidal turbine installations [10]. Aidro has manufactured high-pressure hydraulic manifolds and heat exchangers for oil & gas applications [11]. The Oakridge National Lab (ORNL) has demonstrated a newly designed nuclear reactor core with multi-material and integrated sensor functionality planned in the near future [12]. Optisys manufactures high-performance antennas for critical electromagnetic energy sensor applications for satellites [13]. Thermal management applications—like heat exchangers, fuel cell components, and rocket nozzles—greatly benefit from the design freedom offered by metal AM [14]. The diversity of potential applications for metal AM in the energy sector is just starting to be explored. A better understanding of the potential can be gained by reviewing recent advances in functional metal AM components by classifying the functionality according to geometric and material gradients.

### 1.1.2 Geometrical Gradients in AM

AM lends increased design freedom for the manufacturing of unique functional geometries. Hegab [15] reviewed the design aspects of geometrically graded functional materials. These range from rudimentary component optimization through integrated component design, to flexures, engineered to form compliant mechanisms that are constrained in specific degrees of freedom. The ease by which metamaterials can be manufactured from a highly engineered and well-defined unit cell enables the construction of filters, membranes as well as light and stiff components where the uncertainty from using a stochastic manufacturing method such as foamed materials is taken out of the equation. Said metamaterials can be designed using topology optimization, also known as generative design, to functionally grade a component toward a specific wanted behavior such as thermal conductivity or elastic deformation. The above-mentioned capabilities enable AM to be strategically employed to induce added functionality of a part. For example, how a load-bearing part will buckle and deform and eventually collapse plastically can be precisely be achieved by a combination of integrated component design, compliant mechanisms, and topology optimized metamaterial gradient. The performance of geometrically graded AM parts depends on the voxel size, which in turn determines the limits to physical complexity. Hence, AM processes with smaller voxels, namely powder bed fusion (Section 1.2) and sintering-based hybrid AM (Section 1.5), have the best applications of geometrical gradation.

### 1.1.3 Material Gradients in AM

Traditional manufacturing processes start typically from one feedstock and then convert it into the desired shape. AM processes on account of being a selective consolidation based manufacturing processes offer tremendous flexibility in functional material gradients. Naebe & Shirvanimoghaddam [16] reviewed functionally graded components manufactured through different methods and showcased the importance of AM for expanding the scope and application of functional components. Loh, Pei, Harrison, and Monzón [17] reviewed functionally graded AM components and showed that material gradation was currently only limited by AM process tool design. Rafiee, Farahani, and Therriault [18] reviewed various multi-material additive manufacturing methodologies and found that certain AM processes are better at combining multiple materials than others. Extrusion based additive manufacturing (Section 1.5) is an example where the pointwise deposition of extrudate constitutes the workpiece, and complete control of the extrude through a mixing extruder allows for a multitude of materials to be dynamically blended and deposited [19]. Likewise, direct energy deposition (Section 1.3) allows for graded interfaces due to in-situ powder mixing and hence allows for pointwise grading of the materials while minimizing the waste generated throughout the process [20]. In the case of powder-bed technologies (Section 1.2) and vat photopolymerization AM (VPAM) technologies (Section 1.5), cross contamination of the feedstock materials induces the risk of a waste/component ratio where the waste product is the larger output. This can be remedied by ensuring that the build envelope is packed as densely as possible and improvements to the AM process are necessary to encourage wider adoption of material gradients in these technologies. Solid-state AM is uniquely suited for multiple materials due to the low-temperature processing capability that eliminates typical problems with melting and solidification (Section 1.4).

The following sections focus on one class of AM process each and dive into the potential for functional metal AM components.

## 1.2 Powder Bed Fusion AM

Powder bed fusion (PBF) refers to a class of AM processes. The most common machine tool for metal additive manufacturing is the PBF systems. This type of machine operates by selectively fusing powder following a layer-by-layer deposition principle. The powder is contained in a silo and fed to a re-coater to evenly distributed powder between consolidation operations. The tool distributing the powder can be in the form of a scraper, a roller, or a blade that is designed to both fluidize and distribute the powder. As this tool has a major influence on the distribution of the layer, the powder and the tool have to be carefully matched to the application. Two kinds of PBF technologies are based on the laser (L-PBF) and the electron-beam (EB-PBF) as the power source [21]. The L-PBF process is more versatile since it can operate under a neutral gas atmosphere while EB-PBF requires a vacuum. However, electron beams are faster and can be controlled efficiently through magnetic mirrors and hence offer certain advantages. The high-density parts and small voxel size distinguish powder bed fusion from other metal AM processes for manufacturing functional components. In the past

20 years, PBF technologies (especially L-PBF) have seen a rapid increase across various sectors, such as aerospace, biomedical, tooling, and energy sectors [3]. Several review articles discuss PBF process–structure–property relations and establish applications for Type-I functional materials [22, 23]. A much smaller portion of the literature is focused on more complex functionally gradient metal PBF. Design of functionally gradient additive manufacturing (FGAM) components encompasses distributed site-specific properties with gradual transitions in geometry, chemical compositions, constituents, or microstructures. A lack of guidelines on the selection and distribution of materials has hindered the development of FGAM, thereby limiting the microstructural design and arrangement of transition phases. Although some commercial software packages exist for FGM's and multi-materials 3D printing, they are still far removed from real industrial applications and the ever-increasing demands of novel functionally graded components. FGAM of PBF is a highly underexplored field for metals, especially in the material gradients domain. An overview of the literature on metal PBF functional components is presented in the following sections according to geometrical and material gradients.

### 1.2.1 Geometric Gradients in PBF

Porous metals in various forms, including foams and lattices, have been investigated for many years for a wide variety of applications [24, 25]. These range from lightweight structures to functional applications such as heat exchangers or electrodes. Cellular solids have been investigated to determine relationships between their geometry and their physical properties (thermo-mechanical, acoustic, impact-toughness, etc.). The motivation for much of this research is high-performance applications; for example, in simultaneous structural lightweighting and sound dampening in motor vehicles/aircraft, thus providing high functionality with minimized resource consumption. Recent advances in lattice design have enabled the creation of structures with spatially varying solid volume fraction; these are termed as functionally graded lattices. Gibson-Ashby models have developed a theoretical framework for cellular lattice materials that are well suited for generating topology optimized designs. Powder bed fusion systems are uniquely suitable to manufacture the highly complex lattice structures with struts as small as 100 μm.

Maskery et al. [26] investigated the mechanical properties of graded density Al-Si10-Mg lattice structures manufactured by L-PBF for structural applications and found significant improvement in compressive and impact strength. Choy, Sun, Leong, and Wei [27] showed similar improvements in mechanical properties for Ti-6Al-4V. Li, Hassanin, H., Attallah, Adkins, and Essa [28], and Tan et al. [29] investigated TiNi Shape Memory Alloys (SMA) fabricated by L-PBF. The process conditions that lead to the microstructural characteristics play a major role in the SMA alloy properties. L-PBF of TiNi lattices can enable tailored properties, in addition to the functionality associated with the shape-memory effect. Li et al. [28] used a topology optimized design to manufacture a negative Poissons ratio metamaterial.

Functionally graded metal lattice structures with tailored mechanical response are most commonly used in biomedical applications like implants. The human bone is a functionally graded anisotropic structure, and hence any biocompatible implants need to have similar mechanical performance. A large surface area also helps toward osseointegration and thus leads to fewer implant rejections. A large body of research thus focuses on scaffolds of biocompatible materials like CoCrMo, Titanium alloys, and, more recently, biodegradable alloys of Mg, Zn, and Fe [30]. Hazlehurst, Wang, and Stanford [31] investigated L-PBF of CoCrMo FGMs to imitate bone for a hip implant that is half as light, and twice as flexible compared to a solid metal structure. Ataee, Li, Fraser, Song, and Wen [32] researched high porosity (82–85%) Ti64 scaffolds with varying unit cell dimensions by EB-PBF and found that the ratio of elastic modulus anisotropy in orthogonal directions was comparable to those of a human trabecular bone. Yan, Hao, Hussein, and Young [33] and Yu, Sun, and Bai [34] focused on increasing the surface area by using Triply Periodic Minimal Surfaces (TPMS) in their lattice designs. Zhang, Fang, Leeflang, Zadpoor, and Zhou [35] took an alternate approach through a stepwise topological design based on diamond unit cells to mimic the structure of the femoral diaphysis through L-PBF of Ti-6Al-4V.

### 1.2.2 Material Gradients in PBF

The application of multi-materials in a component expands the already large design space provided by additive manufacturing processes. Directed energy deposition processes (DED) are typically associated with multi-material additive manufacturing in metals due to their intrinsic system-related flexibility of changing to different feedstocks during operation, as further discussed in Section 1.3. However, PBF-based additive manufacturing processes exhibit specific advantages over DED. A prominent example is the possibility to generate more intricate geometries and achieve smaller feature sizes. At the same time, most commercially available PBF systems are limited to a single powder feedstock restricting the deployment of different materials during the build process. This section deals with Type-II, III material gradients in PBF [36].

The most common kind of multi-material PBF components are Type-II involving the use of an existing prefabricated structure of material A and adding a second material B. While the process itself doesn't change, the multi-material components often reduce the manufacturing time thus combining the best of conventional and additive manufacturing. Another Type-II multi-material methodology involves premixed powder compositions that offer significant improvements in desired properties. The review paper by Bandyopadhyay and Heer [20] showcases several examples of Type-II PBF multi-material. More recently, Chen et al. [37] investigated L-PBF of steel–bronze multi-material components and found brittle cracks at the interface. Hinojos et al. [38] used EB-PBF to make IN718/SS316 and SS316/IN718 components and found that the high temperatures in EB-PBF make it less susceptible to solidification cracking. AlMangour, Grzesiak, and Yang [39], Han et al. ([40]), Kun, Beibei, Wenheng, and Cailin ([41]), and Xia et al. [42] worked with premixed titanium composite powders to improve the wear resistance and hardness of L-PBF Ti parts. Recent works successfully demonstrated microstructural control during PBF to make Type-III components. Niendorf et al. [43] used process control to modify the microstructure of SS316 in L-PBF and Koptyug et al. [44] used a similar mechanism in EB-PBF of SS316. Biondani, Bissacco, Mohanty, Tang, and Hansen [45] showcased a hybrid process chain to make Type-II multi-material L-PBF components with a mirror-like finish for molding applications.

Anstaett, Seidel, and Reinhart [46] from Fraunhofer, IGCV showcased a multi-material FGAM part of Copper–Chrome–Zirconia and Tool Steel 1.2790. New systems were used for handling the selective layout of powder within the individual recoated layer for powder-bed systems [47]. Demir and Previtali [48] introduced a prototype L-PBF system with two powder feeders in combination with piezoelectric transducers for in-situ gradual mixing of Fe and Al-12Si powder. Wei, Li, Zhang, and Chueh [49] demonstrated, for the first time, full spatial control in all directions in L-PBF using a custom-made system with multiple powder feeders and a point-by-point micro-vacuum selective material-removing system. Figure 1.3 shows the schematic of an Aurora Labs system with custom-built software for multi-material processing at DTU [36]. The powder feeder systems are accurate within a 1% error by weight of powder dispensed per layer. This allows precise control of the mixing percentage within each layer. An example is presented with a graded Type-III steel sample with a hard phase (SS440C) and a soft phase (SS316L) alternating sinusoidally. Preliminary microstructure and microhardness tests reveal that the component is functionally graded as designed. One of the primary concerns of multi-material PBF is the issue of material waste, due to difficulties in recycling the powder. Aerosint solves this issue through selective powder deposition. At DTU, the issue is addressed by means of tracking software, which automatically adjusts the ratio of the powder materials based on the history of the mixed powder and desired final composition per layer. Bodner et al. [50] used a unique PBF system with integrated liquid dispersed metal powder dispensation. The setup was used to additively manufacture a multilayered structure based on alternating Inconel 625 alloy (IN625) and 316L stainless steel (316L)

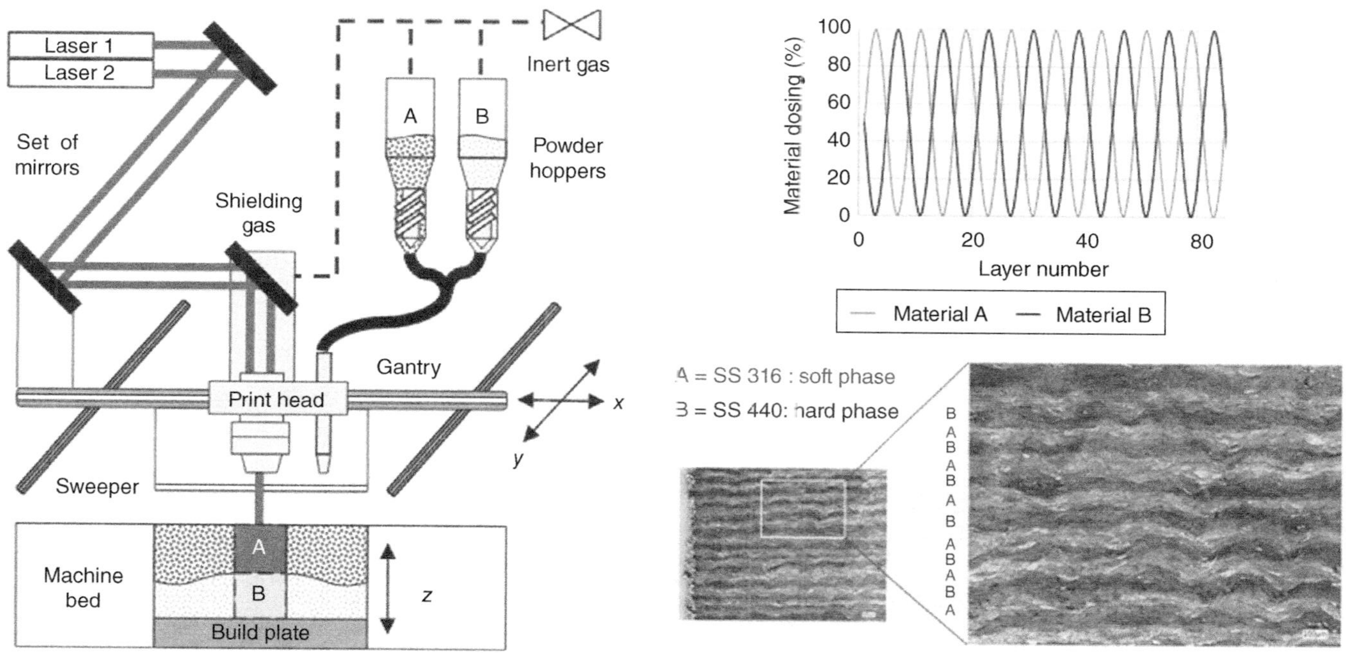


**Figure 1.3** Working schematic of the L-PBF equipment, and functionally graded steel composite components built at The Technical University of Denmark (DTU).

layers on a 316L base plate. New methodologies for applying multi-material powder in PBF are necessary for the wider adoption of FGAM components for critical functional applications.

## 1.3 Direct Material Deposition

### 1.3.1 Powder and Wire Feedstock for Near-Net-Shape AM

Direct Material Deposition (DMD) can be divided into two classes of AM processes based on the feedstock being powder or wire. The class of powder feedstock systems is called Directed Energy Deposition (DED), and with wire, the systems are classified underwire arc additive manufacturing (WAAM) [51]. In both cases, the deposition head consists of a power source (Laser/Wire arc) and a material deposition module to continuously add new material into the process. Often the deposition head is mounted on a five-axis robot, and typically, there is an option to mix multiple materials in-situ. The material flexibility in DMD processes makes them the most suitable for manufacturing functionally graded AM components. Theoretically, any additive process that allows adjustment of feedstock material mid-build is viable. However, directed energy deposition (DMD) processes lend themselves best to functional grading due to the comparatively small feedstock losses as opposed to multi-material PBF, since powder blends cannot be easily separated and recycled for reuse. Figure 1.4 shows a schematic of the DED process, wherein the deposition head is mounted on a five-axis robot. Up to six materials can be used simultaneously in the setup. Laser deposited materials experience a complex thermal history marked by rapid solidification, high cooling rates, steep thermal gradients, and cyclic reheating and cooling from multiple laser passes. This can lead to the formation of nonequilibrium microstructures and significant variations in microstructure from layer to layer, as well as within individual layers. Hence, an online process-control is achieved with the help of an in-axis thermal camera that ensures similar thermal gradients in each new layer.

Hybrid Additive Manufacturing (HAM) combines the cost-saving additive process with the dimensional accuracy of subtractive CNC machining [53]. When a metallic part is additively manufactured by direct material deposition, HAM allows the part to be machined after deposition occurs in the same system. This is largely different compared to the steps taken for conventional CNC techniques, where a block of a single material is placed into a CNC machine and is subjected to extensive subtractive processes to create a final part. A large amount of material waste is generated from the excess material during milling, which translates directly to manufacturing cost and environmental impact. With HAM, however, the initial block material is no longer present because the operator is building the

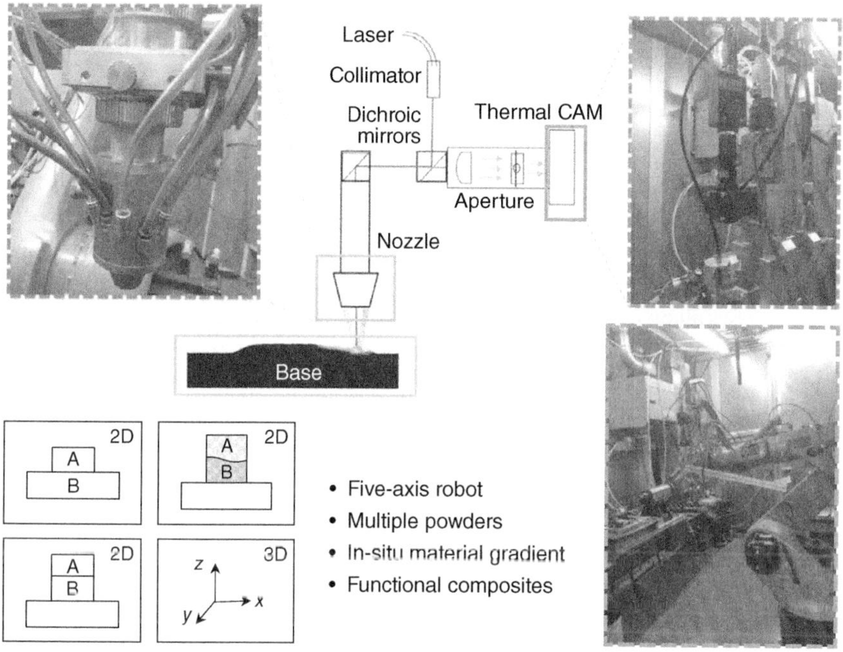


**Figure 1.4** Schematic of powder feedstock based Direct Energy Deposition (DED) system installed at Force technology [52]. Photo credit: Venkata K Nadimpalli.

part near net shape by direct deposition, requiring only surface finishing where necessary and creating little material waste. The presence of the deposition head on a robot-arm leads to a straightforward integration with CNC to convert a typical DMD process into HAM. Several FGM applications are found for DED as compared to WAAM because the near-net-shape resolution of DED is small enough to have a much wider range of complex functional applications in the as-built condition. Further, DED is better at repairing complex components due to the coaxial powder feeding mechanism and faster cooling rate, which allows for material deposition against gravity. The further sections thus focus on DED toward making functionally graded materials.

### 1.3.2 Functional Material Gradients in DED

A promising application of functionally gradient AM is the replacement of problematic dissimilar metal welds with additively manufactured graded transition joints. In the nuclear and aerospace industries, failures related to joints, and interfaces between dissimilar metals are frequent [54]. Typical joining processes produce sharp gradients in composition and properties that ultimately promote failure mechanisms in service. Furthermore, designs are limited by the need to assemble many individual parts composed of different materials in order to meet various services and cost requirements. Hence, for the case of low-volume, high-complexity parts, incorporating functional grading with additive manufacturing presents an attractive alternative. A major advantage of using gradient AM techniques is the ability to fabricate custom compositions, using phase diagrams as a map between desired compositions. In traditional alloy development, multicomponent alloys are made one point at a time until the phase diagram is understood. The gradients in AM processes allow for many compositions along a line to be fabricated in the same part, thus accelerating alloy development. For each gradient alloy, a defined gradient path can be selected for the AM building process, thus unlocking the possibility to reach the desired phase while avoiding the detrimental intermetallics. Table 1.1 below gives an overview of recent research on functional DED components.

Parts can be designed with gradient material properties and can be made by changing powder input. By doing this, the material behavior could be designed across the part by utilizing the changing material/mechanical properties of the material. In an example from Gu, Meiners, Wissenbach, Poprawe [23], a Ni-Cr part was made with a designed negative coefficient of thermal expansion. Ideally, this process could be used to make structures that are piezoelectric and even have a negative Poissons ratio or make a ductile metal with a negative thermal expansion. Compositional gradation also increases the overall properties and integrity of the part since weld-seam stress concentrations weakening joints are reduced. Hofmann et al. [69] performed finite elemental analysis (FEA) and have shown that a gradient transition in an automobile valve stem from steel to Inconel has approximately 10 times less stress concentration at the transitioning zone compared to a traditionally welded joint at the same operating temperatures. Another FGM example is that of LENS-deposited Inconel transitioning to a copper alloy for increased thermal conductivity behaviors in high temperature heat exchangers. The conductivity of the bimetallic structures increased by almost 300% compared to Inconel 718 [70].

The review articles by Bandyopadhyay and Heer [20] list out several applications and research studies conducted on multi-material AM components. Hence, in Table 1.1, the focus was on more recently published work and classifying them according to Type II–IV material gradients. The rapid increase in research on FGAM's and better DED process control, coupled with new software modules for design and simulation of FGAM components, indicates the potential for growth in this section. The ultimate goal is to have integrated materials design freedom at the voxel-scale and the process-control necessary to manufacture the designed FGAM's. Out of all the metal AM processes, DED is the closest toward achieving complete material design freedom.

**Table 1.1** An overview of recently representative work on different types of materials gradients in DED.

| Materials | Type-II | Function | References |
|---|---|---|---|
| 316/IN625 | 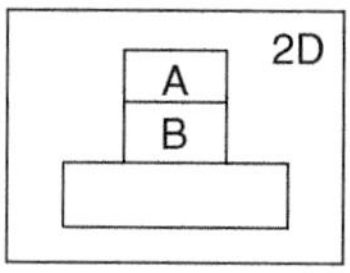 | Optimized to avoid cracks | [55] |
| IN625/Cu | | Bimetallic structure for thermal applications | [56] |
| IN718/SS316L | | Change in hardness and mechanical strength with no cracking | [57] |
| TA15/IN718 with Cu/Nb interlayers | | Interlayers allow the manufacture of traditionally unweldable alloys | [58] |
| | **Type-III** | *Continuous gradients* | |
| IN625/NiCrAlY | 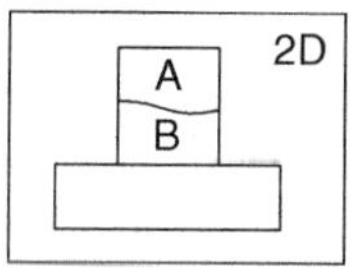 | Functionally graded interface with few/no cracks | [59] |
| 2.25Cr-1Mo-Steel/Alloy 800H | | Graded Ferritic to Austenitic steels with limited Carbon diffusion | [60] |
| Bulk metallic glasses Zr50 alloy (BMG) | | Large gradient BMG structures that are otherwise difficult to produce | [61] |
| Ti/Ti-6Al-4V | | Compositionally graded with phase and microstructural control | [62] |
| Ti-6Al-4V/Invar | | Continuously graded samples with no cracks compared to the discrete interface | [63] |
| | **Type-IV** | *Composites + Continuous gradients* | |
| Ti-$B_4C$ composite | 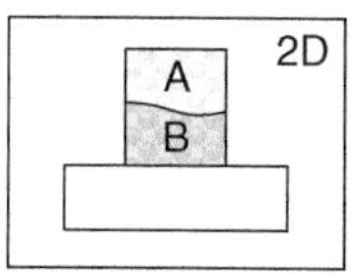 | TiB precipitate control leads to MMC with equiaxed, fine α grains | [64] |
| Cp Ti-NbC composite | | Metal-ceramic composites for high impact, high toughness applications (armor) | [65] |
| Ti-$Al_2O$ MMC w/ nanopowders | | Fine α2 lamellar microstructure within a Ti matrix. High hardness was achieved | [66] |
| TiB-Ti composites | | 3D quasi-continuous precipitate network with high strength | [67] |
| Hybrid Ti-6Al-4V wire + WC powder | | High hardness and wear-resistant composite | ([68]) |

## 1.4 Solid-State Additive Manufacturing

Solid-state AM is a class of processes that use friction and diffusion-based mechanisms to produce mechanical bonding without melting. Solid-state processes like ultrasonic welding and friction-stir welding have established applications [71, 72]. They can also be converted into AM processes by integrating the weld head onto a three or five-axis robot coupled with a CNC mill to make a hybrid AM system. Three such processes are Ultrasonic Additive Manufacturing (UAM), Friction Stir Additive Manufacturing [73], and Cold Spray Additive Manufacturing [74]. FAM and CSAM are relatively new technologies with few research papers in literature exploring hybrid functional components. Yin, Yan, et al. [75] showcased the use of cold spray as a hybrid AM process applied to L-PBF components. UAM, on the other hand, is a well-established and often ignored AM technology that can make unique functional parts. The rest of this section is hence dedicated to functional components made by UAM.

Ultrasonic Additive Manufacturing (UAM) is a solid-state joining manufacturing process that is commonly used in conjunction with a CNC mill to make functional metal components [76]. UAM offers advantages in material properties as compared to traditional metal joining and forming technologies. It uses normal force coupled with low frequency mechanical ultrasonic vibrations to create a solid-state weld between a thin foil and an existing substrate. Ultrasonic welding is an established industrial bonding process for plastics and soft metals. UAM is a hybrid AM process which is essentially layer-by-layer ultrasonic welding combined with CNC machining after each layer, hence providing freeform fabrication capability [77]. Given the right bonding parameters, completely solid-state metallurgically bonded welds can be fabricated. The quality of UAM components depends on several process parameters, including normal force, vibration amplitude, and speed of bonding along with geometrical and environmental factors. For many years, the material systems available to be bonded by the UAM process were limited to softer Aluminum alloys (Al 3003) due to the high energy requirement for other engineering materials. The Fabrisonic UAM systems overcome this hurdle by using a high-power transducer and load cell, which makes it feasible to build components from Copper, Nickel, and Iron-based alloy systems.

UAM involves high-speed relative motion between foils to be joined. Several researchers have studied the mechanism of bonding in ultrasonic metal welding [89]. The bonding process is a function of the three input parameters; the force applied, the velocity of bonding, and the vibration amplitude. It is also dependent on surface roughness, temperature, base plate characteristics, part geometry, and build height, among other factors. The bonding process can be separated into (a) volumetric bonding and (b) surface bonding effects. Volumetric bonding effects include elastic and plastic deformation enhanced through reduced yield stress due to acoustic and thermal softening. Surface bonding effects include interfacial friction and shearing, which break up the oxide layers and bring more nascent metal-to-metal contact. Bond formation by ultrasonic welding requires two conditions to be fulfilled, (a) the generation of clean surfaces with no barrier layers at the atomic scale and (b) direct contact between these clean surfaces. Janaki Ram, Yang, and Stucker [90] suggested that the surface-oxide layers are broken up by the vibrations and are displaced within the vicinity of the interface region. A schematic of the UAM process is shown in the Figure 1.5 with the weld head mounted on a three-axis stage and can be switched with CNC tools.

The additive/subtractive nature and the room temperature processing capability of the process give rise to capabilities such as completely enclosed cooling channels, smart parts with embedded sensors, and composite materials that cannot be produced traditionally. Alloy systems that cannot be typically fusion welded

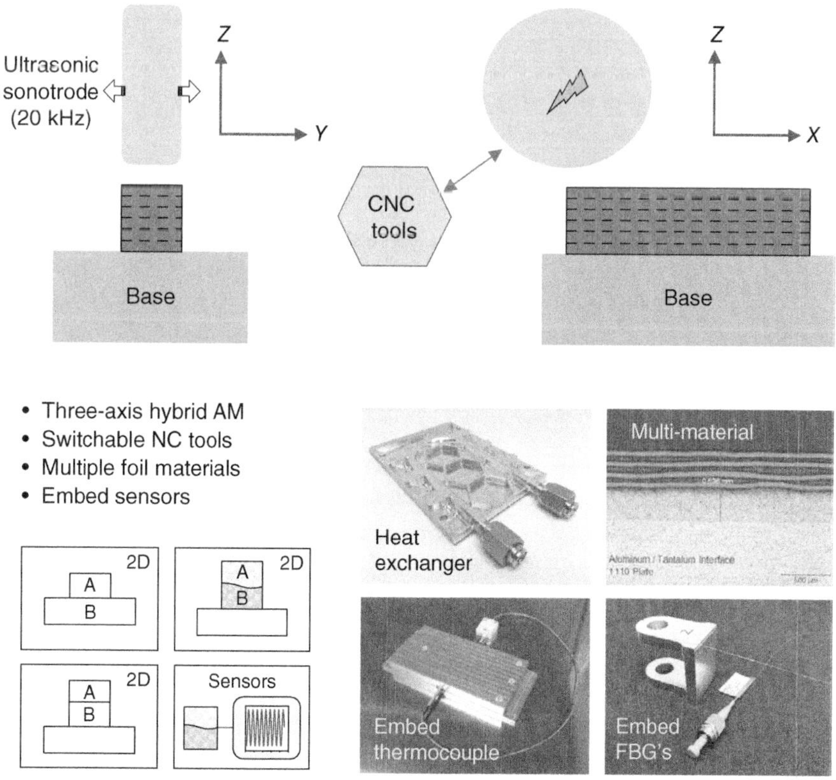


**Figure 1.5** Schematic of the UAM process. FBG's are high-temperature optical fiber Bragg grating (FBG) strain sensors. *Source:* A part of the figure is adapted in accordance with the Creative Commons license and is a copyright of Fabrisonic LLC [91]. © 2020 Fabrisonic LLC.

together can be joined, making multi-material components that do not suffer from common weld defects like embrittlement and solidification cracking. High power UAM can make fully dense bonds within several Al, Cu, Ni, and Fe alloys. However, UAM bonds are prone to delamination and must be treated as anisotropic composites with lower modulus along the build direction. It is also difficult to make high aspect ratio structures with UAM due to the inherent vibration during the process. However, these characteristics could be advantageous for several applications like heat transfer, impact toughness, dissimilar welding, and, most importantly, embedded sensors. An overview of functional UAM applications is given in Table 1.2. UAM is an often-overlooked metal AM process, which has the

**Table 1.2** Overview of functional UAM applications.

| Materials | Type-II | Function | References |
|---|---|---|---|
| Al/Steel joints | 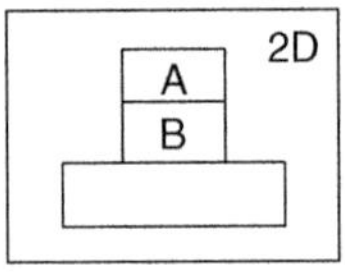  | Al/Fe joints with applications in heat exchangers | [78] |
| Al/Ti | | Strong Al/Ti joints which are difficult with traditional joining methods | [79] |
| Al/Fe/Ni/Ta/Cu | | Demonstrated the multi-material capability with a wide range of metals | [80] |
| Al/Cu | | Interlayers allow the manufacture of traditionally unweldable alloys | [81] |
| | **Type-III** | *Metal matrix composites* | |
| Al/Carbon fiber | 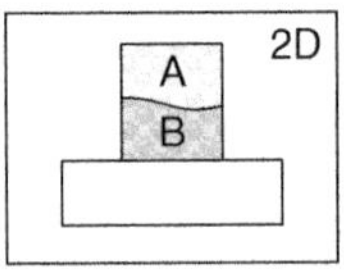  | High impact toughness composite | [82] |
| Al/SiC/TiC | | Fully embedded fibers can only be made by UAM | [83] |
| Al/Metpreg | | Embedded Metpreg MMC makes lightweight and high strength composites | [84] |
| | **Type-IV** | *Integrated sensors* | |
| NiTi, FeGa | 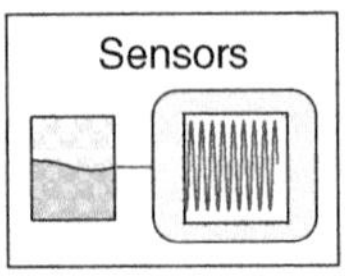  | Shape memory alloys for sensing applications | [85] |
| Al6061 + Ta/ $Eu_2O_3$ | | Fully enclosed neutron absorber components for Oakridge National Lab | [86] |
| Al matrix + optical fiber sensors | | Embedded optical fiber strain sensors for high-temperature sensing (up to 500 °C) | [87] |
| Embed electronics | | Embedded surface mount resistor | [88] |

unique capability to make functional components despite its limitations. Hence, the ongoing research on UAM is bound to find its way into high-impact and niche applications.

## 1.5 Hybrid AM Through Green Body Sintering

The indirect manufacture of metallic and ceramic components via AM employs a process chain by which a blended feedstock is used to build the component. This feedstock contains a powdered filler material and a matrix material that, during additive manufacture, gives the component its as-built strength. This part is known as the green part and is the origin of a multistage process by which the binder matrix is removed. The powdered filler is sintered and densified to form the final component. The process chain can be enabled by a variety of AM technologies and with a variety of process chains for debinding, sintering, and densification. As such, this section deals with the three most common AM methods and the two most common sintering process chains.

### 1.5.1 Common AM Technologies for Green Body Manufacturing

The entry-level technology by which a green body component is commonly manufactured is through filament extrusion [92]. There is a variety of feedstock for extrusion-based AM systems that have been engineered for this purpose. For high throughput industrial application, binder jetting is the fastest build method of all industrially applicable manufacturing methods and is thus competitive in many cases to laser PBF. Since no melt pool and plasma is generated during manufacture, the general tolerances are superior to those of laser PBF. As the green body part shrinks up to 30% during sintering, tolerances are further enhanced. The final method for green body manufacture using AM is by employing vat photopolymerization, where the photopolymer acts as the matrix material and is blended with the powdered filler. This method has a very low throughput yet is still highly industrially relevant, given that vat photopolymerization exhibits a spatial resolution, one order of magnitude better than that of the former methods [93]. Whereas there are no commercially available systems that allow for functionally graded materials employing hybrid AM/sintering process chains, this technology is highly relevant for manufacturing geometrical gradients in materials. Micro components made from metamaterials with topology optimized unit cells can thus be manufactured employing vat photopolymerization and sees usage ranging from medical components through energy applications.

### 1.5.2 CAD Design and Shrinkage Compensation

The hybrid process chain, in its entirety, is as illustrated in Figure 1.6. The CAD body of the component is designed specifically for the process chain. This is done in order to compensate for shrinkage during the sintering step. Shrinkage is typically up to 30% but can be higher for specialty feedstocks. The shrinkage behavior of small-sized parts ($<50 \times 50 \times 50\,mm^3$) is quite uniform and predictable. Larger sized components, however, require accurate multi-physics simulation models for shrinkage behavior. The shrinkage typically is considered to be isotropic for binder jetting and vat photopolymerization [95]. For filament extrusion, the shrinkage is normally anisotropic even for smaller parts with a larger shrinkage in the normal direction to the layers. This is due to a flow phenomenon where the powdered filler has a tendency to concentrate in the very core of the individual extruded strands of filament, leaving a skin layer with a little amount of particles in the skin layer. As such, the powdered filler concentration in the normal direction to the layering of the part exhibits alternating laminae of high and low particles. During sintering, the low concentration zones see more matrix material removal during debinding and thus higher shrinkage.

### 1.5.3 Additive Manufacture

For powder-bed binder jetting and vat photopolymerization, certain industrially available systems are tailored for the hybrid AM process chain [96]. Examples are the Desktop Metal and ExOne (metals) system for binder jetting and the Lithoz

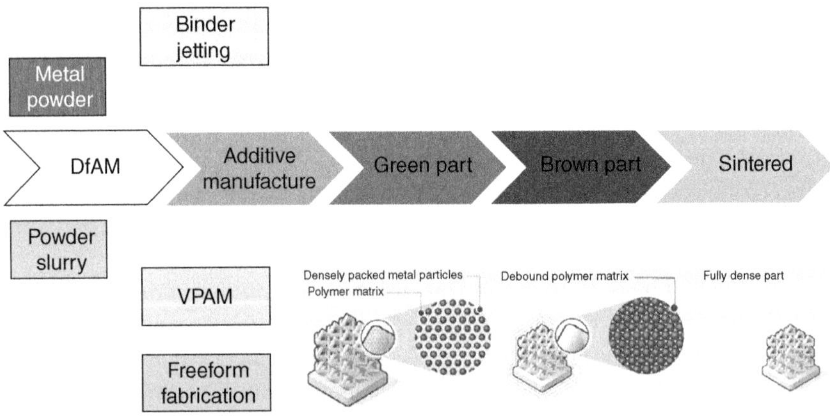


**Figure 1.6** Schematic of the hybrid metal AM process chain through sintering. DfAM refers to Design for AM and VPAM refers to Vat-photo-polymerization AM. *Source:* A part of the figure is adapted in accordance with the Creative Commons license and is a copyright of Holo Inc. [94]. © 2020 John Wiley & Sons.

system for vat photopolymerization (ceramics/metals). Given that these systems are tailored to the process chain, the workflow from CAD body design, through debinding to sintering, is offered as a bundled solution that minimizes the efforts needed for successful adoption. For experimental binder/matrix materials, the readily available industrial systems are less suited, as these are qualified to operate solely in a palette of materials offered by the systems provider. For exotic materials, a custom system is needed that allows for an open parameter setup in order to optimize the build for the feedstock chosen. Such solutions are not readily available and require an initial effort in setting up a custom AM system by the adopter, raising the barrier, therefore tremendously.

For filament extrusion hybrid AM, there is an abundance of capable entry-level systems that allow for a free selection of filament [97]. The powder-loaded filaments for hybrid AM are offered by renowned providers (e.g., BASF) and often accompanied with technical white-papers, aiding the operator of the system in achieving a successful AM build, debind, and sintering. The matrix polymer is often the same polymer as general-purpose feedstock (e.g., polylactic acid), and the powdered filler is often stainless steel (e.g., 316L). Take due notice that some filaments are loaded with metal particles for aesthetic reasons only. Filaments with bronze or copper are often employed to give the additively manufactured part an aesthetic sculpted appearance that will oxidize and wither through time due to exposure to the elements. These materials are not suitable for hybrid AM. The build of a component with a suitable filament for hybrid AM manufacture requires different processing conditions than those of a standard filament. The extrusion temperature is generally increased to aid flowability, and since the extrusion pressure/force is higher due to the altered rheology of the feedstock, a direct drive extruder drive mechanism is favored. Retrofitting a dual traction extruder drive system is oftentimes offered as a drop-in replacement for entry-level filament AM system.

### 1.5.4 Debinding and Sintering

The debinding process serves to remove the matrix material of the green body part. Once removed, the component has reached an intermediate state, commonly known as the brown part. This, at large, is achieved by one of two methods. The most convenient is through a burn-out cycle where the matrix material is debound by decomposing and outgassed from the green body part. This is in particular convenient as it can be integrated into a combined debind and sintering cycle in a programmable oven. This duration of this combined cycle is over multiple hours and in cases up to 2 days. For some specific feedstocks, a separate debinding method is added as an additional link in the process chain. This is introduced out of necessity, to ensure that the component integrity is maintained throughout the sintering cycle. This step involves a chemical dissolving of the binder by exposure to a solvent and is carried out as pertains to one or multiple of the following phenomena; Accumulation of residuals (e.g., soot) as a result of thermal debinding that can form impurities in the sintered component. Mitigation of crack formation during outgassing in case of thermal debinding. The possibility of infiltration of a functional component after debinding (a flux agent or introduction of a matrix component). Solvent-based debinding is rarely applied at a large scale for industrial applications and is primarily seen applied when producing exotic components from novel materials.

### 1.5.5 Functionally Graded Components in Sintered Components

Given the limitations for powder-bed and vat photopolymer systems, functionally graded components that are built from multiple materials are difficult to manufacture by hybrid AM/sintering as feedstock will be cross-contaminated in-situ. As a result, hereof, functional gradients in hybrid AM/sintered components are primarily obtained through geometry. By application of flexure mechanisms, from topology optimization and from graded unit cells in cellular structured metamaterials, it is though possible to derive novel functionality. Select examples are such as negative Poisson's ratio [98], negative thermal expansion [99], acoustic properties [100], and bioinspired structures [101] and bioinspired hierarchical microstructures [102]. The high fidelity of small features that can be manufactured using a hybrid AM/sintering approach, as described in Section 1.5.1, renders these technologies particularly interesting for such topology optimized metamaterials. Extrusion-based hybrid AM/Sintering offers a novel opportunity to sinter multi-material components as a result of the inherent capabilities of this process. This opens a possibility to combine widely different particle loaded filaments to fabricate green bodies that form, for example, metallic/ceramic composites that can be subsequently sintered [103]. This method does set demands for the dissimilar materials to have an overlap in sintering temperature interval and thermal expansion coefficients.

## 1.6 Conclusions

Additive manufacturing facilitates new ways to build functionality into metal components. The unpreceded geometrical freedom that these manufacturing technologies bring forth gives rise to new ways of considering geometry induced functionality through lattice structures, metamaterials, flexures, and topology optimization. Equally contingent is the ability to transition material composition throughout a component, either by-layer in sandwiched gradients or in-layer through a by pointwise material deposition. It is worthwhile to notice that despite AM offering certain new degrees of design freedom, a process inherent design constraints apply and must be thoroughly understood before embarking on the implementation of an AM infrastructure for functionally graded manufacturing of metal components. Some processes such as vat photopolymerization and laser PBF induce issues as relates to keeping process materials separated such that excessive waste is not generated during build operations. For direct energy deposition, this issue is less pronounced as segregated waste powder collection can, to a large extent, remedy cross-contamination. Most favorable is the pointwise deposition by means of multi-material filament extrusion and subsequent sintering. A hybrid approach where no mixed waste product is generated, yet at the cost of having a lower component fidelity as extrusion-based AM, is limited in minimum feature size and minimum layer height. It is crucial that the benefits of AM, including geometrical freedom, digitally driven, highly agile manufacturing is weighted against the process limitations, including low throughput, process-dependent design, and manufacturing paradigms that must be learned. Further, post-processing of AM parts is often performed as an art form by manual fit and finishing. It can be regarded as the secret of additive manufacturing that nonchalantly by the populace is deemed a fully digital process when the underlying modality is that of a high degree of post-processing in order to present a functional component that meets design tolerances.

The overview of recent advances in metal AM is an excellent barometer for the evolution of existing and new applications in the energy sector. The three primary drivers in favor of AM components are geometric/material functionality, lead-time reduction, and sustainable production. The geometric functionality applied toward heat-exchangers, topology-optimized turbine blades, nuclear core-reactors, etc., has the potential to be improved by the addition of material gradient functionality. Nuclear power components, steam/gas turbines components, fuel cell components, oil & gas are the major sectors in which a steady growth of critical functional metal AM components is expected over the coming decades, with an increase in available materials. Metal AM applications driven by lead-time reduction and sustainable production are expected to have an exponential growth in the nuclear, oil & gas sectors specifically for on-demand and on-site production. Consolidated designs produced with AM are increasingly proving to be an efficient solution for the energy industries complex needs. Steadily the capabilities are evolving toward much higher degrees of design freedom, toward better capabilities of multi-material manufacturing in highly dissimilar materials and for the confident and resolute, the capabilities that currently present themselves for introducing a new function to components that have the potential to have an encompassing implication to how new technology can be brought to market.

## Acknowledgment

This work was supported by the Poul Due Jensen Foundation.

## References

**1** Chua, C. K., & Leong, K. F. (2017). *3D Printing and Additive Manufacturing: Principles and Applications. The 5th Edition of Rapid Prototyping: Principles and Applications*. Singapore: World Scientific Publishing Company. doi:10.1142/10200

**2** Gibson, I., Rosen, D. W., & Stucker, B. (2010). *Additive Manufacturing Technologies: Rapid Prototyping To Direct Digital Manufacturing*. New York: Springer. doi:10.1007/978-1-4419-1120-9

**3** Wohlers Report (2020). Analysis. Trends. Forecasts. Feel the pulse of the 3D printing industry. *Wohlers Associates*.

**4** Hofmann, D. C., Kolodziejska, J., Roberts, S., Otis, R., Dillon, R. P., Suh, J. O., … Borgonia, J. P. (2014). Compositionally graded metals: A new frontier of additive manufacturing. *Journal of Materials Research*, 29(17), 1899–1910. doi:10.1557/jmr.2014.208

**5** Sobczak, J. J. J., & Drenchev, L. (2013). Metallic functionally graded materials: A specific class of advanced composites. *Journal of Materials Science and Technology*, 29(4), 297–316. doi:10.1016/j.jmst.2013.02.006

**6** MacDonald, E., & Wicker, R. (2016). Multiprocess 3D printing for increasing component functionality. *Science*, 353(6307), aaf2093-1–aaf2093-8. doi:10.1126/science.aaf2093

**7** GE Additive. (n.d.). New manufacturing milestone: 30,000 additive fuel nozzles. Retrieved from https://www.ge.com/additive/stories/new-manufacturing-milestone-30000-additive-fuel-nozzles

**8** Siemens Global. (n.d.). Additive manufacturing. Energy topics. Retrieved from https://new.siemens.com/global/en/products/energy/topics/additive-manufacturing.html

**9** Oerlikon AM. (n.d.). AM in Power Generation. Energy Market. Retrieved from https://www.oerlikon.com/am/en/markets/power-generationenergy/

**10** Biome Renewables. (n.d.). PowerCone. Retrieved from https://www.biome-renewables.com/powercone

**11** aidro. (n.d.). Additive manufacturing in hydraulics. Retrieved from https://www.aidro.it/3d-metal-printing.html

**12** ORNL. (n.d.). 3D-printed nuclear reactor promises faster, more economical path to nuclear energy. Retrieved from https://www.ornl.gov/news/3d-printed-nuclear-reactor-promises-faster-more-economical-path-nuclear-energy

**13** OPTISYS. (n.d.). High bandwidth antennas. Retrieved from https://www.optisys.tech/blog/tag/Additive-Manufacturing

**14** EOS GmbH. (n.d.). Manufacture heat exchangers additively. Retrieved from https://www.eos.info/en/3d-printing-examples-applications/production-and-industry/electronical-components/heatexchanger-additive-manufacturing

**15** Hegab, H. A. (2016). Design for additive manufacturing of composite materials and potential alloys: A review. *Manufacturing Review*, 3, 1–17. doi:10.1051/mfreview/2016010

**16** Naebe, M., & Shirvanimoghaddam, K. (2016). Functionally graded materials: A review of fabrication and properties. *Applied Materials Today*, 5, 223–245. doi:10.1016/j.apmt.2016.10.001

**17** Loh, G. H., Pei, E., Harrison, D., & Monzón, M. D. (2018). An overview of functionally graded additive manufacturing. *Additive Manufacturing*, 23, 34–44. doi:10.1016/j.addma.2018.06.023

**18** Rafiee, M., Farahani, R. D., & Therriault, D. (2020). Multi-material 3D and 4D printing: A survey. *Advanced Science*, 7(12), 1902307-1–1902307-26. doi:10.1002/advs.201902307

**19** Pascale, D., & Simion, I. (2018). Multi-material 3D printer extruder concept. *Journal of Industrial Design and Engineering Graphics*, 13(1), 25–28.

**20** Bandyopadhyay, A., & Heer, B. (2018). Additive manufacturing of multi-material structures. *Materials Science and Engineering R: Reports*, 129, 1–16.

**21** Murr, L. E., Gaytan, S. M., Ramirez, D. A., Martinez, E., Hernandez, J., Amato, K. N., … Wicker, R. B. (2012). Metal fabrication by additive manufacturing using laser and electron beam melting technologies. *Journal of Materials Science and Technology*, 28(1), 1–14. doi:10.1016/S1005-0302(12)60016-4

**22** DebRoy, T., Wei, H. L., Zuback, J. S., Mukherjee, T., Elmer, J. W., Milewski, J. O., … Zhang, W. (2018). Additive manufacturing of metallic components: Process, structure and properties. *Progress in Materials Science*, 92, 112–224. doi:10.1016/j.pmatsci.2017.10.001

**23** Gu, D. D., Meiners, W., Wissenbach, K., & Poprawe, R. (2012). Laser additive manufacturing of metallic components: Materials, processes and mechanisms. *International Materials Reviews*, 57(3), 133–164. doi:10.1179/1743280411Y.0000000014

**24** Ashby, M. F. (2006). The properties of foams and lattices. *Philosophical Transactions of the Royal Society A: Mathematical, Physical and Engineering Sciences*, 364(1838), 15–30. doi:10.1098/rsta.2005.1678

**25** Simone, A. E., & Gibson, L. J. (1998). The effects of cell face curvature and corrugations on the stiffness and strength of metallic foams. *Acta Materialia*, 46(11), 3929–3935. doi:10.1016/S1359-6454(98)00072-X

**26** Maskery, I., Aboulkhair, N. T., Aremu, A. O., Tuck, C. J., Ashcroft, I. A., Wildman, R. D., & Hague, R. J. M. (2016). A mechanical property evaluation of graded density Al-Si10-Mg lattice structures manufactured by selective laser melting. *Materials Science and Engineering A*, 670, 264–274. doi:10.1016/j.msea.2016.06.013

27 Choy, S. Y. S. Y., Sun, C. N. C.-N., Leong, K. F. K. F., & Wei, J. (2017). Compressive properties of functionally graded lattice structures manufactured by selective laser melting. *Materials and Design*, 131(June), 112–120. doi:10.1016/j.matdes.2017.06.006

28 Li, S., Hassanin, H., Attallah, M. M., Adkins, N. J. E., & Essa, K. (2016). The development of TiNi-based negative Poisson's ratio structure using selective laser melting. *Acta Materialia*, 105, 75–83. doi:10.1016/j.actamat.2015.12.017

29 Tan, C., Li, S., Essa, K., Jamshidi, P., Zhou, K., Ma, W., & Attallah, M. M. M. M. (2019). Laser powder bed fusion of Ti-rich TiNi lattice structures: Process optimisation, geometrical integrity, and phase transformations. *International Journal of Machine Tools and Manufacture*, 141(January), 19–29. doi:10.1016/j.ijmachtools.2019.04.002

30 Carluccio, D., Demir, A. G., Bermingham, M. J., & Dargusch, M. S. (2020). Challenges and opportunities in the selective laser melting of biodegradable metals for load-bearing bone scaffold applications. *Metallurgical and Materials Transactions A: Physical Metallurgy and Materials Science*, 51, 3311–3334. doi:10.1007/s11661-020-05796-z

31 Hazlehurst, K. B., Wang, C. J., & Stanford, M. (2014). An investigation into the flexural characteristics of functionally graded cobalt chrome femoral stems manufactured using selective laser melting. *Materials and Design*, 60, 177–183. doi:10.1016/j.matdes.2014.03.068

32 Ataee, A., Li, Y., Fraser, D., Song, G., & Wen, C. (2018). Anisotropic Ti-6Al-4V gyroid scaffolds manufactured by electron beam melting (EBM) for bone implant applications. *Materials and Design*, 137, 345–354. doi:10.1016/j.matdes.2017.10.040

33 Yan, C., Hao, L., Hussein, A., & Young, P. (2015). Ti-6Al-4V triply periodic minimal surface structures for bone implants fabricated via selective laser melting. *Journal of the Mechanical Behavior of Biomedical Materials*, 51, 61–73. doi:10.1016/j.jmbbm.2015.06.024

34 Yu, S., Sun, J., & Bai, J. (2019). Investigation of functionally graded TPMS structures fabricated by additive manufacturing. *Materials and Design*, 182, 108021. doi:10.1016/j.matdes.2019.108021

35 Zhang, X.-Y., Fang, G., Leeflang, S., Zadpoor, A. A., & Zhou, J. (2019). Topological design, permeability and mechanical behavior of additively manufactured functionally graded porous metallic biomaterials. *Acta Biomaterialia*, 84, 437–452. doi:10.1016/j.actbio.2018.12.013

36 Nadimpalli, V. K., Dahmen, T., Valente, E. H., Mohanty, S., & Pedersen, D. B. (2019, June 3–7). Multi-material additive manufacturing of steels using laser powder bed fusion. Proceedings of the 19th International Conference and

Exhibition European Society for Precision Engineering and Nanotechnology, Conference, EUSPEN 2019, Bilbao.

**37** Chen, J., Yang, Y., Song, C., Zhang, M., Wu, S., & Wang, D. (2019). Interfacial microstructure and mechanical properties of 316L /CuSn10 multi-material bimetallic structure fabricated by selective laser melting. *Materials Science and Engineering A*, 752, 75–85. doi:10.1016/j.msea.2019.02.097

**38** Hinojos, A., Mireles, J., Reichardt, A., Frigola, P., Hosemann, P., Murr, L. E., & Wicker, R. B. (2016). Joining of Inconel 718 and 316 Stainless Steel using electron beam melting additive manufacturing technology. *Materials and Design*, 94, 17–27. doi:10.1016/j.matdes.2016.01.041

**39** AlMangour, B., Grzesiak, D., & Yang, J. M. (2017). In-situ formation of novel TiC-particle-reinforced 316L stainless steel bulk-form composites by selective laser melting. *Journal of Alloys and Compounds*, 706, 409–418. doi:10.1016/j.jallcom.2017.01.149

**40** Han, C., Li, Y., Wang, Q., Cai, D., Wei, Q., Yang, L., … Shi, Y. (2018). Titanium/hydroxyapatite (Ti/HA) gradient materials with quasi-continuous ratios fabricated by SLM: Material interface and fracture toughness. *Materials and Design*, 141, 256–266. doi:10.1016/j.matdes.2017.12.037

**41** Kun, C., Beibei, H. H., Wenheng, W., & Cailin, Z. (2017). The formation mechanism of TiC reinforcement and improved tensile strength in additive manufactured Ti matrix nanocomposite. *Vacuum*, 143, 23–27. doi:10.1016/j.vacuum.2017.05.029

**42** Xia, M., Liu, A., Hou, Z., Li, N., Chen, Z., & Ding, H. (2017). Microstructure growth behavior and its evolution mechanism during laser additive manufacture of in-situ reinforced (TiB+TiC)/Ti composite. *Journal of Alloys and Compounds*, 728, 436–444. doi:10.1016/j.jallcom.2017.09.033

**43** Niendorf, T., Leuders, S., Riemer, A., Brenne, F., Tröster, T., Richard, H. A., & Schwarze, D. (2014). Functionally graded alloys obtained by additive manufacturing. *Advanced Engineering Materials*, 16(7), 857–861. doi:10.1002/adem.201300579

**44** Koptyug, A., Popov, V. V., Botero Vega, C. A., Jiménez-Piqué, E., Katz-Demyanetz, A., Rännar, L. E., & Bäckström, M. (2020). Compositionally-tailored steel-based materials manufactured by electron beam melting using blended pre-alloyed powders. *Materials Science and Engineering A*, 771(July 2019), 138587-1–138587-11. doi:10.1016/j.msea.2019.138587

**45** Biondani, F. G., Bissacco, G., Mohanty, S., Tang, P. T., & Hansen, H. N. (2020). Multi-metal additive manufacturing process chain for optical quality mold generation. *Journal of Materials Processing Technology*, 277, 116451.

**46** Anstaett, C., Seidel, C., & Reinhart, G. (2017). Fabrication of 3D multi-material parts using laser-based powder bed fusion. Proceedings of the 28th Annual International Solid Freeform Fabrication Symposium.

**47** Aerosint. (n.d.). Selective powder deposition for AM. Retrieved from https://aerosint.com/

**48** Demir, A. G., & Previtali, B. (2017). Multi-material selective laser melting of Fe/Al-12Si components. *Manufacturing Letters*, 11, 8–11.

**49** Wei, C., Li, L., Zhang, X., & Chueh, Y.-H. (2018). 3D printing of multiple metallic materials via modified selective laser melting. *CIRP Annals*, 67(1), 245–248.

**50** Bodner, S. C., van de Vorst, L. T. G., Zalesak, J., Todt, J., Keckes, J. F., Maier-Kiener, V., … Keckes, J. (2020). Inconel-steel multilayers by liquid dispersed metal powder bed fusion: Microstructure, residual stress and property gradients. *Additive Manufacturing*, 32, 101027-1–101027-11. doi:10.1016/j.addma.2019.101027

**51** Wang, J., Pan, Z., Ma, Y., Lu, Y., Shen, C., Cuiuri, D., & Li, H. (2018). Characterization of wire arc additively manufactured titanium aluminide functionally graded material: Microstructure, mechanical properties and oxidation behaviour. *Materials Science and Engineering A*, 734, 110–119. doi:10.1016/j.msea.2018.07.097

**52** FORCE Technology. (n.d.). Large-scale 3D printing facility. Retrieved from https://forcetechnology.com/en/all-industry-facilities/large-scale-3d-printing-facility

**53** Stavropoulos, P., Foteinopoulos, P., Papacharalampopoulos, A., & Bikas, H. (2018). Addressing the challenges for the industrial application of additive manufacturing: Towards a hybrid solution. *International Journal of Lightweight Materials and Manufacture*, 1(3), 157–168. doi:10.1016/j.ijlmm.2018.07.002

**54** Lundin, C. D. (1982). Dissimilar metal welds: Transition joints literature review. *Welding Journal (Miami, Fla)*, 61(2), 58-s–63-s.

**55** Chen, N., Khan, H. A., Wan, Z., Lippert, J., Sun, H., Shang, S.-L., … Li, J. (2020). Microstructural characteristics and crack formation in additively manufactured bimetal material of 316L stainless steel and Inconel 625. *Additive Manufacturing*, 32, 101037-1–101037-16. doi:10.1016/j.addma.2020.101037

**56** Anderson, R., Terrell, J., Schneider, J., Thompson, S., & Gradl, P. (2019). Characteristics of bi-metallic interfaces formed during direct energy deposition additive manufacturing processing. *Metallurgical and Materials Transactions B: Process Metallurgy and Materials Processing Science*, 50(4), 1921–1930. doi:10.1007/s11663-019-01612-1

**57** Li, P., Gong, Y., Xu, Y., Qi, Y., Sun, Y., & Zhang, H. (2019). Inconel-steel functionally bimetal materials by hybrid directed energy deposition and thermal milling: Microstructure and mechanical properties. *Archives of Civil and Mechanical Engineering*, 19(3), 820–831. doi:10.1016/j.acme.2019.03.002

**58** Shang, C., Wang, C., Xu, G., Li, C., & You, J. (2019). Laser additive manufacturing of TA15: Inconel 718 bimetallic structure via Nb/Cu multi-interlayer. *Vacuum*, 169(July), 108888. doi:10.1016/j.vacuum.2019.108888

**59** Savitha, U., Srinivas, V., Jagan Reddy, G., Gokhale, A. A., & Sundararaman, M. (2018). Additive laser deposition of YSZ on Ni base superalloy for thermal barrier application. *Surface and Coatings Technology*, 354, 257–267. doi:10.1016/j.surfcoat.2018.08.089

**60** Zuback, J. S., Palmer, T. A., & DebRoy, T. (2019). Additive manufacturing of functionally graded transition joints between ferritic and austenitic alloys. *Journal of Alloys and Compounds*, 770, 995–1003. doi:10.1016/j.jallcom.2018.08.197

**61** Lu, Y., Huang, Y., & Wu, J. (2018). Laser additive manufacturing of structural-graded bulk metallic glass. *Journal of Alloys and Compounds*, 766, 506–510. doi:10.1016/j.jallcom.2018.06.259

**62** Liu, Y., Liang, C., Liu, W., Ma, Y., Liu, C., & Zhang, C. (2018). Dilution of Al and V through laser powder deposition enables a continuously compositionally Ti/Ti6Al4V graded structure. *Journal of Alloys and Compounds*, 763, 376–383. doi:10.1016/j.jallcom.2018.05.289

**63** Bobbio, L. D., Otis, R. A., Borgonia, J. P., Dillon, R. P., Shapiro, A. A., Liu, Z.-K., & Beese, A. M. (2017). Additive manufacturing of a functionally graded material from Ti-6Al-4V to Invar: Experimental characterization and thermodynamic calculations. *Acta Materialia*, 127, 133–142. doi:10.1016/j.actamat.2016.12.070

**64** Nartu, M. S. K. K. Y., Mantri, S. A., Pantawane, M. V., Ho, Y.-H., McWilliams, B., Cho, K., … Banerjee, R. (2020). In situ reactions during direct laser deposition of Ti-B<inf>4</inf>C composites. *Scripta Materialia*, 183, 28–32. doi:10.1016/j.scriptamat.2020.03.021

**65** Traxel, K. D., & Bandyopadhyay, A. (2020). Naturally architected microstructures in structural materials via additive manufacturing. *Additive Manufacturing*, 34, 101243-1–101243-14. doi:10.1016/j.addma.2020.101243

**66** Lanfant, B., Bär, F., Mohanta, A., & Leparoux, M. (2019). Fabrication of metal matrix composite by laser metal deposition-a new process approach by direct dry injection of nanopowders. *Materials*, 12(21), 3584-1–3584-16. doi:10.3390/ma12213584

**67** Hu, Y., Cong, W., Wang, X., Li, Y., Ning, F., & Wang, H. (2018). Laser deposition-additive manufacturing of TiB-Ti composites with novel three-dimensional quasi-continuous network microstructure: Effects on strengthening and toughening. *Composites Part B: Engineering*, 133, 91–100. doi:10.1016/j.compositesb.2017.09.019

**68** Li, F., Gao, Z., Li, L., & Chen, Y. (2016). Microstructural study of MMC layers produced by combining wire and coaxial WC powder feeding in laser direct metal deposition. *Optics and Laser Technology*, 77, 134–143. doi:10.1016/j.optlastec.2015.09.018

**69** Hofmann, D. C., Roberts, S., Otis, R., Kolodziejska, J., Dillon, R. P., Suh, J.-O., … Borgonia, J.-P. (2014). Developing gradient metal alloys through radial deposition additive manufacturing. *Scientific Reports*, 4, 5357-1–5357-8. doi:10.1038/srep05357

**70** Heer, B., & Bandyopadhyay, A. (2018). Compositionally graded magnetic-nonmagnetic bimetallic structure using laser engineered net shaping. *Materials Letters*, 216, 16–19. doi:10.1016/j.matlet.2017.12.129

**71** Akinlabi, E. T., & Akinlabi, S. A. (2014). Friction stir welding of dissimilar metals. In M.-K. Besharati-Givi & P. Asadi (Eds.), *Advances in Friction-Stir Welding and Processing*. Cambridge: Woodhead Publishing. doi:10.1533/9780857094551.241

**72** Domack, M. S., & Baughman, J. M. (2005). Development of nickel-titanium graded composition components. *Rapid Prototyping Journal*, 11(1), 41–51. doi:10.1108/13552540510573383

**73** Dilip, J. J. S., & Ram, G. D. J. (2013). Microstructure evolution in aluminum alloy AA 2014 during multi-layer friction deposition. *Materials Characterization*, 86, 146–151.

**74** Yin, S., Cavaliere, P., Aldwell, B., Jenkins, R., Liao, H., Li, W., & Lupoi, R. (2018). Cold spray additive manufacturing and repair: Fundamentals and applications. *Additive Manufacturing*, 21, 628–650.

**75** Yin, S., Yan, X., Chen, C., Jenkins, R., Liu, M., & Lupoi, R. (2018). Hybrid additive manufacturing of Al-Ti6Al4V functionally graded materials with selective laser melting and cold spraying. *Journal of Materials Processing Technology*, 255, 650–655. doi:10.1016/j.jmatprotec.2018.01.015

**76** Nadimpalli, V. K., & Nagy, P. B. (2018). Designing an in-situ ultrasonic nondestructive evaluation system for ultrasonic additive manufacturing. *AIP Conference Proceedings*, 1949, 020005-1–020005-9. doi:10.1063/1.5031502

**77** Nadimpalli, V. K., Yang, L., & Nagy, P. B. (2018). In-situ interfacial quality assessment of Ultrasonic Additive Manufacturing components using ultrasonic NDE. *NDT and E International*, 93, 117–130. doi:10.1016/j.ndteint.2017.10.004

**78** Sridharan, N., Wolcott, P., Dapino, M., & Babu, S. S. S. S. (2017). Microstructure and mechanical property characterisation of aluminium–steel joints fabricated using ultrasonic additive manufacturing. *Science and Technology of Welding and Joining*, 22(5), 373–380. doi:10.1080/13621718.2016.1249644

**79** Wolcott, P. J. J., Sridharan, N., Babu, S. S. S., Miriyev, A., Frage, N., & Dapino, M. J. J. (2016). Characterisation of Al–Ti dissimilar material joints fabricated using ultrasonic additive manufacturing. *Science and Technology of Welding and Joining*, 21(2), 114–123. doi:10.1179/1362171815Y.0000000072

**80** Stucker, B. E., Obielodan, J. O., Ceylan, A., & Murr, L. E. (2010). Multi-material bonding in ultrasonic consolidation. *Rapid Prototyping Journal*, 16(3), 180–188. doi:10.1108/13552541011034843

**81** Kumar, S., & Kruth, J.-P. (2010). Composites by rapid prototyping technology. *Materials and Design*, 31(2), 850–856. doi:10.1016/j.matdes.2009.07.045

**82** Guo, H., Gingerich, M. B., Headings, L. M., Hahnlen, R., & Dapino, M. J. (2019). Joining of carbon fiber and aluminum using ultrasonic additive manufacturing (UAM). *Composite Structures*, 208, 180–188. doi:10.1016/j.compstruct.2018.10.004

**83** Yang, Y., Janaki Ram, G. D. D., & Stucker, B. E. E. (2009). Bond formation and fiber embedment during ultrasonic consolidation. *Journal of Materials Processing Technology*, 209(10), 4915–4924. doi:10.1016/j.jmatprotec.2009.01.014

**84** Obielodan, J., & Stucker, B. (2014). A fabrication methodology for dual-material engineering structures using ultrasonic additive manufacturing. *International Journal of Advanced Manufacturing Technology*, 70(1–4), 277–284. doi:10.1007/s00170-013-5266-5

**85** Dapino, M. J. (2014). Smart structure integration through ultrasonic additive manufacturing. ASME 2014 Conference on Smart Materials, Adaptive Structures and Intelligent Systems, SMASIS 2014, 2. 10.1115/SMASIS20147710.

**86** Hehr, A., Wenning, J., Terrani, K., Babu, S. S., & Norfolk, M. (2017). Five-axis ultrasonic additive manufacturing for nuclear component manufacture. *JOM*, 69(3), 485–490. doi:10.1007/s11837-016-2205-6

**87** Petrie, C. M. C. M., Sridharan, N., Subramanian, M., Hehr, A., Norfolk, M., & Sheridan, J. (2019). Embedded metallized optical fibers for high temperature applications. *Smart Materials and Structures*, 28(5), 055012-1–055012-33. doi:10.1088/1361-665X/ab0b4c

**88** Bournias-Varotsis, A., Friel, R. J., Harris, R. A., & Engstrøm, D. S. (2018). Ultrasonic Additive Manufacturing as a form-then-bond process for embedding electronic circuitry into a metal matrix. *Journal of Manufacturing Processes*, 32, 664–675. doi:10.1016/j.jmapro.2018.03.027

**89** Sriraman, M. R., Babu, S. S., & Short, M. (2010). Bonding characteristics during very high power ultrasonic additive manufacturing of copper. *Scripta Materialia*, 62(8), 560–563. doi:10.1016/j.scriptamat.2009.12.040

**90** Janaki Ram, G. D., Yang, Y., & Stucker, B. E. (2006). Effect of process parameters on bond formation during ultrasonic consolidation of aluminum alloy 3003. *Journal of Manufacturing Systems*, 25(3), 221–238. doi:10.1016/S0278-6125(07)80011-2

**91** Fabrisonic. (n.d.). Embedding sensors and electronics. Retrieved from https://fabrisonic.com/applications/

**92** Gonzalez-Gutierrez, J., Cano, S., Schuschnigg, S., Kukla, C., Sapkota, J., & Holzer, C. (2018). Additive manufacturing of metallic and ceramic components by the material extrusion of highly-filled polymers: A review and future perspectives. *Materials*, 11(5), 840-1–840-36. doi:10.3390/ma11050840

**93** Pedersen, D. B., Andersen, S. A., & Hansen, H. N. (2019). *Measurements in Additive Manufacturing* (pp. 369–397). USA: Springer. doi:10.1007/978-981-10-4938-5_13

**94** Holo Additive Manufacturing. (n.d.). PureForm Technology. Retrieved from https://holoam.com/technology/

**95** Salcedo, E., Baek, D., Berndt, A., & Ryu, J. E. (2018). Simulation and validation of three dimension functionally graded materials by material jetting. *Additive Manufacturing*, 22, 351–359. doi:10.1016/j.addma.2018.05.027

**96** Sufiiarov, V., Polozov, I., Kantykov, A., & Khaidorov, A. (2020). Binder jetting additive manufacturing of 420 stainless steel: Densification during sintering and effect of heat treatment on microstructure and hardness. Materials Today: Proceedings.

**97** Thompson, Y., Gonzalez-Gutierrez, J., Kukla, C., & Felfer, P. (2019). Fused filament fabrication, debinding and sintering as a low cost additive manufacturing method of 316L stainless steel. *Additive Manufacturing*, 30, 100861.

**98** Larsen, U. D., Signund, O., & Bouwsta, S. (1997). Design and fabrication of compliant micromechanisms and structures with negative Poisson's ratio. *Journal of Microelectromechanical Systems*, 6(2), 99–106.

**99** Takezawa, A., Kobashi, M., & Kitamura, M. (2015). Porous composite with negative thermal expansion obtained by photopolymer additive manufacturing. *APL Materials*, 3(7), 76103.

**100** Andersen, P. R., Henríquez, V. C., & Aage, N. (2019). Shape optimization of micro-acoustic devices including viscous and thermal losses. *Journal of Sound and Vibration*, 447, 120–136.

**101** Wu, J., Aage, N., Westermann, R., & Sigmund, O. (2017). Infill optimization for additive manufacturing: Approaching bone-like porous structures. *IEEE Transactions on Visualization and Computer Graphics*, 24(2), 1127–1140.

**102** Martin, J. J., Fiore, B. E., & Erb, R. M. (2015). Designing bioinspired composite reinforcement architectures via 3D magnetic printing. *Nature Communications*, 6, 8641-1–8641-7. doi:10.1038/ncomms9641

**103** Abel, J., Scheithauer, U., Janics, T., Hampel, S., Cano, S., Müller-Köhn, A., … Moritz, T. (2019). Fused Filament Fabrication (FFF) of metal-ceramic components. *JoVE (Journal of Visualized Experiments)*, 143, e57693.

# 2

# Additive Manufacturing of Functional Ceramics

*José Fernando Valera-Jiménez[1], Juan Ramón Marín-Rueda[1,2], Juan Carlos Pérez-Flores[1], Miguel Castro-García[1], and Jesús Canales-Vázquez[1]*

[1] *3D-ENERMAT, Materials for Energy & 3D Printing Lab, Renewable Energy Research Institute, Universidad de Castilla-La Mancha, Albacete, Spain*
[2] *Print3D Solutions, Albacete, Spain*

## 2.1 Introduction

Until the second part of the twentieth century, ceramics were residually used in Industry and there were very few applications beyond porcelains, clays, glass, bricks, and abrasives. Over the last decades, this field has undergone phenomenal progress and expanded to a vast number of applications ranging from electronics or semiconductors to biomaterials, including energy production and storage devices, filtration, magnetism, etc., and alternatives to traditional niches such as high temperature structural applications.

One of the key issues behind such exponential growth is the development of advanced processing techniques allowing the production of high quality ceramics controlling both the geometry and the microstructure. Nevertheless physical properties such as hardness, wearing resistance, mechanical strength, high temperature durability also imply complicated processing routes and as a consequence shaping ceramics, more specifically advanced functional ceramics, has become a very relevant issue.

Undoubtedly, processing and available fabrication techniques have been one of the most relevant limiting factors to widespread ceramics and find niches in technical applications and although techniques such as tape casting, injection molding, CVD, spray coating, screen-printing have facilitated enormously the use

*3D Printing for Energy Applications*, First Edition. Edited by Albert Tarancón and Vincenzo Esposito.

of ceramics outside the traditional applications mentioned above, there are still difficulties in recreating complex geometrics and/or reduced number of replicas.

As happens in the case of polymers or metals, additive manufacturing may well address those issues and consequently this has become a hot research topic and a growing number of applications are ongoing. Since the first patents back in the 1980s additive manufacturing has been mostly confined to laboratories, which has prevented a widespread use until the very last 5–10 years. A good example of such constraints is the case of stereolithography. The first filed patent dates back to 1984 by C.W. Hull, which required the use of rather costly systems c. $30,000 [1]. Until the first decade of the 21st century, the cost of these high precision systems was in the range of tens of thousand dollars, whereas nowadays some commercial kits are readily available on the market for a few hundred dollars due to the expiry of patent protection periods. Such accessibility is clearly behind the spectacular growth in the number of users, not restricted to specialists anymore, and also potential applications. A clear indication of this can be found when analyzing the number of research articles in specialized journals. Materials and 3D printing are clearly becoming a hot topic and the same trend can be observed in the case of Ceramics and 3D Printing (Figure 2.1).

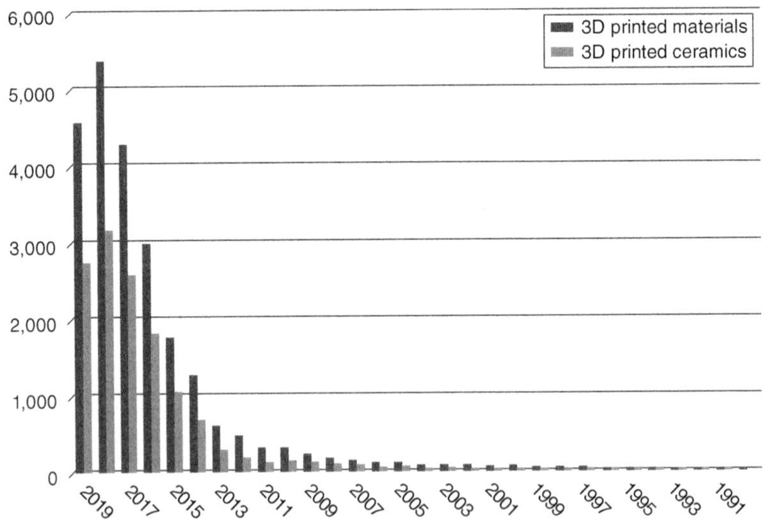


**Figure 2.1** Evolution in number of research articles in materials 3D printing and, particularly, ceramics. Search terms: "3D printing" or "additive manufacturing" for 3D printed materials and "3D printing" or "additive manufacturing" and "ceramic" for 3D printed ceramics. Search categories: "Materials Science Multidisciplinary," "Engineering Manufacturing," "Engineering Mechanical," "Engineering Electrical Electronic," "Physics Applied," "Engineering Biomedical," "Nanoscience Nanotechnology," "Materials Science Biomaterials," "Chemistry Multidisciplinary," and "Polymer Science." Search date: November 4, 2019. *Source:* Based on Materials and 3D printing are clearly becoming a hot topic and the same trend can be observed in the case of Ceramics & 3D Printing. https://3dprintingindustry.com/news/evolution-3d-printing-past-present-future-90605/.

### 2.1.1 Why 3D Printing of Technical Ceramics?

From a mere economic point of view, analyzing the market forecasts for both technical ceramics and additive manufacturing points to an emerging field that can be considered as a business opportunity for investors. Indeed, in 2018, the global advanced ceramics market size was valued at USD over 70 billion and is expected to exhibit a compound annual growth rate (CAGR) of 9.1% during the 2019–2024 period. Such high expectations are a consequence of the rising demand from various industries such as heavy machinery, electronics, energy, automotive, cutting tools, and defense [2]. In Europe, the production value exceeds €30 billion, though 8% corresponds to technical ceramics.

On the other hand, additive manufacturing is undergoing remarkable progress in recent times and there are several reports analyzing the ceramics additive manufacturing markets for the 2017–2028 period (Figure 2.2), which conclude that the ceramics 3D printing market is expected to generate overall revenues for over USD 3.6 billion, driven by CAGRs over 25% in end-use part production. Such high expectations are largely due to the development of both 3D printing technologies adapted to ceramics and an ever growing portfolio of available materials.

### 2.1.2 Materials and Applications

Availability of materials devoted to additive manufacturing technologies shows a parallel development to the recent progress of these technologies. Although initial

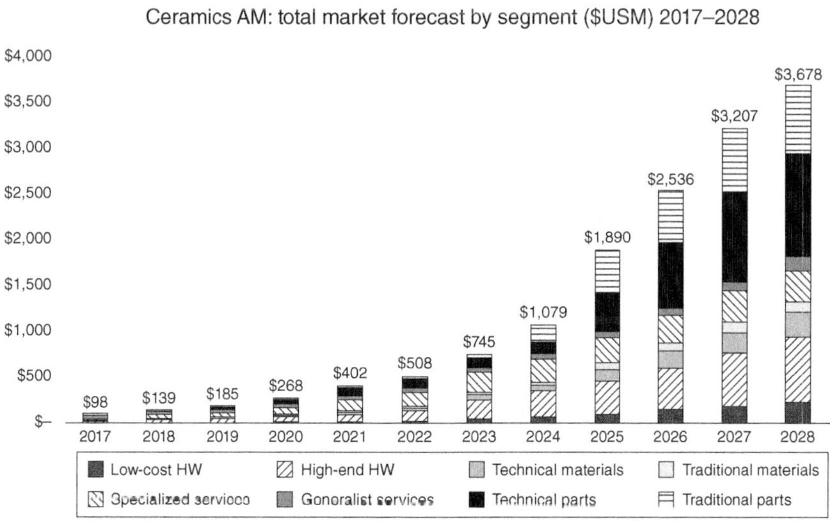


**Figure 2.2** Total market forecast for ceramics additive manufacturing for the 2017–2028 period. Figures are in $USM. *Source:* From Ref. [3]. © 2020 SmarTech Market Publishing LLC.

applications of 3D printing were restricted mostly to polymers and metals/alloys, ceramics have also become relevant in this area.

This is a key issue as metals and, particularly, ceramics may benefit from the implementation of additive manufacturing as complex geometries could be produced, resulting in competitive advantages, which can be considered as "unfair" as the state of the art processing techniques cannot produce them. This is perhaps the key issue behind the use of additive manufacturing technologies for the production of functional ceramics: the possibility of producing geometries that imply an extra value that cannot be achieved by the state of the art, despite the individual properties of the materials may be lower in 3D printed objects compared to bulk properties.

In the case of metals and alloys, this may not appear as a decisive argument toward the implementation in the large scale of 3D Printing since subtractive machining may be used to produce a wide range of geometries in a relatively large number of alloys. One may argue that there are still situations where subtractive techniques are not adequate or whether additive manufacturing implies a more rational use of materials, minimizing waste. In the case of ceramics, the arguments favoring additive manufacturing are even stronger as the combination of hardness and brittleness impedes an extensive use of subtractive manufacturing and therefore 3D printing appears as a viable alternative to expand and improve the applications of technical ceramics in fields ranging from Biomedicine (from the development of dentures/dental covers to bone implants), devices for generation and storage of energy, such as batteries, electrolysers or solid oxide fuel cells, and novel catalysts with application in the production of biodiesel or photoinduced removal of organic contaminants. In most of these applications, high aspect ratio structures are required and conventional ceramic processing techniques are rather restrictive in the geometry choice and 3D printing may help to produce complex geometries in the search of higher efficiencies. The case of Biomedicine is even clearer as there exists a combination of needs for customization and very complex geometries. Similarly, in the case of electroceramics, 3D printed devices and components fabricated with zirconia or alumina may result advantageous in microwave and radiofrequency (RF) applications [4]. Thus, functional ceramics can be used to create a new generation of devices and additive manufacturing may provide means to produce ceramic-based devices with non-conventional geometries that offer those unfair competitive advantages.

## 2.2 Ceramics 3D Printing Technologies

The ISO/ASTM 529000:2015 defines seven additive manufacturing technologies according to the layer fabrication method, that is, Extrusion, Material Jetting, Binder Jetting, Photopolymerization, Powder Bed Fusion, Sheet Lamination, and Directed Energy Deposition.

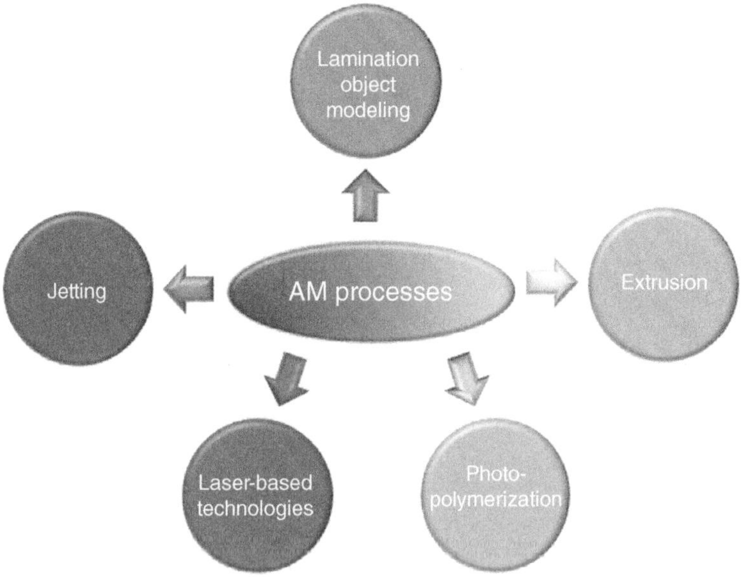


**Figure 2.3** Additive manufacturing technologies as a function of the physical principle.

In this chapter, 3D printing technologies have been classified as a function of the physical principle behind them (e.g., Binder and Material Jetting are both based in spraying/jetting through a printhead). According to this, we have identified five 3D printing technologies that can be used for the production of technical ceramics: Extrusion-based, Photopolimerization, Laser-based, Jetting, and Lamination Object Modeling (LOM) (Figure 2.3).

### 2.2.1 Lamination Object Modeling (LOM)

LOM can be considered as an evolution of tape-casting processes, which consists on stacking ceramic tapes and then laser cutting them to the targeted geometry (Figure 2.4) as first demonstrated in ceramics by Griffith and co-workers [5, 6]. A variation of LOM, computer aided manufacturing of laminated engineering materials (CAM-LEM), considers the opposite approach, that is, cutting and then stacking [7, 8].

LOM was originally used in the production of metal and ceramic parts [9], and despite a certain limitation in the geometries that can be produced, this technology has been extensively used to produce 3D-structured components over the last 20 years [10–12]. LOM has been applied mostly in the field of structural ceramics, particularly when processing reinforced ceramic matrix composites, such as

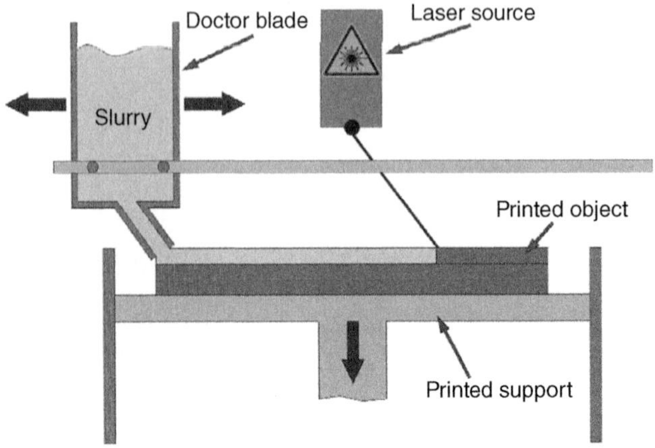


**Figure 2.4** Schematic representation of a LOM system.

SiC [8, 13, 14], $Si_3N_4$ [15], and $Ti_3SiC_2$ [16]. Nevertheless, as a consequence of the potential layer-wise integration of further materials and the opportunity for embedding different functionalities, LOM may be used in the production of heaters and sensors among others. Indeed, LOM has been considered for the fabrication of telescoping actuators [17–19]; however, some other additive manufacturing technologies are more adequate for such functionalities [20]. A similar case can be found for customized bone implants, which were produced by LOM using hydroxyapatites [21], but cannot compete with other technologies as discussed below. In all cases, LOM demonstrates an enormous potential to produce relatively simple geometries, exhibiting high densities and low shrinkage rates compared to the green state, though as the general tendency in functional materials is towards miniaturization and high aspect ratio structures, LOM is not usually considered in this field. The potential application of LOM for functional ceramics will be rather restricted compared to the rest of additive manufacturing technologies described next.

### 2.2.2 Ceramics Extrusion

Extrusion is one of the most used technologies for ceramics shaping. A mixture of ceramics with organics/polymers is forced through an opening, which defines the cross-section of an elongated material. This apparently simple principle can be used to further fabricate complex geometries using different additive manufacturing technologies via direct extrusion (i.e., robocasting/direct ink writing) or producing filaments that can be used in Fused Deposition Modelling (FDM) systems (Figure 2.5).

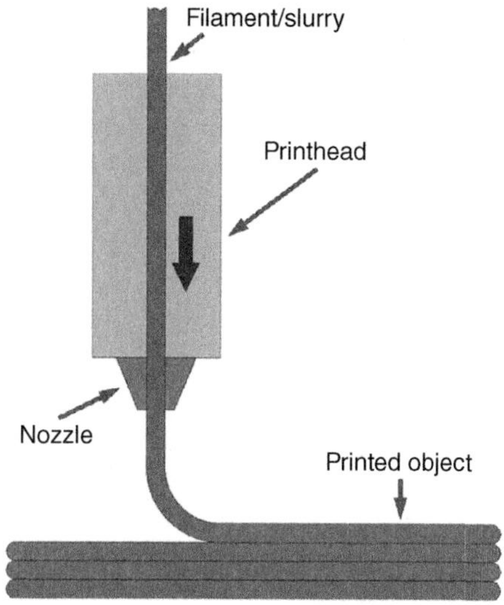


**Figure 2.5** Schematic representation of extrusion-based systems.

#### 2.2.2.1 Robocasting/Direct Ink Writing

The production of ceramics via robocasting or direct ink writing (DIW) is undoubtedly the most common choice of extrusion-based technologies and there are a large number of applications described in the literature ranging from buildings to microdevices for energy generation/storage. The reason behind such a wide variety of applications is that this technology is a simple (in theory) and direct evolution of conventional extrusion, where the extrusion head (nozzle) is coupled to an *xyz* stage that controls the movement of the system in the three axes and therefore offering the opportunity of creating 3D objects with complex geometries. Moreover, the 3D printing material is an ink or slurry containing a high load of the ceramic [22–27] that passes through the printhead and consolidates/cures rapidly to produce a green body that after an adequate thermal treatment leads to the ceramic object. Obviously, this is perhaps the most developed additive manufacturing technique in ceramics as the process is simpler and quicker than other technologies such is the case of photopolymerization, which requires the use of photosensitive resins (restricted choice) that in some cases are rather toxic. Furthermore and as will be explained later, the production of stable photosensitive resins with a high load of ceramic is far from being trivial and there are a number of parameters related to energy dispersion and/or absorption that must be taken into account. Among the limitations of robocasting, it should be noted that surface finish and lateral resolution are rather poor compared to photopolymerization techniques and also some geometries cannot be afforded when depositing ceramic slurries. In any case, hybridization of robocasting and stereolithography (SLA) may allow thc production of more complex geometries and this has already been applied in fields such as catalysis [28]. A further relevant issue regarding this technique is the preparation of stable slurries, particularly in the long-term, which in turn affects largely storage [29].

Generally speaking, the ceramic ink must exhibit some requirements [30]: (a) reversible shear-thinning behavior to viscosity in the 10–100 Pa s range at high shear rate to facilitate material extrusion and retention of geometries, (b) perfect particle dispersion, avoiding the formation of agglomerates that may collapse the nozzle, (c) High modulus ($G'$) with $\tau_y > 200$ Pa to allow structural self-support and fabrication of high aspect ratio structures.

A vast array of ceramics exhibiting complex geometries and controlled porosity have been fabricated using direct extrusion techniques with applications ranging from filters [31] and thermal barriers [32, 33] to scaffolds for tissue engineering [34–37], devices for energy generation and storage [38–48], electronics components [49–52], or monoliths for catalysis [53–55]. The development of piezoelectrics was among the first target applications for robocasting, and in 2001 Tuttle et al. [49] were able to produce PZT monoliths that could withstand repeated cycles of saturated polarization switching under fields of 30 kV/cm and also poled ferroelectric-to-antiferroelectric phase transformations. As the production of printable inks is perhaps the key issue in this technology, once the proof of concept has been demonstrated, this could be extended to some other materials and hence explore novel functionalities and applications, such as relaxor ferroelectrics like lead magnesium niobate (PMN) or lead nickel niobate (PNN), commonly used in the fabrication of multilayer ceramic capacitors, actuators, transducers, and motors [56, 57]. Due to environmental concerns, the development of lead-free piezoelectrics have become a major target and therefore, it was a question of time that additive manufacturing could be considered for fabricating freeform monoliths of other ceramics. Indeed, dense bulk $BaTiO_3$ has been produced using polyvinylidene fluoride (PVDF) and *N*,*N*-Dimethylformamide (DMF) solutions, in the search for piezoelectric and dielectric properties in commercial applications. Despite the rather simple structures proven and further optimization of the process required [58], high performing piezoelectric and dielectric properties, 200/pCN and 4,730 at 103 Hz respectively, were reported, demonstrating the potential commercial application for sensor and energy fields, such as automobile, biomedical, and aerospace [59]. Further work on the production of dielectrics, such as alumina platelets for MEMS applications [60] or in some other lead-free piezoelectrics, has been reported [61], which constitutes a clear example of the potential use of extrusion-based techniques in this area.

ZnO, which finds application in electronic devices due to its piezoelectric, pyroelectric, and optical properties, has also been prepared via robocasting to demonstrate the optoelectronic functionality of printed alumina-doped ZnO (AZO) patterns using their photosensitivity properties upon different illumination conditions [62].

Some other electroceramics have been produced via this technology, including the superconducting YBCO [63] or ferrite-based ($NiFe_2O_4$ and $BaFe_{12}O_{19}$) soft and hard magnetic structures, with densities of approximately 95% after sintering above 1,200 °C and exhibiting excellent magnetic properties with saturation magnetization as high as 48.39 and 64.94 emu/g for the $NiFe_2O_4$ and $BaFe_{12}O_{19}$ samples at room temperature [64].

Robocasting has also been considered for energy-related applications such is the case of electrodes for batteries and supercapacitors [39, 65] or transition metal dichalcogenides [66] that exhibit superior capacitance and energy density in proof of concept devices as compared to the state of the art microsupercapacitors, and enhanced performance compared to hybrid supercapacitor-battery devices. Nevertheless, perhaps one of the most relevant developments on this technology applied to energy storage is the development of an entire Li-microbattery based on $La_4Ti_5O_{14}$ and $LiFePO_4$ electrodes, delivering up to 1.5 mA h/cm$^2$ at a stable 1.8 V at discharge rates below 5C [38]. Although the electrochemical performance was somewhat low compared to conventional batteries, this work became a clear proof of concept of the potential use of robocasting/direct ink writing for the production of energy storage devices and, in recent times, a growing number of reports have appeared in the literature [42, 44, 48, 67], including some references to other devices such as Solid Oxide Fuel Cells (SOFCs) [46, 68].

After the number of applications described above, one may believe the technology is mature and ready for extensive use. However, there is still a lot of work on-going to produce freeform functional ceramics exhibiting the best properties. For instance, inks do not depend exclusively on having an adequate mixture of solvents (organic or aqueous), plasticizers, and dispersants to meet the rheological conditions for robocasting. Morphology and particle size distribution of the ceramic also play an important role in the properties of the sintered 3D printed object [69]. A further example of how important the raw material is can be found in the production of 3D printed structures of dielectrics at room temperature [70]. Lithium molybdate is a non-toxic dielectric, which has been considered for applications, such as microwaves, battery electrode, corrosion inhibition, and moisture sensor. As this compound is water soluble and exhibits a fairly low sintering temperature, it is suitable for robocasting with no need of sintering process. The dielectric properties (both dielectric permittivity and dielectric loss) are extraordinarily close to those measured for lithium molybdate antennas fabricated by room temperature fabrication at 540 °C [70].

At this point, it must be highlighted that a number of the systems described in the literature refers to composites rather than just ceramics. This is the case of the development of piezoelectric inks of 10 wt% barium titanate in a polyvinylidene fluoride (PVDF) matrix [71] or the co-extrusion of such inks with Ag inks to produce piezoelectric sensors in just one step with resolutions down to the millimetre range, which are particularly interesting for wearables [72].

#### 2.2.2.2 Fused Deposition of Ceramics

Fused deposition modeling is perhaps the most commonly used 3D printing technology worldwide due to a combination of low cost, large-size capabilities, simple use (does not require special infrastructures, desktop systems, and possibility of DIY systems), and a relatively large range of commercially available thermoplastic materials. It was first patented by Scott Crump [73] and soon after exploited by Stratasys.

When analyzing the different available technologies to produce ceramics and metal alloys by additive manufacturing, fused deposition modeling is perhaps a non-conventional choice, as these materials (apart from glasses) do not exhibit vitreous transitions and therefore modeling via softening of the raw materials upon heating cannot be performed. Consequently, ceramics must be mixed within a polymer matrix exhibiting thermoplastic properties to produce adequate filaments for FDM, which after a thermal treatment to remove the organics, yields sintered bodies. This is not trivial, as the resulting filament must exhibit a combination of elasticity and hardness in order to be extruded through the printer nozzle. In addition, the debinding process must be controlled to avoid the generation of cracks and other defects in the final ceramic. In the literature, there exist some few examples of the so-called Fused Deposition of Ceramics (FDC), most of them from the joint team including the University of Rutgers (Prof. Danforth and Prof. Safari) and Washington State University (Prof. Bandyopadhyay) in collaboration with Stratasys [74–77]. This research group developed a polymer formulation based on four components: base binder, tackifier, wax, and plasticizer [76]. This polymer could be mixed with up to 60%v/v of ceramic load to produce high quality filaments that could be fed onto a 1650 Modeller Stratasys 3D printer. Using a specially modified printhead, and after debinding and sintering, they were able to produce high quality functional freeform ceramics of alumina, mullite, fused silica, silicon nitride, apatite, titania, PZT, PNM for various applications [74–79], mostly related to dielectrics. The actual composition of the binder has not been published, though the authors mention the use of a combination of polyolefines as base binder with relatively high softening temperatures. Although the so-3D printed objects produced exhibit very high quality and acceptable finish surface, the potential application or upgrade for mass-scale implementation of this technique appears somehow limited due to the relatively high temperatures required at the 3D printers hot-end, as this may impede the use of conventional low-cost 3D printing kits. Indeed, FDC is based on the extrusion of semisolid slurries through a nozzle and most of the conventional ("non-industrial") systems cannot operate beyond 250 °C. Additionally, most of these examples require the use of sintering aids or fillers to facilitate the production of highly dense ceramics. In any case, as ceramic processing cannot be considered for individual implementation (not a field for home-made devices), industrial production centers may not be limited for such constraints.

On the other hand, care must be taken when developing new materials as powder chemical composition and particle size distribution may affect largely the preparation of the green bodies. Depending on the ceramic to be printed, the organic base and the dispersants must be adjusted/changed for each composition. In other words, every ceramic system requires a thorough preliminary study to produce high quality filaments for FDM processing and this is a similar situation as that described for other extrusion-based technologies such as robocasting/direct ink writing. Despite these limitations, there are still some nice examples in the literature describing the potential use of FDC to process electroceramics with application in photonic bandgap structures or transducers [80–82]. It should be noted that the typical layer thickness in this research from the late 1990s and early 2000s was larger than the state of the art FDM commercial systems operating with thermoplastics such as PLA or ABS, where layer thickness may be well below 100 μm, but the electromagnetic measurements displayed the presence of a bandgap between 17.1 and 23.3 GHz, showing a good agreement with the predicted values.

For over a decade, there have been no reports on the use of this technology for the production of functional ceramics. However, some very recent reports explore the use of FDM in ceramics for aeronautics and also for the generation/storage of energy in devices, such as ceramic fuel cells or solid state batteries [83–85]. According to these research works, it is possible to produce filaments for the FDM processing of any ceramic or metal with the possibility of adjusting the porosity in the ceramics after debinding and sintering, and therefore it is highly likely that more examples of the potential use of this technology for a wide range of application will appear in the literature in the next few years. Indeed, a couple of European companies are developing filaments in a range of ceramic compositions for various applications [86, 87].

So far, only examples of the production of fully ceramic components have been discussed, though perhaps this is a rather residual research line compared to the FDM fabrication of composites containing ceramics. Indeed, in recent times, FDM technology has found an important niche of applications in Biomedicine by using composite materials combining biopolymers and (bio)ceramics, particularly to produce objects in which the geometrical restrictions are not very severe. The use of these biocomposites has become a hot topic in research, considering that the sintering stage is avoided and therefore there exists a wide range of polymer–ceramic combinations that may be used straightaway. Obviously, one of the most relevant applications of FDM-produced bioceramics is the production of scaffolds for bone and/or tissue regeneration [88–92].

Similarly, the fabrication of ceramic-based composites via FDM has also attracted the interest of research groups, which have developed freeform BTO/PVDF sensors exhibiting improved piezoelectric properties compared to the state of the art due to the homogeneous dispersion and alleviation of agglomeration for BTO particles achieved upon the production of the filaments [93]. Very recently, a similar approach has been used for the production of 3D printed functional opto-electronic devices [94].

### 2.2.3 Photopolymerization

This group involves several technologies which consist in the polymerization of a resin via UV light, originally from a laser SLA. SLA is considered as the first 3D printing technology, developed by C.W. Hull [95] in the mid-eighties. The printing process of ceramics may be described as follows: the monomer as liquid resin is mixed up with a catalyst initiator and the ceramic powder (50–80 wt%) and the resulting slurry is introduced in a vat. Then the UV source (laser, LED–LCD, etc.) interacts with the resin causing local polymerization only in the illuminated areas. This process is repeated successively to create the 3D object over a platform as depicted in Figure 2.6. Among the advantages of this process, we could highlight the high lateral resolution and high surface finish and the excellent mechanical properties of the printed objects. Among the disadvantages, the number of available printable ceramics is very restricted, the control upon the final porosity depends largely on the thermal post-processing and the stability of the ceramic

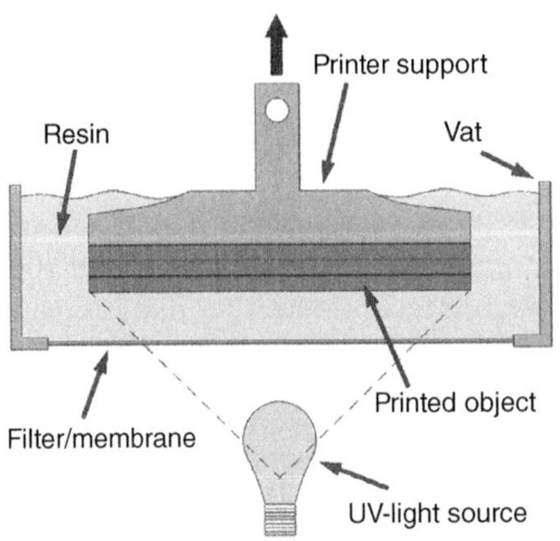


**Figure 2.6** Schematic representation of a photopolymerization-based 3D printing system.

pastes in the long-term and the cost of the printers compared to the simpler extrusion-based systems.

Although the original light source was a laser, in recent times there have appeared alternatives such as the use of LCDs giving rise to the so-called Digital-Light Processing (DLP) technique, which allows projecting UV light layer-by-layer, which may accelerate the fabrication process although the resolution may be compromised.

As already mentioned, ceramics fabrication via photopolymerization involves mixing photosensitive resins with powder and other organic additives. Therefore, a thermal treatment is required to eliminate the organic matter (debinding) and promote sintering [96–98]. Despite the concept may appear as simple, the ceramic slurry must exhibit some requirements such as high solid loading to avoid the structure collapsing upon debinding and sintering, which must be combined with shear thinning behavior, i.e. viscosities typically below 3,000 mPa s, although nowadays it is possible to operate with suspensions exhibiting in the 1–10 of Pa s range at a shear rate of 1,000/s [99]. The introduction of ceramic particles certainly causes an increase in the viscosity. If this exceeds 3,000 mPa s, the slurry may not flow adequately to cover the printed layers. This can be adjusted by adding organic dispersing agents and similar, but too high organic fractions may lead to problems during post-thermal treatment. Related to this, it should be noted that the ink must also show stability to prevent particles precipitation. Morphology and particle size distribution of the ceramic must be precisely controlled due to their impact on light scattering [100], which can make the process far more complicated. A systematic study by Badev et al. [101] revealed that the refractive index ratio between the ceramic particles and the organic content, in addition to the viscosity, are the parameters controlling the polymerization rate (i.e., conversion) in the suspensions. The rate decreases with increasing the index ratio due to light scattering and absorption, which is worse for larger particle sizes. The conducting nature of the ceramic is another important parameter as in the case of working with semiconductors, UV irradiation may be partially absorbed. Finally, temperature is another key parameter as it directly affects the photopolymerization process and during the photocuring process, the resin may be heated up hence the last layers would be different to the first.

The rather complex process leading to high performance ceramic pastes explains the commercial success of companies, such as 3D Ceram, LitHoz, or Prodways, which are specialists in the production of top quality freeform ceramics via SLA/DLP. Nevertheless, there are some important limitations such is the case of the rather narrow choice of commercially available materials: alumina, ZTA, stabilized zirconias, silicon nitride, hydroxyapatites, and few more. This, in addition to the relatively high cost of the printing systems, may limit the access of photopolymerization technologies to application niches that require different materials [102].

Nevertheless, over the last couple of decades, there have been many attempts to develop novel materials for SLA/DLP, which in all cases imply the development of photosensitive resins with a high ceramic loading [103–112] and exhaustive characterization before and after the 3D printing process [113–120]. In many cases, a very popular option is the fabrication of DIY systems mostly based on UV-emitting DLP projectors. Using this home-made approach, several research groups have been able to produce freeform ceramic objects exhibiting quite good quality [102, 121–125], with lateral resolutions in the 50–100 μm range.

In any case, as both the quality of the ceramics and the mechanical properties exceed those of the extrusion-based techniques, the fabrication of functional ceramics via photopolymerization techniques is currently a hot topic in research and the number of potential applications is growing considerably, particularly as the number of available ceramic printable slurries increases. For instance, using a microstereolithography system, it has been possible to fabricate $SiO_2$ photonic crystals for terahertz-wave applications, exhibiting the formation of a common band gap in all directions between 490 and 510 GHz [126, 127]. As happened in the case of extrusion-based techniques, the fabrication of dielectrics via photopolymerization has been explored rather extensively [109, 128, 129], though the production and characterization of functional components is more recent. For example, a piezoelectric ultrasonic transducer based on $BaTiO_3$ has been developed exhibiting a piezoelectric constant and relative permittivity of 160/pCN and 1,350 respectively, with potential application for both sensing and energy harvesting. With the DLP-printed system, the authors performed a 6.28 MHz ultrasonic scan to visualize the structure of a porcine eyeball [130]. A PZT-based ultrasound transducer array was also developed by SLA, showing piezoelectric and dielectric constants of 212–345/pCN and 760–1,390 respectively, which were somewhat lower than those measured on uniaxially pressed disks [131]. Further examples of the potential use of ceramics SLA include the production of alumina photonic crystals for resonators, that is, 3D-electromagnetic bandgap antennas [132–135].

Nevertheless, the reader should note that as discussed in the case of extrusion-based technologies, the field of sensors, actuators and all sorts of dielectrics/piezoelectrics is also strongly driven towards the development of polymer/ceramic composites, see for example [136]. Similar relevant examples appear in the literature regarding the application of photopolymerization techniques to produce sensors [135, 137], heat exchangers [138], and catalytic reactors [139] or SOFC components [102, 123, 140].

In the last few years, due to the advances in microfabrication and nanotechnology there is a strong demand for high resolution 3D printing systems. In this context, the two-photon polymerization (2PP) can be considered as an upgrade of

SLA systems, where the resin absorbs two photons simultaneously (in the near infrared and green) in a very restricted location. This results in the polymerization of submicron volumes within the resin, hence boosting the capacity to reproduce details in the 200 nm range, which is 1–2 orders of magnitude better than conventional photopolymerization systems. This approach has been used to produce nanostructured lattices of (a) $SiO_2$ with the ability to tailor the width of the complete photonic band gap [141], (b) TiN deposited by ALD over a polymer nanolattice and able to withstand up to 1.75 GPa [142], and (c) $Al_2O_3$ [143], exhibiting outstanding optical, mechanical properties or even allowing striking ductile-like fracture in ceramic structures.

### 2.2.4 Laser-Based Technologies

Additive manufacturing can be performed by using a high energy laser that sinters/melts directly the ceramic particles in a powder bed as depicted in Figure 2.7.

Laser-based technologies can be classified according to the physical process used to consolidate the initial ceramic powder. Following this criterion, particles may sinter (Selective Laser Sintering, SLS) or melt (Selective Laser Melting, SLM). Both technologies have undergone limited progress as a result of the very high melting point of most of functional ceramics [118]. SLM is commonly devoted to metals and alloys, although it can be used in ceramics and this technology is often considered as the only additive manufacturing technique capable of producing ready-to-use fully dense ceramics in a single step showing mechanical properties close to those of conventional manufacturing technologies. A limitation related to the production of

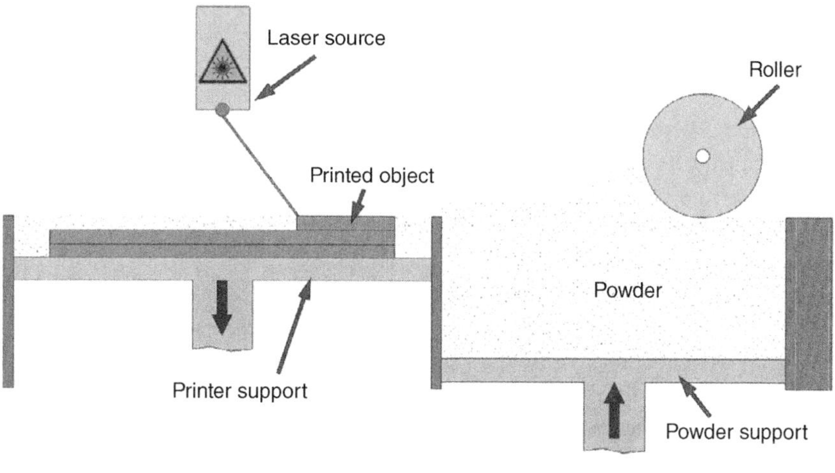


**Figure 2.7** Schematic representation on a laser-powder bed system.

fully dense monoliths via SLM is the likely formation of cracks as consequence of thermal shock, which obviously has a negative impact on the final quality [144]. There are many examples described in the literature reporting the addition of organic fillers to the ceramic powder, hence facilitating particle aggregation and sintering, but few of them focused on technical ceramics and therefore only SLS will be considered for the production of freeform functional ceramics.

As previously mentioned, one of the most important challenges of SLS technology is the production of dense monoliths. This may be achieved by combining SLS with infiltration of liquid phases and/or hot pressing. Using this strategy, there are quite a few examples in the literature reporting high quality ceramic monoliths produced via SLS such as zirconias [145–147], alumina [148–150], Si carbide [151], TiC-$Al_2O_3$ composites [152], $Si_3N_4$ [153, 154], aluminosilicates [155], apatites [156], or phosphates [157–159]. These last have become quite popular as they are biocompatible and the commercial systems have a moderate cost, such is the case of Z-Corp systems in the €30–50k range.

As occurs in most of the additive manufacturing technologies, the number of functional materials investigated under working conditions for device applications is rather limited. In SLS, the laser–ceramic interaction may cause local inhomogeneities (e.g., at the grain boundaries), with a very negative impact on the properties. Nevertheless, dielectrics such as PZT have been successfully fabricated [160]. In most of cases, to achieve full density, it is required to combine SLS with some other manufacturing techniques. Guo et al. [161] used a combination of gel-casting and SLS to produce piezoelectrics exhibiting almost identical properties to those produced by gel-casting and sintering, with special focus on the use of adequate dispersants when conditioning the initial raw powder. Occasionally, laser sintering gives rise to unexpected advantages as described for several lead-free piezoelectrics by Du et al. [162, 163]. 97.5% dense lead-free tungsten bronze $Sr_{1.86}Ca_{0.14}NaNb_5O_{15}$ (SCNN) was prepared successfully by $CO_2$ laser sintering. Compared with specimens conventionally sintered in high temperature furnaces, the dielectric permittivity at room temperature and 1 MHz of SCNN ceramics is enhanced from 1,312 to 1,419, the electromechanical coupling coefficient and piezoelectric strain constants of the laser sintered sample increases significantly from 17 to 27% and 60 to 93 pC/N, respectively. A similar behavior has been observed in other dielectric systems [164, 165] explained in terms of a combination of different crystallographic orientation and somewhat higher densities.

Some other dielectrics have been produced via SLS, exhibiting excellent properties which demonstrate the feasibility of this technology to produce high performance components for electronic applications such as sensors, actuators or ferroelectric memories [166]. Analogously, bismuth germanates have been prepared by SLS for their potential use as scintillator devices for radiation detection [167].

On the other hand, the high energy of the $CO_2$ laser may cause degradation in the response of certain systems as is the case of $CaCu_3Ti_4O_{12}$ (CCTO). A systematic study of the properties of conventionally sintered CCTO versus laser sintered reveals that the high energy may locally affect the oxidation state of Cu and lead to the formation of $TiO_2$ precipitates, hence causing a marked drop in the dielectric permittivity. However, the dissipation values were also reduced in the case of laser sintering and consequently this could be a potential route to fabricate ceramic capacitors [168].

Another important example is the fabrication of $BaTiO_3$ via SLS, which is possible, though the fabrication of thin layers was not good enough for electronic applications [169]. This is a clear indication that there is still plenty of room for further improvement in this technology.

Outside piezo- and ferroelectrics, the examples of SLS-fabricated functional ceramics are scarce. Agarwala et al. [170] combined sol–gel synthesis of $YBa_2Cu_3O_{7-\delta}$ (YBCO) and laser sintering plus Ag infiltration to produce superconductors. The results showed almost no changes in the structure of YBCO compared to conventional sintering, and the $T_c$ onset is at 88–90 K.

### 2.2.5 Jetting

The last group of technologies discussed in the present chapter refers to jetting technologies, usually referred as inkjet. This follows the basic principles of desktop inkjet printers, were the ink passes through nozzles by using a piezoelectric control system that regulates drop formation. As for ceramics, printheads may jet directly inks containing ceramics or, alternatively, they may jet a binder-based ink onto a powder bed. In both cases, the printhead recreates the pattern of the 3D object layer by layer as in other additive manufacturing technologies and after the printing process, it is necessary to perform an adequate thermal treatment for debinding and sintering (Figure 2.8).

Ceramics inkjet has been clearly associated with tile decoration technology as the introduction of this technology in this industrial sector helped to overcome some of the limitations of the traditional screen-printing processing lines, such as higher resolution, capability to reproduce complex patterns using the entire surface of the tile, minimization of tile fracture as this is a non-contact technology, possibility to work with non-flat substrates, wider color palette, optimized use of materials, size reduction of the production systems. Since the commercialization of the first industrial system by Kerajet in the 1990s [171], this industrial sector has undergone an exponential increase and indeed in 2015 the market penetration of this technology was 85% [172, 173].

In other words, this is a well-established technology in the case of ceramics, although initially was mostly oriented toward thin film fabrication. Any ceramic

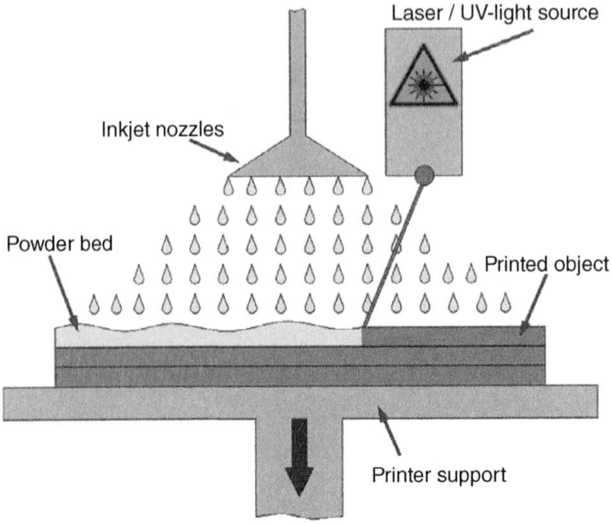


**Figure 2.8** Schematic representation of a jetting-powder bed system.

materials could be produced with this technology a priori, as far as inks containing a high loading of well-dispersed particles are produced. Obviously, particles must be smaller than nozzle diameter (typically bellow 0.5 μm), although some relevant companies like Ferro have started production large particle aqueous inks [174]. Good particle dispersion is also a must, as if the particles agglomerate they would also clog-up the nozzles. The key issues for ceramic inkjet are fine powders [175] and adequate solvent mixture to produce inks with suitable rheological properties. The addition of ceramics clearly modifies them and that is the main reason why binder jetting is preferred as the control over the rheology is much simpler and issues related to particle agglomeration/sedimentation are avoided.

Regarding the rheological properties of the inks, the critical parameters to ensure good quality ink jetting are density ($\rho$), dynamic viscosity ($\eta$), and surface tension ($\gamma$), which can be grouped in one single formula, defined as the Ohnesorge number:

$$\mathrm{oh} = \frac{\sqrt{\mathrm{We}}}{\mathrm{Re}} = \frac{\eta}{(\gamma \rho a)^{1/2}}$$

For 1/Oh values in the 1–10 range, the ink is considered as printable [176–178]. This usually implies that viscosities must be below 30 mPa s [179], which in turn implies that the maximum ceramic loading should not exceed 20–30% in volume. Such low solid contents necessarily result in large shrinkage rates upon sintering and may also lead to the formation of inhomogeneities in the final ceramic after organic matter removal (the so-called coffee stains), which are related to defects and loss of quality [180]. The use of solvent mixtures that mitigate gradients in surface tension during evaporation seems a good strategy to achieve high quality results [181].

As this is one of the most extensively researched additive manufacturing technologies in electroceramics due to the outstanding capacity of producing high quality multi-layer thin-film devices, the number of potential applications and results demonstrating the possibilities of inkjet printing exceed those of the already described 3D printing strategies [182–197].

Apart from the initial reports on inkjet printing of dielectric materials such as $BaTiO_3$ or PZT [198–200], functional devices have also been produced and measured rendering very good performances. Sr-substituted barium titanate has been used to produce tunable capacitors with up to 30% tuning and a loss tangent ($\tan \delta$) less than 0.002 at 1 MHz [201]. PZT-based piezoelectric ceramic thin films for microelectromechanical system (MEMS) applications fabricated via inkjet have also been reported, showing dielectric constant and loss tangent of 23 and 0.0013, respectively, at microwave frequencies, that is, very close to bulk material values. Such combination of high dielectric constant and low loss tangent suggests the potential use of inkjet to fabricate a new generation of microwave devices including antennae, lenses, and filters [202].

Contrary to other additive manufacturing technologies, the number of electroceramics fabricated and tested is much larger. Thus, one of the target applications is the production of thin film capacitors as they can be considered as multilayered devices and are among the essential building blocks for electronic components such as thin-film capacitors and field-effect transistors for transparent electronics. Multiple layers of $Ta_2O_5$ doped with Al and Si exhibiting smooth, flat surface free of defects has been produced via ink-jet, rendering performances comparable with about-three-times-thicker (135 nm) spin-coated films of the same composition [203].

Further examples of transparent electronics include ZnO-based systems [204–206], although compositional changes may have an impact on the rheological properties and hence the microstructure may be negatively affected. This is particularly relevant in the case of thin films as the presence of defects or surface roughness results in performances well below the state of the art [207].

As inkjet technology has been repeatedly demonstrated as valid in this research field, the next stage includes the production of environmentally friendly products. This would include the development of inks, which minimize the production of VOCs, which is the case of aqueous-based inks for optoelectronic applications, for example, Al-oxide phosphate thin layers developed by Meyers et al. [208, 209], although the optoelectronic device was fabricated combining inkjet and sputtering.

Energy and environmental applications are another research field where inkjet-printed functional ceramics may become highly relevant [102], particularly as thin film technologies offer superior performances in these multimaterial devices. Reports on the development of components and devices for energy storage such as Li batteries or SOFCs keep on growing [190, 191, 195, 210–214].

Nevertheless, 3D printing of multimaterial devices must not be restricted to just one technology or one type of materials. One of the most important fabrication advantages of additive manufacturing comes from the possibility of hybridizing both technologies and materials. Examples of materials hybridization may be found for inkjet technology in the development of dielectrics [215] or even fully functional optoelectronic devices where there is only one ceramic component [216]. This is further evidence of the versatility to produce complex multimaterial functional devices and also of the advanced state of development of ink-jet technology.

## References


**1** 3 Printing Industry. The evolution of 3D printing: Past, present and future. Retrieved from https://3dprintingindustry.com/news/evolution-3d-printing-past-present-future-90605/

**2** *Advanced Ceramics Market Size, Share & Trends Analysis Report By Material (Alumina, Titanate), By Product (Monolithic, Coatings), By Application, By End Use, By Region, and Segment Forecasts*, 2019–2025.

**3** SmarTech Markets Publishing LLC. (2018, September). *Ceramics additive manufacturing markets 2017–2028* (ID: 4649506 Report). Crozet, VA: 3DR Holdings.

**4** Oh, Y., Bharambe, V., Mummareddy, B., Martin, J., McKnight, J., Abraham, M. A., … Adams, J. J. (2019). Microwave dielectric properties of zirconia fabricated using NanoParticle Jetting™. *Additive Manufacturing*, 27, 586–594.

**5** Griffin, C., Daufenbach, J., & McMillin, S. (1994). Desktop manufacturing: LOM vs. pressing. *American Ceramic Society Bulletin*, 73(8), 109–113.

**6** Griffin, C., Daufenbach, J., & McMillin, S. (1994). Solid freeform fabrication of functional ceramic components using a laminated object manufacturing technique. *Solid Freeform Fabrication*, 17, 17–24.

**7** Klosterman, D., Chartoff, R., Graves, G., Osborne, N., Lightman, A., Han, G., … Rodrigues, S. (1998). *Automated fabrication of monolithic ceramics and ceramic matrix composites (CMCs) using a novel rapid prototyping method*. Proceedings of the 21st Annual Conference on Composites, Advanced Ceramics, Materials, and Structures. The American Ceramic Society, Cocoa Beach, FL.

**8** Klosterman, D., Chartoff, R., Osborne, N., Graves, G., Lightman, A., Han, G., … Rodrigues, S. (1999). Development of a curved layer LOM process for monolithic ceramics and ceramic matrix composites. *Rapid Prototyping Journal*, 5(2), 61–71.

**9** Agarwala, M., Klosterman, D., Osborne, N., & Lightman, A. (1999). *Hard metal tooling via SFF of ceramics and powder metallurgy*. Solid Freeform Symposium of The University of Texas, Austin, TX.

**10** Williams, C. B. (2008). *Design and development of a layer-based additive manufacturing process for the realization of metal parts of designed mesostructure* (PhD thesis). Georgia Institute of Technology, Atlanta, GA.

**11** Gomes, C. M., Oliveira, A. P. N., Hotza, D., Travitzky, N., & Greil, P. (2008). LZSA glass-ceramic laminates: Fabrication and mechanical properties. *Journal of Materials Processing Technology*, 206, 194–201.

**12** Cui, X., Ouyang, S., Yu, Z., Wang, C., & Huang, Y. (2003). A study on green tapes for LOM with water-based tape casting processing. *Materials Letters*, 57, 1300–1304.

**13** (a) Klosterman, D. A., Chartoff, R., Graves, G., Osborne, N., & Priore, B. (1998). Interfacial characteristics of composites fabricated by laminated object manufacturing. *Composites Part A: Applied Science and Manufacturing*, 29A, 1165–1174.(b) Windsheimer, H., Travitzky, N., Hofenauer, A., & Greil, P. (2007). Laminated object manufacturing of preceramic-paper-derived Si-SiC composites. *Advanced Materials*, 19(24), 4515–4519.

**14** Travitzky, N., Windsheimer, H., Fey, T., & Greil, P. (2008). Preceramic paper-derived ceramics. *Journal of the American Ceramic Society*, 91(11), 3477–3492.

**15** Gomes, C. M., Gutbrod, B., Travitzky, N., Fey, T., & Greil, P. (2010). Preceramic paper derived fibrillar ceramics. *Ceramic Transactions*, 210, 421–426.

**16** Krinitcyn, M., Fu, Z., Harris, J., Kostikov, K., Pribytkov, G. A., Greil, P., & Travitzky, N. (2017). Laminated object manufacturing of in-situ synthesized MAX-phase composites. *Ceramics International*, 43(12), 9241–9245.

**17** Bender, B. A., Rayne, R. J., & Jessen, T. L. (2001). Laminated object manufacturing of functional ceramics. In M. Singh & T. Jessen (Eds.), *Ceramic engineering and science proceedings* (pp. 127–134). Hoboken, NJ: The American Society, Wiley.

**18** Bender, B. A., Rayne, R. J., & Wu, C. C. (2000). Solid freeform fabrication of a telescoping actuator via laminated object manufacturing. *Ceramic Engineering and Science Proceedings*, 21(4), 143–150.

**19** Pan, M.-J., Leung, A., Wu, C., & Bender, B. (2004). Optimizing the performance of telescoping actuators through rapid prototyping and finite element modelling. *Ceramic Transactions*, 150, 53–62.

**20** Xu, Y., Wu, X., Guo, X., Kong, B., Zhang, M., Qian, X., … Sun, W. (2017). The boom in 3D-printed sensor technology. *Sensors*, 17, 1166–1202.

**21** Steidle, C., Klosterman, D., Chartoff, R., Graves, G., & Osborne, N. (1999). *Automated fabrication of custom bone implants using rapid prototyping*. 44th International SAMPE Symposium and Exhibition, 1866–1877 of Society for the Advancement of Material and Process Engineering, Longbeach, CA.

22 Cesarano, J., III, Segalman, R., & Calvert, P. (1998). Robocasting provides moldless fabrication from slurry deposition. *Ceramic Industry*, 148, 94–102.

23 Chrisey, D. B. (2000). The power of direct writing. *Science*, 289, 879–881.

24 Smay, J. E., Cesarano, J., III, & Lewis, J. A. (2002). Colloidal inks for directed assembly of 3-D periodic structures. *Langmuir*, 18, 5429–5437.

25 Smay, J. E., Gratson, G., Shepard, R. F., Cesarano, J., III, & Lewis, J. A. (2002). Directed colloidal assembly of 3D periodic structures. *Advanced Materials*, 14, 1279–1283.

26 Lewis, J. A., & Gratson, G. M. (2004). Direct writing in three dimensions. *Materials Today*, 7, 32–39.

27 Lewis, J. A., Smay, J. E., Stuecker, J., & Cesarano, J. (2006). Direct ink writing of three-dimensional ceramic structures. *Journal of the American Ceramic Society*, 89(12), 3599–3609.

28 de Hazan, Y., Thänert, M., Trunec, M., & Misak, J. (2012). Robotic deposition of 3D nanocomposite and ceramic fiber architectures via UV curable colloidal inks. *Journal of the European Ceramic Society*, 32(6), 1187–1198.

29 Feilden, E. (2017). *Additive manufacturing of ceramics and ceramic composites via robocasting* (PhD thesis). Imperial College, London.

30 Peng, E., Zhang, D., & Ding, J. (2018). Ceramic robocasting: Recent achievements, potential and future developments. *Advanced Materials*, 30(47), 1802404–18002418. doi:10.1002/adma.201802404

31 Stuecker, J. N., Miller, J. E., Ferrizz, R. E., Mudd, J. E., & Cesarano, J. (2004). Advanced support structures for enhanced catalytic activity. *Industrial & Engineering Chemistry Research*, 43(1), 51–55.

32 Cai, K., Román, B., Smay, J. E., Zhou, J., Osendi, M. I., Belmonte, M., & Miranzo, P. (2012). Geometrically complex silicon carbide structures fabricated by robocasting. *Journal of the American Ceramic Society*, 95(8), 2660–2666.

33 Bandyopadhyay, A., & Heer, B. (2018). Additive manufacturing of multi-material structures. *Materials Science and Engineering R: Reports*, 129, 1–16.

34 Michna, S., Wu, W., & Lewis, J. A. (2005). Concentrated hydroxyapatite inks for direct-write assembly of 3-D periodic scaffolds. *Biomaterials*, 26, 5632–5639.

35 Miranda, P., Saiz, E., Gryn, K., & Tomsia, A. P. (2006). Sintering, robocasting of beta-tricalcium phosphate scaffolds for orthopaedic applications. *Acta Biomaterialia*, 2, 457–466.

36 Martínez-Vázquez, F. J., Perera, F. H., Miranda, P., Pajares, A., & Guiberteau, F. (2010). Improving the compressive strength of bioceramic robocast scaffolds by polymer infiltration. *Acta Biomaterialia*, 6, 4361–4368.

37 Shao, H., Liu, A., Ke, X., Sun, M., He, Y., Yang, X., … Gou, Z. (2017). 3D robocasting magnesium-doped wollastonite/TCP bioceramic scaffolds with improved bone regeneration capacity in critical sized calvarial defects. *Journal of Materials Chemistry B*, 5, 2941–2951.

38 Sun, K., Wei, T. S., Ahn, B. Y., Seo, J. Y., Dillon, S. J., & Lewis, J. A. (2013). 3D printing of interdigitated li-ion microbattery architectures. *Advanced Materials*, 25(33), 4539–4543.

39 Fu, K., Wang, Y., Yan, C., Yao, Y., Chen, Y., Dai, J., … Hu, L. (2016). Graphene oxide-based electrode inks for 3D-printed lithium-ion batteries. *Advanced Materials*, 28(13), 2587–2594.

40 McOwen, D. W., Xu, S., Gong, Y., Wen, Y., Godbey, G. L., Gritton, J. E., … Wachsman, E. D. (2018). 3D-Printing electrolytes for solid-state batteries. *Advanced Materials*, 30, e1707132.

41 Zekoll, S., Marriner-Edwards, C., Hekselman, A. K. O., Kasemchainan, J., Kuss, C., Armstrong, D. E. J., … Bruce, P. G. (2018). Hybrid electrolytes with 3D bicontinuous ordered ceramic and polymer microchannels for all-solid-state batteries. *Energy & Environmental Science*, 11, 185.

42 Cheng, M., Jiang, Y., Yao, W., Yuan, Y., Deivanayagam, R., Foroozan, T., … Shahbazian-Yassar, R. (2018). Elevated-temperature 3D printing of hybrid solid-state electrolyte for Li-ion batteries. *Advanced Materials*, 30, 1800615–1800624.

43 Li, J., Liang, X., Liou, F., & Park, J. (2018). Macro-/micro-controlled 3D lithium-ion batteries via additive manufacturing and electric field processing. *Scientific Reports*, 8, 1846.

44 Gao, X., Sun, Q., Yang, X., Liang, J., Koo, A., Lia, W., … Sun, X. (2019). Toward a remarkable Li-S battery via 3D printing. *Nano Energy*, 56, 595–603.

45 Ambrosini, A., & Pumera, M. (2016). 3D-printing technologies for electrochemical applications. *Chemical Society Reviews*, 45, 2740–2755.

46 Kuhn, M., Napporn, T., Meunier, M., Vengallatore, S., & Therriault, D. (2007). Direct-write microfabrication of single-chamber micro solid oxide fuel cells. *Journal of Micromechanics and Microengineering*, 18(1), Article 015005.

47 Liu, C., Cheng, X., Li, B., Chen, Z., Mi, S., & Lao, C. (2017). Fabrication and characterization of 3D-printed highly-porous 3D LiFePO4 electrodes by low temperature direct writing process. *Materials*, 10(8), 934.

48 Wei, T. S., Ahn, B. Y., Grotto, J., & Lewis, J. A. (2018). 3D printing of customized li-ion batteries with thick electrodes. *Advanced Materials*, 30(16), 1703027. doi:10.1002/adma.201703027

49 Tuttle, B. A., Smay, J. E., Cesarano, J., Voigt, J. A., Scofield, T. W., Olson, W. R., & Lewis, J. A. (2001). Robocast Pb(Zr0.95Ti0.05)O3 ceramic monoliths and composites. *Journal of the American Ceramic Society*, 84(4), 872–874.

50 San Marchi, C., Kouzeli, M., Rao, R., Lewis, J. A., & Dunand, D. (2003). Alumina–aluminum interpenetrating-phase composites with three-dimensional periodic architecture. *Scripta Materialia*, 49(9), 861–866.

51 Young, M., Rao, R., Almer, J., Haeffner, D., Lewis, J. A., & Dunand, D. (2009). Load partitioning in Al2O3–Al composites with three-dimensional periodic architecture. *Acta Materialia*, 57(8), 2362–2375.

52 Smay, J. E., Tuttle, B., & Cesarano, J. (2018). Robocasting of three-dimensional piezoelectric structures. In A. Safari & E. K. Ardogan (Eds.), *Piezoelectric and acoustic materials for transducer applications* (pp. 305–318). Springer.

53 Parra-Cabrera, C., Achille, C., Kuhn, S., & Ameloot, R. (2018). 3D printing in chemical engineering and catalytic technology: Structured catalysts, mixers and reactors. *Chemical Society Reviews*, 47, 209–230.

54 Elkoro, A., & Casanova, I. (2018). 3D printing of structured nanotitania catalysts: A novel binder-free and low-temperature chemical sintering method, 3D print. *Additive Manufacturing*, 5, 220–226.

55 Elkoro, A., Soler, L., Llorca, J., & Casanova, I. (2019). 3D printed microstructured Au/TiO2 catalyst for hydrogen photoproduction. *Applied Materials Today*, 16, 265–272.

56 Şakar-Deliormanlı, A., Çelik, E., & Polat, M. (2008). Rheological behaviour of PMN gels for solid freeform fabrication. *Colloids and Surfaces A: Physicochemical and Engineering Aspects*, 326, 159–166.

57 Qiu, J., Tani, J., Yanada, N., Kobayashi, Y., & Takahashi, H. (2004). Fabrication of Pb(Nb, Ni)O3-Pb(Zr, Ti)O3 piezoelectric ceramic fibers by extrusion of a sol-powder mixture. *Journal of Intelligent Material Systems and Structures*, 15(8), 643–653.

58 Renteria, A., Fontes, H., Diaz, J. A., Regis, J. E., Chavez, L. A., Tseng, T.-L., … Lin, Y. (2019). Optimization of 3D printing parameters for BaTiO3 piezoelectric ceramics through design of experiments. *Materials Research Express*, 6(8), 085706. doi:10.1088/2053-1591/ab200e

59 Kim, H., Renteria-Marquez, A., Islam, M. D., Chavez, L. A., Garcia Rosales, C. A., Ahsan, M. A., … Lin, Y. (2019). Fabrication of bulk piezoelectric and dielectric BaTiO3 ceramics using paste extrusion 3D printing technique. *Journal of the American Ceramic Society*, 102(6), 3685–3694.

60 Feilden, E., Giovannini, T., Ferraro, C., Vandeperre, L., Saiz, E., & Giuliani, F. (2017). Micromechanical strength of individual Al2O3 platelets. *Scripta Materialia*, 131, 55–58.

61 Nan, B., Olhero, S., Pinho, R., Vilarinho, P. M., Button, T. W., & Ferreira, J. M. F. (2019). Direct ink writing of macroporous lead-free piezoelectric Ba0.85Ca0.15Zr0.1Ti0.9O3. *Journal of the American Ceramic Society*, 102(6), 3191–3203.

62 Torres-Arango, M. A., Abidakun, O. A., Korakakis, D., & Sierros, K. A. (2017). Tuning the crystalline microstructure of Al-doped ZnO using direct ink writing. *Flexible and Printed Electronics*, 2(3), 035006.

63 Wei, X., Nagarajan, R. S., Peng, E., Xue, J., Wang, J., & Ding, J. (2016). Fabrication of YBa2Cu3O7−x (YBCO) superconductor bulk structures by extrusion freeforming. *Ceramics International*, 42(14), 15836–15842.

64 Peng, E., Wei, X., Herng, T. S., Garbe, U., Yu, D., & Ding, J. (2017). Ferrite-based soft and hard magnetic structures by extrusion free-forming. *RSC Advances*, 7, 27128–27138.

**65** Rocha, V. G., García-Tuñón, E., Markoulidis, F., Feilden, E., D'Elia, E., Ni, N., … Saiz, E. (2017). Multi-material 3D printing of graphene-based electrodes for electrochemical energy storage using thermo-responsive inks. *ACS Applied Materials & Interfaces*, 9(42), 37136–37145.

**66** Grotta, C. (2018). *3D printing of transition metal dichalcogenides* (PhD thesis). Imperial College, London, UK.

**67** Wei, M., Zhang, F., Wang, W., Alexandridis, P., Zhou, C., & Wu, G. (2017). 3D direct writing fabrication of electrodes for electrochemical storage devices. *Journal of Power Sources*, 354, 134–147.

**68** Peng, E., Wei, X., Garbe, U., Yu, D., Brunet, E., Liu, A., & Ding, J. (2018). Robocasting of dense yttria-stabilized zirconia structures. *Journal of Materials Science*, 53(1), 247–273.

**69** Renteria, A., Diaz, J. A., He, B., Renteria-Marquez, I. A., Chavez, L. A., Regis, J. E., … Lin, Y. (2019). Particle size influence on material properties of BaTiO3 ceramics fabricated using freeze-form extrusion 3D printing. *Materials Research Express*, 6(11), 115211. doi:10.1088/2053-1591/ab4a36

**70** Väätäjä, M., Kähäri, H., Ohenoja, K., & Sobocinski, M. (2018). 3D printed dielectric ceramic without a sintering stage. *Scientific Reports*, 8, 15955.

**71** Bodkhe, S., Turcot, G., Gosselin, F. P., & Therriault, D. (2017). One-step solvent evaporation-assisted 3D printing of piezoelectric PVDF nanocomposite structures. *ACS Applied Materials & Interfaces*, 9, 20833–20842.

**72** Bodkhe, S., Noonan, C., Gosselin, F. P., & Therriault, D. (2018). Coextrusion of multifunctional smart sensors. *Advanced Engineering Materials*, 20(10), 1800206.

**73** Crump, S. S. (1989). Apparatus and method for creating three-dimensional objects. *US Patent US5121329A*.

**74** Agarwala, M. K., Bandyopadhyay, A., Weeren, R., Whalen, P., Safari, A., & Danforth, S. C. (1996). Fused deposition of ceramics: Rapid fabrication of structural ceramic components. *Ceramic Bulletin*, 75(11), 60–65.

**75** Agarwala, M. K., Jamalabad, V. R., Langrana, N. A., Safari, A., Whalen, P. J., & Danforth, S. C. (1996). Structural quality of parts processed by fused deposition. *Rapid Prototyping Journal*, 2(4), 4–19.

**76** McNulty, T. F., Mohammadi, F., Bandyopadhyay, A., Danforth, S. C., & Safari, A. (1998). Development of a binder formulation for fused deposition of ceramics (FDC). *Rapid Prototyping Journal*, 4(4), 144–150.

**77** Cheeseman, B., Ruan, X., Safari, A., Danforth, S. C., & Chou, T. (2000). Modeling and optimization of novel actuators produced by solid freeform fabrication. *MRS Proceedings*, 625(159). doi:10.1557/PROC-625-159

**78** Hall, A., Allahverdi, M., Akdogan, E. K., & Safari, A. (2005). Piezoelectric/electrostrictive multimaterial PMN-PT monomorph actuators. *Journal of the European Ceramic Society*, 25, 2991–2997.

**79** Onagoruwa, S., Bose, S., & Bandyopadhyay, A. (2001). *Fused deposition of ceramics (FDC) and composites*. Proceedings of Solid Freeform Fabrication Conference of The University of Austin, TX, 224–231.

**80** Allahverdi, M., Danforth, S. C., Jafari, M., & Safari, A. (2001). Processing of advanced electroceramic components by fused deposition technique. *Journal of the European Ceramic Society*, 21, 1485–1490.

**81** Pilleux, M. E., Safari, A., Allahverdi, M., Chen, Y., Lu, Y., & Jafari, M. A. (2002). 3-D photonic bandgap structures in the microwave regime by fused deposition of multimaterials. *Rapid Prototyping Journal*, 8(1), 46–52.

**82** Chen, Y., Bartzos, D., Lu, Y., Niver, E., Pilleux, M., Allahverdi, M., … Safari, A. (2001). Simulation, fabrication, and characterization of 3-D alumina photonic bandgap structures. *Microwave and Optical Technology Letters*, 30(5), 305–307.

**83** Marín Rueda, J. R., Yagüe-Alcaraz, V., López-López, J. J., Sánchez-Bravo, G., & Canales-Vázquez, J. (2017). Procedure for 3D printing of ceramics via FDM, PCT/ES2017/070202.

**84** Valera-Jiménez, J. F., Pérez-Flores, J. C., Marín Rueda, J. R., Castro-García, M., & Canales-Vázquez, J. (2019). *3D-printing of electrodes for Li-batteries via fused deposition modelling*. Proceedings of XIV Reunión Nacional de Electrocerámica, Castellon, Spain.

**85** Yagüe-Alcaraz, V. (2017). *Fabricación y caracterización de pilas de combustible de óxidos sólidos (SOFC) y componentes mediante modelado por deposición fundida (FDM)* (Master thesis). EII-Albacete, University of Castilla-La Mancha, Albacete, Spain.

**86** nanoe. (2019). Retrieved from www.nanoe.fr

**87** Print3D Solutions. (2019) Especialistas en Impresión 3D y asesoria técnica. Retrieved from www.print3dsolutions.net

**88** Sa, M. W., Nguyen, B. N. B., Moriarty, R. A., Kamalitdinov, T., Fisher, J. P., & Kim, J. Y. (2018). Fabrication and evaluation of 3D printed BCP scaffolds reinforced with ZrO2 for bone tissue applications. *Biotechnology and Bioengineering*, 115(4), 989–999.

**89** Hutmacher, D. W., Schantz, T., Zein, I., Ng, K. W., Teoh, S. H., & Tan, K. C. (2001). Mechanical properties and cell cultural response of polycaprolactone scaffolds designed and fabricated via fused deposition modeling. *Journal of Biomedical Materials Research: An Official Journal of The Society for Biomaterials, The Japanese Society for Biomaterials, and The Australian Society for Biomaterials and the Korean Society for Biomaterials*, 55(2), 203–216.

**90** Cao, T., Ho, K.-H., & Teoh, S.-H. (2003). Scaffold design and in vitro study of osteochondral coculture in a three-dimensional porous polycaprolactone scaffold fabricated by fused deposition modeling. *Tissue Engineering*, 9(4), 103–112.

**91** Tsang, V. L., & Bhatia, S. N. (2004). Three-dimensional tissue fabrication. *Advanced Drug Delivery Reviews*, 56(11), 1635–1647.

**92** Xu, N., Ye, X., Wei, D., Zhong, J., Chen, Y., Xu, G., & He, D. (2014). 3D artificial bones for bone repair prepared by computed tomography-guided fused deposition modeling for bone repair. *ACS Applied Materials & Interfaces*, 6(17), 14952–14963.

**93** Kim, H., Fernando, T., Li, M., Lin, Y., & Tseng, T. B. (2018). Fabrication and characterization of 3D printed BaTiO3/PVDF nanocomposites. *Journal of Composite Materials*, 52(2), 197–206.

**94** Loke, G., Yuan, R., Rein, M., Khudiyev, T., Jain, Y., Joannopoulos, J., & Fink, Y. (2019). Structured multimaterial filaments for 3D printing of optoelectronics. *Nature Communications*, 10, 4010–4020.

**95** Hull, C. W. (1984). Apparatus for production of three-dimensional objects by stereolithography. *US Patent US4575330A*.

**96** Griffith, M. L., & Halloran, J. W. (1996). Freeform fabrication of ceramics via stereolithography. *Rapid Prototyping Journal*, 79(10), 2601–2608.

**97** Chu, T. M., & Halloran, J. W. (2000). Curing of highly loaded ceramics in acrylates. *Journal of the American Ceramic Society*, 83(10), 2075–2080.

**98** Doreau, F., Chaput, C., & Chartier, T. (2000). Stereolithography for manufacturing ceramic parts. *Advanced Engineering Materials*, 2(8), 493–496.

**99** Chen, Z., Li, Z., Li, J., Liu, C., Lao, C., Fu, Y., … He, Y. (2019). 3D printing of ceramics: A review. *Journal of the European Ceramic Society*, 39, 661–687.

**100** Griffith, M. L., & Halloran, J. W. (1997). Scattering of ultravio-let radiation in turbid suspensions. *Journal of Applied Physics*, 81(6), 2538–2546.

**101** Badev, A., Abouliatim, Y., Chartier, T., Lecamp, L., Lebaudy, P., Chaput, C., & Delage, C. (2011). Photopolymerization kinetics of a polyether acrylate in the presence of ceramic fillers used in stereolithography. *Journal of Photochemistry and Photobiology A: Chemistry*, 222(1), 117–122.

**102** Ruiz-Morales, J. C., Tarancón, A., Canales-Vázquez, J., Méndez-Ramos, J., Hernández-Afonso, L., Acosta-Mora, P., … Fernández-González, R. (2017). Three dimensional printing of components and functional devices for energy and environmental applications. *Energy & Environmental Science*, 10, 846–859.

**103** Griffith, M. L., & Halloran, J. W. (1994). Ultraviolet curable ceramic suspensions for stereolithography of ceramics. *Manufacturing Science and Engineering*, 68–72, 529–534.

**104** Griffith, M. L., Chu, T.-M., Wagner, W., & Halloran, J. W. (1995). *Fabrication of ceramic objects by solid freeform fabrication (SFF)*. Solid Freeform Fabrication Conference (SFF'95) of The University of Austin, TX, (pp. 31–38).

**105** Griffith, M. L. (1995). *Stereolithography of ceramics* (PhD thesis). University of Michigan, Ann Arbor, USA.

**106** Hinczewski, C., Corbel, S., & Chartier, T. (1998). Ceramic suspensions suitable for Stereolithography. *Journal of the European Ceramic Society*, 18, 583–590.

**107** Hinczewski, C., Corbel, S., & Chartier, T. (1998). Stereolithography for the fabrication of ceramic three-dimensional parts. *Rapid Prototyping Journal*, 4(3), 104.

**108** Provin, C., Monneret, S., Le Gall, H., & Corbel, S. (2003). Three-dimensional ceramic microcomponents made using microstereolithography. *Advanced Materials*, 15(12), 994–997.

**109** Dufaud, O., Marchal, P., & Corbel, S. (2002). Rheological properties of PZT suspensions for stereolithography. *Journal of the European Ceramic Society*, 22, 2081–2092.

**110** Dufaud, O., & Corbel, S. (2003). Oxygen diffusion in ceramic suspensions for stereolithography. *Chemical Engineering Journal*, 92(1–3), 55–62.

**111** Dufaud, O. (2002). *Prototypage rapide de composites céramiques fonctionnels* (PhD thesis). Institut National Polytechnique de Lorraine, Lorraine.

**112** Dufaud, O., & Corbel, S. (2002). Stereolithography of PZT ceramic suspensions. *Rapid Prototyping Journal*, 8(2), 83–90.

**113** Tomeckova, V., & Halloran, J. (2010). Predictive models for photopolymerization of ceramic suspensions. *Journal of the European Ceramic Society*, 30(14), 2833–2840.

**114** Tomeckova, V., & Halloran, J. (2010). Cure depth for photopolymerization of ceramic suspensions. *Journal of the European Ceramic Society*, 30(15), 3023–3033.

**115** Tomeckova, V., & Halloran, J. (2010). Critical energy for photopolymerization of ceramic suspensions. *Journal of the European Ceramic Society*, 30(16), 3273–3282.

**116** Halloran, J. W., Tomeckova, V., Gentry, S., Das, S., Cilino, P., Yuan, D., … Long, D. (2011). Photopolymerization of powder suspensions for shaping ceramics. *Journal of the European Ceramic Society*, 31(14), 2613–2619.

**117** Bae, C., & Halloran, J. W. (2011). Influence of residual monomer on cracking in ceramics fabricated by stereolithography. *International Journal of Applied Ceramic Technology*, 8(6), 1289–1295.

**118** Travitzky, N., Bonet, A., Dermeik, B., Fey, T., Filbert-Demut, I., Schlier, L., … Greil, P. (2014). Additive manufacturing of ceramic-based materials. *Advanced Engineering Materials*, 16(6), 729–754.

**119** Guo, N., & Leu, M. C. (2013). Additive manufacturing: Technology, applications and research needs. *Frontiers of Mechanical Engineering*, 8(3), 215–243.

**120** Wu, H., Liu, W., He, R., Wu, Z., Jiang, Q., Song, X., … Wu, S. (2017). Fabrication of dense zirconia-toughened alumina ceramics through a stereolithography-based additive manufacturing. *Ceramics International*, 43, 968–972.

**121** Mitteramskogler, G., Gmeiner, R., Felzmann, R., Gruber, S., Hofstetter, C., Stampfl, J., … Laubersheimer, J. (2014). Light curing strategies for lithography-based additive manufacturing of customized ceramics. *Additive Manufacturing*, 1–4, 100–118.

**122** Ferrage, L., Bertrand, G., Lenormand, P., Grossin, D., & Ben-Nissan, B. (2016). A review of the additive manufacturing (3DP) of bio ceramics: Alumina, zirconia (PSZ) and hydroxyapatite. *Journal of the Australian Ceramic Society*, 1, 42–56.

**123** Hernández-Rodríguez, E. M., Acosta-Mora, P., Méndez-Ramos, J., Borges Chinea, E., Esparza Ferrera, P., Canales-Vázquez, J., … Ruiz-Morales, J. C. (2014). Prospective use of the 3D printing technology for the microstructural engineering of solid oxide fuel cell components. *Boletín de la Socieedad Española de Cerámica y Vidrio*, 53(5), 213–216.

**124** Varghese, G., Moral, M., Castro-García, M., López-López, J. J., Marín-Rueda, J. R., Yagüe-Alcaraz, V., … Canales-Vázquez, J. (2018). Fabrication and characterisation of ceramics via low-cost DLP 3D printing. *Boletin de la Sociedad Espanola de Ceramica y Vidrio*, 57(1), 9–18.

**125** Hernández-Afonso, L., Fernández-González, R., Esparza, P., Borges, M. E., Díaz-González, S., Canales-Vázquez, J., & Ruiz-Morales, J. C. (2017). Ceramic-based 3D printed supports for photocatalytic treatment of wastewater. *Journal of Chemistry*, 2017, 7602985. doi:10.1155/2017/7602985

**126** Chen, W., Kirihara, S., & Miyamoto, Y. (2007). Fabrication and measurement of micro three-dimensional photonic crystals of SiO2 ceramic for terahertz wave applications. *Journal of the American Ceramic Society*, 90(7), 2078–2081.

**127** Kirihara, S., & Niki, T. (2015). Three-dimensional stereolithography of alumina photonic crystals for terahertz wave localization. *International Journal of Applied Ceramic Technology*, 12(1), 32–37.

**128** Jang, J. H., Wang, S., Pilgrim, S. M., & Schulze, W. A. (2000). Preparation and characterization of barium titanate suspensions for stereolithography. *Journal of the American Ceramic Society*, 83(7), 1804–1806.

**129** Dufaud, O., & Corbel, S. (2001). Stéréolithographie pour la fabrication de composites céramiques piézoélectriques. *Nano et Micro Technologies*, 1(2), 177–191.

**130** Chen, Z., Song, X., Lei, L., Chen, X., Fei, C., Chiu, C. T., … Zhou, Q. (2016). 3D printing of piezoelectric element for energy focusing and ultrasonic sensing. *Nano Energy*, 27, 78–86.

**131** Chen, Y., Bao, X., Wong, C.-M., Cheng, J., Wu, H., Song, H., … Wu, S. (2018). PZT ceramics fabricated based on stereolithography for an ultrasound transducer array application. *Ceramics International*, 44(18), 22725–22730.

**132** Chartier, T., Dupas, C., Lasgorceix, M., Brie, J., Champion, E., Delhote, N., & Chaput, C. (2015). Additive manufacturing to produce complex 3D ceramic parts. *Journal of Ceramic Science & Technology*, 6(2), 95–104.

**133** Delhote, N., Baillargeat, D., Verdeyme, S., Delage, C., & Chaput, C. (2007). Ceramic layer-by-Layer stereolithography for the manufacturing of 3-D millimeter-wave filters. *IEEE Transactions on Microwave Theory and Techniques*, 55, 548–554.

**134** Brakora, K. F., Halloran, J., & Sarabandi, K. (2007). Design of 3-D monolithic MMW antennas using ceramic stereolithography. *IEEE Transactions on Antennas and Propagation*, 55, 790–797.

**135** Nguyen, N. T., Delhote, N., Ettorre, M., Baillargeat, D., Le Coq, L., & Sauleau, R. (2010). Design and characterization of 60-GHz integrated Lens antennas fabricated through ceramic stereolithography. *IEEE Transactions on Antennas and Propagation*, 58, 2757–2762.

**136** Yang, Y., Chen, Z., Song, X., Zhu, B., Hsiai, T., Wu, P.-I., ... Shung, K. K. (2016). Three dimensional printing of high dielectric capacitor using projection based stereolithography method. *Nano Energy*, 22, 414–421.

**137** Leigh, S. J., Purssell, C., Bowen, J., Hutchins, D. A., Covington, J. A., & Billson, D. (2011). A miniature flow sensor fabricated by micro-stereolithography employing a magnetite/acrylic nanocomposite resin. *Sensors and Actuators A: Physical*, 168(1), 66–71.

**138** Scheithauer, U., Schwarzer, E., Moritz, T., & Michaelis, A. (2018). Additive manufacturing of ceramic heat exchanger: Opportunities and limits of the lithography-based ceramic manufacturing (LCM). *Journal of Materials Engineering and Performance*, 27(1), 14–20.

**139** Scheithauer, U., Schwarzer, E., Ganzer, G., Kornig, A., Becker, W., Reichelt, E., ... Moritz, T. (2015). Micro-reactors made by lithography-based ceramic manufacturing (LCM). *Additive Manufacturing and Strategic Technologies in Advanced Ceramics: Ceramic Transactions*, 258, 31–41.

**140** Masciandaro, S., Torrell, M., Leone, P., & Tarancón, A. (2019). Three-dimensional printed yttria-stabilized zirconia self-supported electrolytes for solid oxide fuel cell applications. *Journal of the European Ceramic Society*, 39(1), 9–16.

**141** Tétreault, N., von Freymann, G., Deubel, M., Hermatschweiler, M., Pérez-Willard, F., John, S., ... Ozin, G. A. (2006). New route to three-dimensional photonic bandgap materials: Silicon double inversion of polymer templates. *Advanced Materials*, 18(4), 457–460.

**142** Meza, L. R., & Greer, J. R. (2014). Mechanical characterization of hollow ceramic nanolattices. *Journal of Materials Science*, 49(6), 2496–2508.

**143** Meza, L. R., Das, S., & Greer, J. R. (2014). Strong, lightweight, and recoverable three-dimensional ceramic nanolattices. *Science*, 345(6202), 1322–1326.

**144** Zocca, A., Colombo, P., Gomes, C. M., & Günster, J. (2015). Additive manufacturing of ceramics: Issues, potentialities, and opportunities. *Journal of the American Ceramic Society*, 98, 1983–2001.

**145** Bertrand, P., Bayle, F., Combe, C., Goeuriot, P., & Smurov, I. (2007). Ceramic components manufacturing by selective laser sintering. *Applied Surface Science*, 254, 989.

146 Shahzad, K., Deckers, J., Zhang, Z., Kruth, J. P., & Vleugels, J. (2014). Additive manufacturing of zirconia parts by indirect selective laser sintering. *Journal of the European Ceramic Society*, 34(1), 81–89.

147 Ferrage, L., Bertrand, G., & Lenormand, P. (2018). Dense yttria-stabilized zirconia obtained by direct selective laser sintering. *Additive Manufacturing*, 21, 472–478.

148 Wu, Y., Du, J., Choy, K. L., & Hench, L. L. (2007). Laser densification of alumina powder beds generated using aerosol assisted spray deposition. *Journal of the European Ceramic Society*, 27, 4727.

149 Deckers, J., Shahzad, K., Vleugels, J., & Kruth, J. P. (2012). Isostatic pressing assisted indirect selective laser sintering of alumina components. *Rapid Prototyping Journal*, 18, 409.

150 Tang, H. H., Chiu, M. L., & Yen, H. C. (2011). Slurry-based selective laser sintering of polymer-coated ceramic powders to fabricate high strength alumina parts. *Journal of the European Ceramic Society*, 31, 1383.

151 Löschau, W., Lenk, R., Scharek, S., Teichgräber, M., Nowotny, S., & Richter, C. (2000). Prototyping of complex-shaped parts and tools of Si. *Industrial Ceramics*, 20, 95.

152 Slocombe, A., & Li, L. (2001). Selective laser sintering of TiC–Al2O3 composite with self-propagating high-temperature synthesis. *Journal of Materials Processing Technology*, 118, 173.

153 Minasyan, T., Liu, L., Aghayan, M., Kollo, L., Kamboj, N., Aydinyan, S., & Hussainova, I. (2018). A novel approach to fabricate Si3N4 by selective laser melting. *Ceramics International*, 44(12), 13689–13694.

154 Minasyan, T., Liu, L., Holovenko, Y., Aydinyan, S., & Hussainova, I. (2019). Additively manufactured mesostructured MoSi2-Si3N4 ceramic lattice. *Ceramics International*, 45(8), 9926–9933.

155 Chen, A. N., Li, M., Xu, J., Lou, C. H., Wu, J. M., Cheng, L. J., … Li, C. H. (2018). High-porosity mullite ceramic foams prepared by selective laser sintering using fly ash hollow spheres as raw materials. *Journal of the European Ceramic Society*, 38(13), 4553–4559.

156 Goodridge, R. D., Wood, D. J., Ohtsuki, C., & Dalgarno, K. W. (2007). Biological evaluation of an apatite–mullite glass-ceramic produced via selective laser sintering. *Acta Biomaterialia*, 3, 221.

157 Bose, S., Vahabzadeh, S., & Bandyopadhyay, A. (2013). Bone tissue engineering using 3D printing. *Materials Today*, 16(12), 496–504.

158 Vorndran, E., Klarner, M., Klammert, U., Grover, L. M., Patel, S., Barralet, J. E., & Gbureck, U. (2008). 3D powder printing of β-tricalcium phosphate ceramics using different strategies. *Advanced Engineering Materials*, 10(12), B67–B71.

159 Klammert, U., Gbureck, U., Vorndran, E., Rödiger, J., Meyer-Marcotty, P., & Kübler, A. C. (2010). 3D powder printed calcium phosphate implants for reconstruction of cranial and maxillofacial defects. *Journal of Cranio-Maxillofacial Surgery*, 38(8), 565–570.

**160** Gureev, D. M., Ruzhechko, R. V., & Shishkovskii, I. V. (2000). Selective laser sintering of PZT ceramic powders. *Technical Physics Letters*, 26(3), 262–264.

**161** Guo, D., Li, L.-T., Cai, K., Gui, Z.-l., & Nan, C.-W. (2004). Rapid prototyping of piezoelectric ceramics via selective laser sintering and gelcasting. *Journal of the American Ceramic Society*, 87(1), 17–22.

**162** Du, X. Y., Ji, L. F., & Jiang, Y. J. (2009). Laser sintering of lead-free Sr1.86Ca0.14 NaNb5O15 piezoelectric ceramics. *Journal of Optoelectronics Laser*, 20(8), 1037–1040.

**163** Du, X. Y., Ji, L. F., & Jiang, Y. J. (2010). Dielectric and piezoelectric properties of CO2 laser sintered Sr1.86Ca0.14NaNb5O15 piezoelectric ceramic. *Ferroelectrics*, 402(1), 163–167.

**164** Ji, L. F., & Jiang, Y. J. (2007). Investigations of dielectric enhancement in (Ta2O5)1-x(TiO2)x ceramics prepared by laser-sintering technique. *Applied Physics A: Materials Science & Processing*, 87(4), 733–738.

**165** Ji, L. F., Jiang, Y. J., Wang, W., & Yu, Z. L. (2004). Enhancement of the dielectric permittivity of Ta2O5 ceramics by CO2 laser irradiation. *Applied Physics Letters*, 85, 1577–1579.

**166** Macedo, Z. S., Lente, M. H., Eiras, J. A., & Hernandes, A. C. (2004). Dielectric and ferroelectric properties of Bi4Ti3O12 ceramics produced by a laser sintering method. *Journal of Physics: Condensed Matter*, 16, 2811–2818.

**167** Macedo, Z. S., Silva, R. S., Valerio, M. E. G., Martinez, A. L., & Hernandes, A. C. (2004). Laser-sintered bismuth germanate ceramics as scintillator devices. *Journal of the American Ceramic Society*, 87(6), 1076–1081.

**168** Jesus, L. M., Santos, J. C. A., Sampaio, D. V., Barbosa, L. B., Silva, R. S., & M'Peko, J.-C. (2016). Polymeric synthesis and conventional versus laser sintering of CaCu3Ti4O12 electroceramics: (Micro)structures, phase development and dielectric properties. *Journal of Alloys and Compounds*, 654(5), 482–490.

**169** Basile, N., Gonon, M., Petit, F., & Cambier, F. (2012). Interaction between laser beam and BaTiO3 powders in selective laser sintering treatments. *Journal of the European Ceramic Society*, 32(12), 3303–3311.

**170** Agarwala, M. K., Bourell, D. L., Manthiram, A., Birmingham, B. R., & Marcus, H. (1995). High Tc dual phase Ag-YBa2Cu3O7−x composites prepared by selective laser sintering and infiltration. *Journal of Materials Science*, 30(2), 459–464.

**171** Hutchings, I. (2010). *Ink-jet printing for the decoration of ceramic tiles: Technology and opportunities* (pp. 1–16). Castellón, Spain: Actas de QUALICER.

**172** Baraldi, L. (2014). Digital tiles. *Ceramic World Review*, 105, 62–69.

**173** Baraldi, L. (2015). 100% digital lines. *Ceramic World Review*, 1, 66–68.

**174** FERRO. (2019). Retrieved from https://www.ferro.com/products/product-category/inks-for-digital-ceramic-and-glass-printing/inks-for-digital-ceramic-tile-printing/digital-ceramic-inks/water-based-inks-for-digital-ceramic-tile-printing

**175** Güngör, G. L., Kara, A., Blosi, M., Gardini, D., Guarini, G., Zanelli, C., & Dondi, M. (2015). Micronizing ceramic pigments for inkjet printing: Part I. Grindability and particle size distribution. *Ceramics International*, 41(5), 6498–6506.

**176** Fromm, J. E. (1984). Numerical-calculation of the fluid-dynamics of drop-on-demand jets. *IBM Journal of Research and Development*, 28, 322–333.

**177** Reis, N., & Derby, B. (2000). Ink jet deposition of ceramic suspensions: Modelling and experiments of droplet formation. *MRS Symposium Proceedings*, 624, 65–70.

**178** Derby, B. (2015). Additive manufacture of ceramics components by inkjet printing. *Engineering*, 1(1), 113–123.

**179** Derby, B. (2011). Inkjet printing ceramics: From drops to solid. *Journal of the European Ceramic Society*, 31(14), 2543–2550.

**180** Dou, R., Wang, T., Guo, Y., & Derby, B. (2011). Inkjet printing of zirconia: Coffee staining and line stability. *Journal of the American Ceramic Society*, 94(11), 3787–3792.

**181** Zhang, Y., Yang, S., Chen, L., & Evans, J. R. G. (2008). Shape changes during the drying of droplets of suspensions. *Langmuir*, 24(8), 3752–3758.

**182** Pan, Z., Wang, Y., Huang, H., Ling, Z., Dai, Y., & Ke, S. (2015). Recent development on preparation of ceramic inks in ink-jet printing. *Ceramics International*, 41(10), 12515–12528.

**183** Blazdell, P. F., & Evans, J. R. G. (2000). Application of a continuous ink jet printer to solid freeforming of ceramics. *Journal of Materials Processing Technology*, 99, 94–102.

**184** Noguera, R., Lejeune, M., & Chartier, T. (2005). 3D fine scale ceramic components formed by ink-jet prototyping process. *Journal of the European Ceramic Society*, 25, 2055–2059.

**185** Özkol, E., Ebert, J., & Telle, R. (2010). An experimental analysis of the influence of the ink properties on the drop formation for direct thermal inkjet printing of high solid content aqueous 3Y-TZP suspensions. *Journal of the European Ceramic Society*, 30(7), 1669–1678.

**186** Blazdell, P. F., Evans, J. R. G., Edirisinghe, M. J., Shaw, P., & Bin-stead, M. J. (1995). The computer aided manufacture of ceramics using multilayer jet printing. *Journal of Materials Science Letters*, 14, 1562.

**187** Xiang, Q. F., Evans, J. R. G., Edirisinghe, M. J., & Blazdell, P. F. (1997). Solid freeforming of ceramics using a drop-on-demand jet printer. *Proceedings of the Institution of Mechanical Engineers, Part B*, 211, 211.

**188** Blazdell, P. F., & Evans, J. R. G. (1999). Preparation of ceramic inks for solid freeforming using a continuous jet printer. *Journal of Materials Synthesis and Processing*, 7, 349.

**189** Wang, C., Tomov, R. I., Kumar, R. V., & Glowacki, B. A. (2011). Inkjet printing of gadolinium-doped ceria electrolyte on NiO-YSZ substrates for solid oxide fuel cell applications. *Journal of Materials Science*, 46(21), 6889–6896.

**190** Sukeshini, A. M., Cummins, R., Reitz, T. L., & Miller, R. M. (2009). Ink-jet printing: A versatile method for multilayer solid oxide fuel cells fabrication. *Journal of the American Ceramic Society*, 92, 2913–2919.

**191** Esposito, V., Gadea, C., Hjelm, J., Marani, D., Hu, Q., Agersted, K., … Jensen, S. H. (2015). Fabrication of thin yttria-stabilized-zirconia dense electrolyte layers by inkjet printing for high performing solid oxide fuel cells. *Journal of Power Sources*, 273, 89–95.

**192** Sukeshini, A. M., Gardner, P., Meisenkothen, F., Jenkins, T., Miller, R., Rottmayer, M., & Reitz, T. L. (2011). Aerosol jet printing and microstructure of SOFC electrolyte and cathode layers. *ECS Transactions*, 35, 2151.

**193** Shimada, H., Ohba, F., Li, X., Hagiwara, A., & Ihara, M. (2012). Electrochemical behaviors of nickel/yttria-stabilized zirconia anodes with distribution controlled yttrium-doped barium zirconate by ink-jet technique. *Journal of the Electrochemical Society*, 159, F360.

**194** Sukeshini, A. M., Cummins, R., Reitz, T. L., & Miller, R. M. (2009). Ink-jet printing: A versatile method for multilayer solid oxide fuel cells fabrication. *Electrochemical and Solid-State Letters*, 12, B176.

**195** Yashiro, N., Usui, T., & Kikuta, K. (2010). Application of a thin intermediate cathode layer prepared by inkjet printing for SOFCs. *Journal of the European Ceramic Society*, 30, 2093–2098.

**196** Mott, M., Song, J.-H., & Evans, J. R. G. (1999). Microengineering of ceramics by direct ink-jet printing. *Journal of the American Ceramic Society*, 82(7), 1653–1658.

**197** Song, J. H., Edirisinghe, M. J., & Evans, J. R. G. (1999). Formulation and multilayer jet printing of ceramic inks. *Journal of the American Ceramic Society*, 82, 3374–3380.

**198** Ding, X., Li, Y., Wang, D., & Yin, Q. (2004). Fabrication of BaTiO3 dielectric films by direct ink-jet printing. *Ceramics International*, 30, 1885–1887.

**199** Tseng, W. J., Lin, S.-Y., & Wang, S.-R. (2006). Particulate dispersion and freeform fabrication of BaTiO3 thick films via direct inkjet printing. *Journal of the American Ceramic Society*, 16(4), 537–540.

**200** Ding, X., Li, Y., Wang, D., & Yin, Q. (2003). Preparation of (BaxSrx-1TiO3) sols used for ceramic inkjet printing. *Materials Science and Engineering B*, 99, 502–505.

**201** Kaydanova, T., Miedaner, A., Perkins, J. D., Curtis, C., Alleman, J. L., & Ginley, D. S. (2007). Direct-write inkjet printing for fabrication of barium strontium titanate based tunable circuits. *Thin Solid Films*, 515(7–8), 3820–3824.

**202** Martin, C. R., & Aksay, I. A. (2004). Submicrometer-scale patterning of ceramic thin films. *Journal of Electroceramics*, 12(1–2), 53–68.

**203** Matavž, A., Frunză, R. C., Drnovšek, A., Bobnar, V., & Maličab, B. (2016). Inkjet printing of uniform dielectric oxide structures from sol–gel inks by adjusting the solvent composition. *Journal of Materials Chemistry C*, 4, 5634–5641.

**204** Lee, H., Chang, Y. J., Herman, G. S., & Chang, C. H. (2007). A general route to printable high-mobility transparent amorphous oxide semiconductors. *Advanced Materials*, 19, 843–847.

**205** Shen, W., Zhao, Y., & Zhang, C. (2005). The preparation of ZnO based gas-sensing thin films by ink-jet printing method. *Thin Solid Films*, 483, 382–387.

**206** Kim, G. H., Kim, H. S., Shin, H. S., Ahn, B. D., Kim, K. H., & Kim, H. J. (2009). Inkjet-printed InGaZnO thin film transistor. *Thin Solid Films*, 517, 4007–4010.

**207** Tellier, J., Malic, B., Kuscer, D., Trefalt, G., & Kosec, M. (2011). Ink-jet printing of In2O3/ZnO two-dimensional structures from solution. *Journal of the American Ceramic Society*, 94(9), 2834–2840.

**208** Meyers, S. T., Anderson, J. T., Hong, D., Hung, C. M., Wager, J. F., & Keszler, D. A. (2007). Solution-processed aluminum oxide phosphate thin-film dielectrics. *Chemistry of Materials*, 19(16), 4023–4029.

**209** Meyers, S. T., Anderson, J. T., & Keszler, D. A. (2008). Aqueous inorganic inks for low-temperature fabrication of ZnO TFTs. *Journal of the American Chemical Society*, 130(51), 17603–17609.

**210** Kolchanov, D. S., Mitrofanov, I., Kim, A., Koshtyal, Y., Rumyantsev, A., Sergeeva, E., … Maximov, M. Y. Inkjet printing of Li-rich cathode material for thin-film lithium-ion micro-batteries. *Energy Technology*. doi:10.1002/ente.201901086

**211** Li, C., Shi, H., Ran, R., Su, C., & Shao, Z. (2013). Thermal inkjet printing of thin-film electrolytes and buffering layers for solid oxide fuel cells with improved performance. *International Journal of Hydrogen Energy*, 38, 9310–9319.

**212** Gadea, C., Hanniet, Q., Lesch, A., Marani, D., Jensen, S. H., & Esposito, V. (2017). Aqueous metal–organic solutions for YSZ thin film inkjet deposition. *Journal of Materials Chemistry C*, 5, 6021–6029.

**213** Rosa, M., Gooden, P. N., Butterworth, S., Zielke, P., Kiebach, R., Xu, Y., … Esposito, V. (2019). Zirconia nano-colloids transfer from continuous hydrothermal synthesis to inkjet printing. *Journal of the European Ceramic Society*, 39(1), 2–8.

**214** Li, C., Chen, H., Shi, H., Tadé, M. O., & Shao, Z. (2015). Green fabrication of composite cathode with attractive performance for solid oxide fuel cells through facile inkjet printing. *Journal of Power Sources*, 273, 465–471.

**215** Mikolajęk, M., Reinheimer, T., Bohn, N., Kohler, C., Hoffmann, M. J., & Binder, J. R. (2019). Fabrication and characterization of fully inkjet printed capacitors based on ceramic/polymer composite dielectrics on flexible substrates. *Scientific Reports*, 9, 13324.

**216** Singh, M., Haverinen, H. M., Dhagat, P., & Jabbour, G. E. (2010). Inkjet printing – process and its applications. *Advanced Materials*, 22, 673–685.

# 3

# 3D Printing of Functional Composites with Strain Sensing and Self-Heating Capabilities

*Xin Wang and Jihua Gou*

*Composite Materials and Structures Laboratory, Department of Mechanical and Aerospace Engineering, University of Central Florida, Orlando, FL, USA*

## 3.1 Introduction

Digital printing is the "process of joining materials to make objects from 2D/3D model data, usually layer by layer," which was first described in 1986 by Charles Hull [1]. This technology creates objects by adding materials to reduce waste while reaching satisfactory geometric accuracy [2]. It begins with a meshed computer model that can be created by acquired image data or structures built in computer-aided design (CAD) software. Conventional fabrication techniques such as molding, casting and machining create polymer composites with complex geometry through material removal processes [3]. While the manufacturing processes of composites using these methods are well-controlled and understood, the ability to control the geometry and properties of fabricated composites is limited. Digital printing technique is able to fabricate complex composite structures without the typical waste. The size and geometry of composites can be precisely controlled with the assistance of computer-aided design. Additionally, the properties could also be controlled and adjusted due to the flexibility of digital printing process. Through optimizing the processing parameter and materials design, the functional performance of composites could be improved. Therefore, digital printing of polymer composites has found many potential applications, including in aerospace industries for creating complex lightweight structures [4], architectural industries for structural models [5], art fields for artifact replication or education [6], and medical fields for printing patient-specific tissues and organs [7].

*3D Printing for Energy Applications*, First Edition. Edited by Albert Tarancón and Vincenzo Esposito.

A variety of printing techniques have been employed to fabricate polymer composites, such as fused deposition modeling (FDM), selective laser sintering (SLS), inkjet 3D printing, stereolithography, and 3D plotting. The selection of fabrication technique depends on the starting materials, requirements of processing speed and resolution, costs and performance requirements of final products. This chapter introduces the inkjet printing and fused deposition modeling technique used to fabricate functional carbon nanotube (CNT) reinforced polymer composites. FDM printers work with controlled extrusion of thermoplastic filaments. In the FDM process, filaments are melt into a semi-liquid state at the nozzle and extruded layer by layer onto a build platform where layers are fused together and then solidified to final parts. The inkjet printing is a popular technique due to its ability to print fine and easily controllable patterns, and the advantages of noncontact, solution saving, and scalability. It recreates a digital image by propelling droplets of ink on to various types of substrates. This chapter also gives some examples of functional nanocomposites fabricated by digital printing technique including strain sensors, heating elements, and shape memory composites.

## 3.2 Carbon Nanotube Reinforced Functional Polymer Nanocomposites

### 3.2.1 Strain Sensing of CNT Reinforced Polymer Nanocomposites

The electrical characteristic of CNTs coupled with their high mechanical strength makes them a promising material for piezo-resistive strain sensors [8–10]. In CNT strain sensors, there is a reversible correlation between the mechanical deformation and the electrical resistance, which is called piezo-resistive effect [11]. When the strain sensors are stretched out, reorientation and repositioning of CNTs within the CNT network increases the base resistance of the strain sensors. Upon relaxation of the strain, reestablishment of the CNT network induced by the elastic force of the substrate restores the original resistance.

Traditional strain sensors were fabricated through non-digital technique, such as spray coating [11] and molding techniques [12], which could only produce strain sensors with simple geometry. For example, a strain sensor made of single-walled carbon nanotube (SWCNT) was developed through spray coating process [13]. The batch-fabricated SWCNT strain sensors showed a linear relationship between resistance changes and externally applied strains. The sensitivity of the strain sensor was measured to be approximately 30 times higher than that of commercial foil-type strain sensors. Stretchable and flexible strain sensors have attracted considerable attention for human motion monitoring and analysis. Several studies have been conducted on the flexible strain sensors based on carbon nanotubes and elastomers. For instance, CNTs were spray-deposited on a PDMS substrate to prepare a strain sensor, which could accommodate strain up to 150% to monitor large-scale human motions [14]. A CNT-based strain sensor produced by compression molding method exhibited a high sensitivity with a gauge factor of 153 at 30% strain [15]. Recent work on the fabrication of nanocomposites with printing technology offers a promising method to fabricate strain sensors with controllable sensitivity and designed geometry to monitor complex human motion [16–18]. The additive feature and precise print path control of digital printing enable the continuous fabrication of arbitrary sensor configuration with low cost.

### 3.2.2 Resistive Heating of CNT Reinforced Polymer Nanocomposites

In the case of electrical heating materials, electrical energy can be converted into heat energy through the Ohmic joule heating [19]. Usually, electrical heating elements are made by metallic materials and used for the plane heater, wire heater, water heater, and so on [20]. However, metal-based electrical heating devices have several drawbacks, such as easy oxidization, emission of electromagnetic waves, and high manufacturing costs. Therefore, polymer composites containing CNTs have attracted interest for preparing electrical heating materials due to their light weight, easy processing, and high energy efficiency [21]. When a certain voltage was applied to the polyethylene/CNT composite films, the surface temperature of the film reaches the equilibrium value within less than 100 s [22]. Incorporating CNTs into shape memory polymers (SMPs) can also make them into conductive composites and the electrically resistive heating of these composites can be a desirable stimulus to activate their shape memory effect without an external heater. As a typical smart material, SMP can be deformed into a temporary shape and then recover to its original (or permanent) shape upon exposure to the heat stimuli [23–25]. Therefore, besides acting as an electric heater, due to their superior structural versatility, CNT/SMPs can also be used in many other fields, including smart tooling [26], deployable medical device [27], and morphing aircraft [28]. Many efforts have been proposed focusing on the electrical actuation of CNT/SMP composites. Electro-active multi-walled carbon nanotubes filled polyurethane shape memory composites showed more than 98% shape recovery ratio and a rapid recovery time of 9 s under a voltage of 40 V [29]. The CNT nanopaper/SMP composites showed even better performance. Owing to the excellent electrical conductivity of CNT nanopaper, their composites showed a full shape recovery at a low voltage of 8.4 V [30]. However, non-digital fabricated nanocomposites can only show global actuation, using 3D printing technology, selective localized actuation can be realized for smart skin or morphing structure applications.

## 3.3 Printing Strategies

### 3.3.1 Spray Deposition Modeling and Fused Deposition Modeling

All of the nanocomposites in this chapter were fabricated with help of home-made spray deposition modeling printer. Spray deposition modeling (SDM) is a digital fabrication process in which the CNT ink droplets are supplied in the form of a stream to any desired location since the nozzle is connected to a computer controlled $x$ and $y$ axis. SDM is similar to an ink jet printer, but SDM has the ability to model over the same area with precision accuracy. Thermal jet nozzle (Figure 3.1a) is used in SDM machine and it contains a thin film resistor in the nozzle. In order to eject a droplet, the thin film resistor is heated by passing current through it. This causes the ink in the nozzle to vaporize, creating a bubble and a large increase in pressure, which forces ink droplets out of the nozzle. A thermal jet nozzle with a 12 nozzle outlet array and 235 μm pore size is used in SDM. This allows for a large enough pore size to keep the nozzles form clogging while creating particles that are small enough to ensure the high resolution. Spray deposition modeling uses computer numeric controls (CNC) to move the nozzles into position and allows the thermal jet nozzle to lay down the material properly. This digital processing ensures the created composite is geometrically accurate. This process is also additive, which enables the composite to be built layer by layer. Each layer and location could have a different composition of materials. The functionality could be realized by the material design. The fused deposition modeling printer used in this work is a commercially available Mbot printer (Figure 3.1b). The quality of printed parts can be controlled by altering printing parameters, such as layer thickness, printing orientation, raster width, raster angle, and air gap.

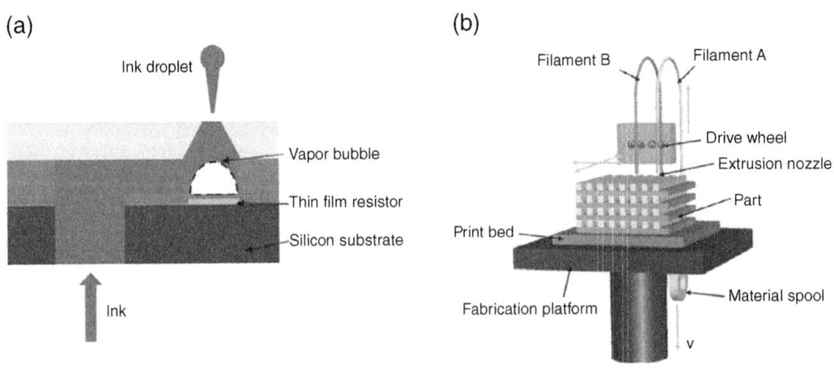


**Figure 3.1** Schematic of a typical (a) thermal jet nozzle setup (b) FDM printer setup.

### 3.3.2 Printing of Highly Flexible Carbon Nanotube/Polydimethylsilicone Strain Sensor

Multi-walled carbon nanotubes are commonly used as printing materials. The carbon nanotube powder can be dispersed in water to form a suspension for digital printing. Generally, 0.1 g CNT materials are dispersed in 150 ml water using a sonication tip. The surfactant Triton X-100 can be added to the solvent to help dispersion. Owing to their high elasticity and fast response speed to strain, polydimethylsilicone (PDMS) are commonly used as a flexible substrate for composite strain sensors [31]. Acid treatment ($H_2O_2$:$H_2SO_4$ = 1:3, $H_2O_2$ 30%; $H_2SO_4$ 98%) is needed to increase the hydrophilicity of PDMS substrate for aqueous ink printing. The prepared CNT ink is printed on the PDMS substrate using the SDM printer. After the printing process, copper wires are attached to the end of CNT patterns and finally another layer of PDMS can be cured on top of printed patterns to obtain the composite strain sensors. Figure 3.2 schematically depicts the digital fabrication process of composite strain sensors carried out through deposition CNT on PDMS substrate using SDM printing process.

### 3.3.3 Printing of Carbon Nanotube/Shape Memory Polymer Nanocomposites

CNT inks can be printed on two types of substrates to produce CNT/SMP nano composites. One type of the substrate is SMP thin film. SMP pellets can be dissolved in dimethylformamide at 70 °C and after that, the hot solution can be poured into petri dishes to produce a thin film substrate. Another substrate is a honeycomb-structured substrate. Shape memory polymer filaments are fabricated through single-screw extruder. FDM 3D printer is used to print the CAD designed honeycomb-structured substrate by extruding the filaments. Figure 3.3 schematically depicts the digital fabrication process of shape memory polymer nanocomposites carried out through deposition CNT on SMP substrates using SDM printing process.

## 3.4 Strain Sensing of Printed Nanocomposites

To control the thickness of CNT layer in composites, different numbers of printing cycles are performed. Figure 3.4a shows the CNT layers that fabricated with different numbers of printing cycles from 10 to 50 with an interval of 10 layers. It shows that the CNT layer became darker as printing cycle increases, indicating the increasing of thickness of CNT layer. It should be noticed that when the

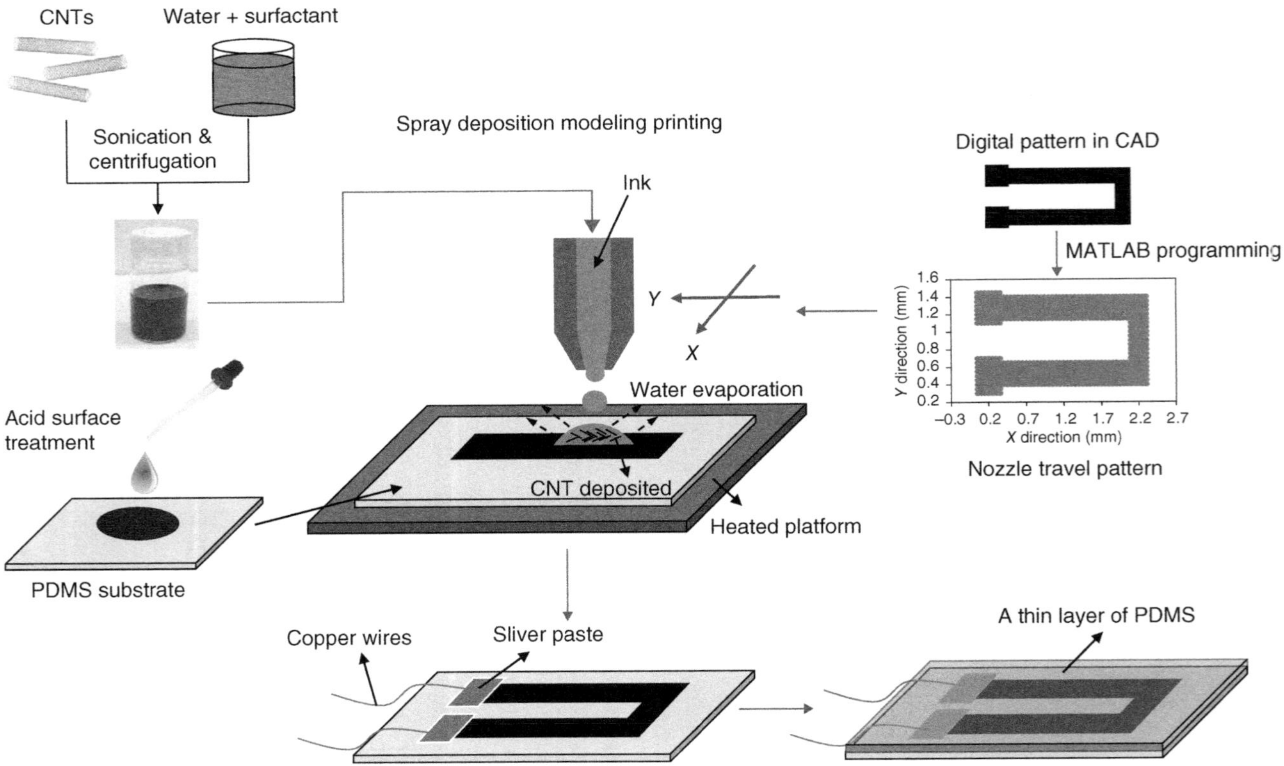


**Figure 3.2** Schematic of the digital fabrication of composite strain sensors through the SDM technique.

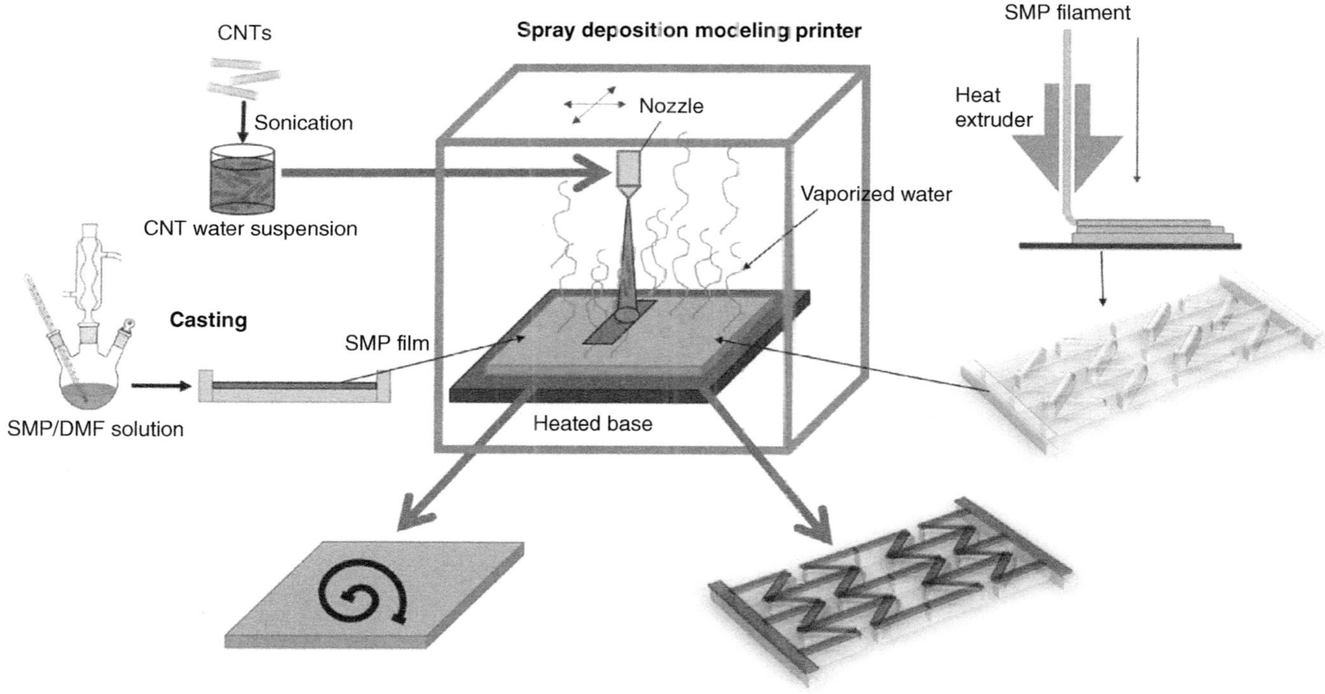


**Figure 3.3** Schematic of the digital fabrication of CNT/SMP nanocomposites.

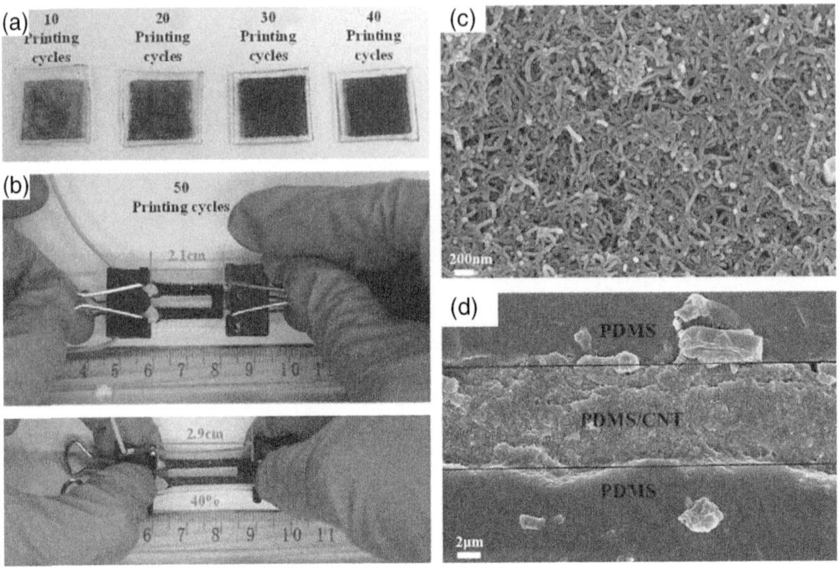


**Figure 3.4** (a) Optical images of CNT layers printed with 10, 20, 30, 40 cycles; (b) photograph of the strain sensor subjected to 40% strain; (c) cross-section SEM images of pristine CNT layers on PDMS substrate with 50 printing cycles; (d) cross-section SEM images of CNT/PDMS composite.

number of printing cycle is small, the deposited CNT layer is not perfectly uniform because of the "coffee ring" effect with CNT dispersion [32], but as the printing cycle increases, more homogenous CNT layer is formed. The fabricated strain sensors are shown in Figure 3.4b, which indicating the excellent stretchability.

To identify the morphology of the printed pristine CNT layer on PDMS substrate, cross-section SEM images of CNT layers are shown in Figure 3.4c. It clearly shows that CNTs were uniformly deposited on PDMS substrate to form a dense and continuous conductive network without the agglomeration of CNTs, owing to the homogenous dispersion of CNTs in water as well as the optimized and effective SDM printing process. It also shows that the PDMS resin fully impregnates throughout CNT networks indicating a good interfacial bonding between matrix and fillers. The "sandwich" layered structure with one CNT embedded composite layer was also confirmed from Figure 3.4d.

The thickness effect of CNT layers on the strain sensing of the composite sensors is investigated by comparing the sensors made from different printing cycles of CNT layer. The strain sensors with an area of $15 \times 50\,\text{mm}^2$ and thickness of 0.5 mm are prepared for sensing testing, and are named as CP$x$, where $x$ denotes the printing cycles of CNT layer. These composite strain sensors are characterized with regard to the applied strain under tension, and the strain sensors are stretched up to 45% strain until the plastic deformation is found. Figure 3.5a shows the representative resistance change–strain data. All the sensors exhibit similar piezoresistive responses to the strain, and with increasing of strain, the electrical

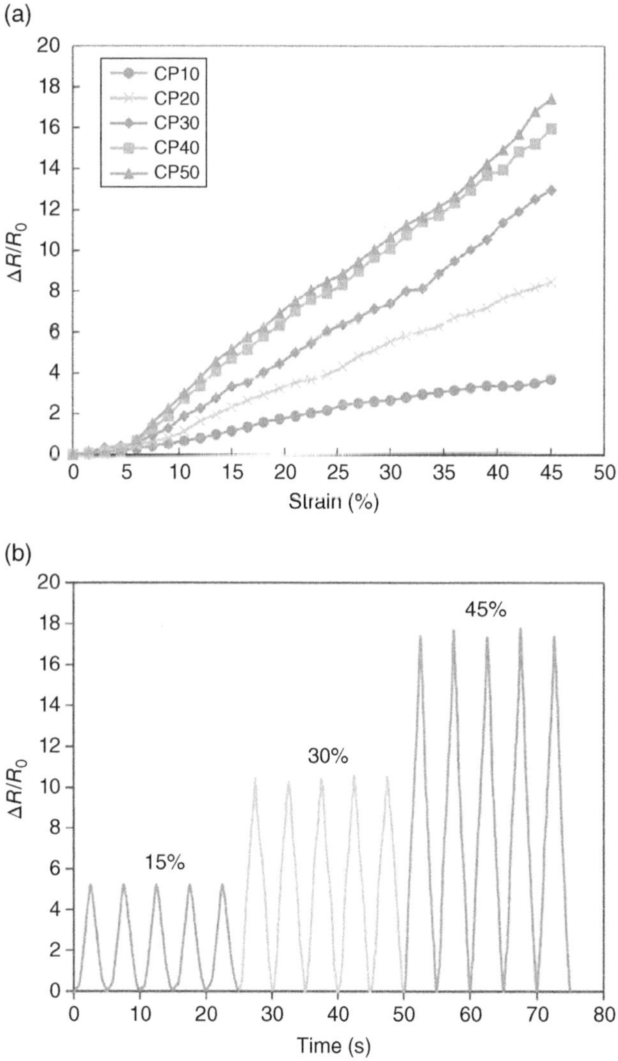


**Figure 3.5** (a) Resistance change–strain relationship of composite sensors with different CNT layer thickness. (b) Pieoresistive response of CP50 under different strains.

resistance of all the sensors increases. The strain-dependent responses of the sensors are found to have two regions. In low strain region (less than 10%), the stretching of CNT network is just initiated in the loading direction, and there are still many CNT junctions due to the compression of transverse direction, so the resistance change is slow at the beginning of the stretching. The linear increasing of electrical resistance as the strain increases becomes stable after the initiation stage. In particular, the sensitivity of the fabricated sensors could be modulated by simply tailoring the printing cycles of printed CNT layer to accommodate diverse sensing requirements. The resistance of composite strain sensors with more printed CNT layers increases significantly, indicating that the sensors with more CNT layers are more sensitive to applied strain. This is possibly explained by the conducting mechanism in the CNT network [33]. At unstretched state, strain sensors with thicker CNT layer have a higher density CNT network and more CNT joints, thus the contact resistance between CNTs dominantly determines the overall resistance. When the strain is applied, the cracking occurs in the CNT network resulting in the disconnection of most CNT joints so that the connection mode transfers to the tunneling state where a large increase of electrical resistance exists. However, for sensors with thinner CNT layer, the CNT joint is few at the original state and the conducting mechanism transition is less visible so that they are insensitive to the external stimulus.

Generally, the gauge factor [$K = (\Delta R/R_0)/\varepsilon$] is calculated to determine the sensitivity of composite strain sensors [34], where $R_0$ is the resistance of carbon nanotube strain sensor without strain, $\Delta R$ is the resistance change under strain, $\varepsilon$ is the strain, and $K$ is the gauge factor. For all the sensors in this chapter, the electrical resistance increases linearly with the strain ($R^2 > 0.98$), and the gauge factors are almost constant after the initiation stage. This good linearity is of vital importance in precise sensing of strains. Among these composite sensors, CP50 sensors have the highest gauge factor of 35.75. Moreover, these CNT/PDMS strain sensors are capable of measuring tensile strain up to 45%, which is more desirable for monitoring human motions compared to commercially available metallic strain gauges. Generally, the joint movement of human body results in less than 50% strain [35], thus the printed composite strain sensors hold a great promise for monitoring human motions. To examine the response repeatability and durability of the strain sensors, repeated stretching and releasing cycles are applied to strain sensors. As shown in Figure 3.5b, sensors under three different strain all present excellent resistance recoverability and reproducibility.

The printed nanocomposite strain sensors can be used for wearable electronics applications because they are flexible to be easily attached to human body. Their capability of monitoring human motion is demonstrated in this chapter. Human body deformation is generally multidirectional, which cannot be accurately monitored by single-directional strain sensors. Due to the flexibility of the SDM fabrication, multiaxial strain sensors can be fabricated to monitor complicated strain. A rosette-shaped strain sensor is fabricated with an intersecting angle of 120°. By using this type of strain sensor, the magnitude and direction of the principal strain could be measured. The strain data obtained from the correlated piezoresistive signal are compared with strain data calculated from measuring the 3D position of reflective markers attached to the strain sensor during wrist flexion movements using a motion capture. As shown in Figure 3.6a, the fabricated rosette-shaped strain sensor is mounted on the tester's wrist. Figure 3.6b shows the strain recorded by the motion capture of each individual strain sensors during the cyclic flexion and extension motions of the wrist. Figure 3.6c shows the corresponding electrical resistance changes of each individual sensors. Maximum strains calculated from the signal changes of three individual sensors are 6.29, 2.45, and 0.82%, respectively, which is only slightly different from the strain recorded by motion capture, indicating the accuracy of the strain sensors to monitor human movement. In particular, the strain tested using the sensor that is in the same direction as the wrist extensors is larger than the strain tested by the angled sensor panels, which is reasonable considering that our testing motion is wrist flexion and extension movements. Additionally, using this multidirectional strain sensor, the strain along other directions could be identified. According to the defined coordinate system, the principal strain and their angle of orientation are calculated to be 6.43% and 8.4° by solving the strain transformation equations using strains obtained from three sensors in the rosette configuration. It indicates that during wrist flexion, the direction of the skin stretch is not completely identical with wrist flexion due to the complexity of human motion. When this rosette-type strain sensor is subjected to any strain state, the principal strain and its orientation could be calculated through the strain data obtained from the three independent strain sensor, enabling to monitor diverse human motions.

## 3.5 Electric Heating Performance Analysis

Due to the flexibility of the SDM fabrication method, CNT/SMP composite with different heating patterns could be obtained. In Figure 3.7a, three patterned CNT/SMP composites are fabricated as (I) a rectangular area, (II) a semi-circle area, and (III) a spiral line area of printed CNT layers. Contact pads at the end of the patterns are designed for attaching the wires using silver paste. Their thermal images are taken when certain voltages are applied. It is found that the surface temperature does not show many variations in central locations with printed CNT layers, which confirms the uniformity of fabricated CNT/SMP composites. The temperature on the edge of composites is lower due to the heat exchange between the heated

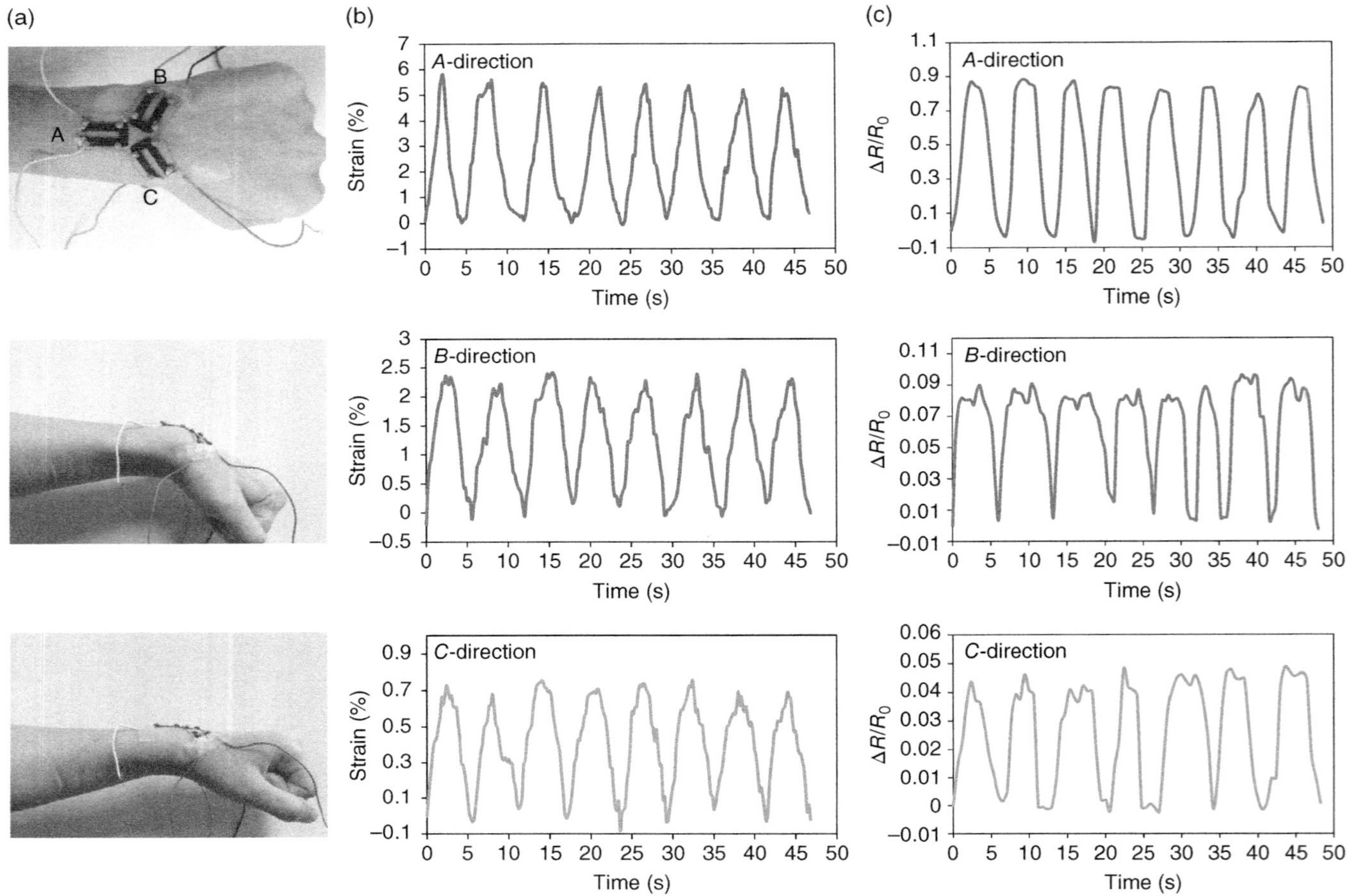


**Figure 3.6** (a) Photographs of the rosette-type strain sensor attached to tester's wrist. (b) Strain recorded by camera of each individual strain sensor. (c) Electrical resistance changes of each individual sensor.

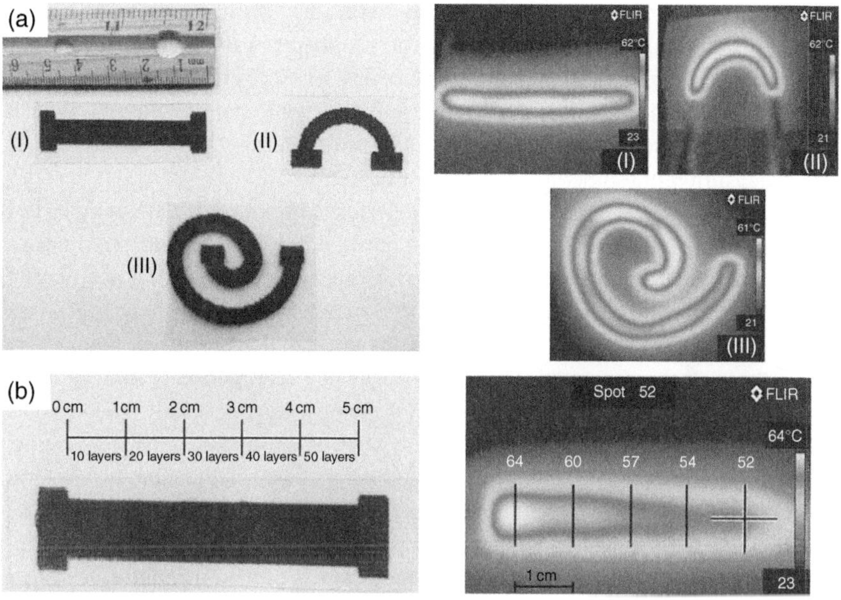


**Figure 3.7** (a) Left: CNT/SMP composites that have (I) rectangular, (II) hemi-sphere area, and (III) spiral line-shaped CNT layers. Right: Corresponding temperature distribution thermal images. (b) Left: CNT/SMP composites that have different numbers of printed CNT layers in five districts. Right: Corresponding temperature distribution thermal images.

composite and the unheated atmosphere. Using digital controlled SDM method, composites that have arbitrary heating pattern could be manufactured, which indicating that electric heating could be accurately applied at desired locations.

The printed CNT/SMP composite is not only designed for areas needing even temperature distribution, but also designed for areas needing thermal gradient. Figure 3.7b shows the CNT/SMP composite that is fabricated with different numbers of printed CNT layers in five districts. In the five consecutive districts from left to right, the composite contains 10, 20, 30, 40, and 50 printed CNT layers at each district, respectively. These five districts could act a series circuit with five resistors. In a series circuit, the heating power is proportional to the resistance and the square of resistor current. Since the electrical current is the same for each resistor and thus, the differences in heating power between each resistor are determined by their resistance. The district with more printed CNT layers has lower resistance and lower heating power, thus has a lower surface temperature. Under a constant voltage of 30 V, the highest surface temperature observed at the location with 10 printed CNT layers is 64 °C, whereas the lowest surface temperature observed at the location with 50 printed CNT layers is 52 °C. Consequently, through designing the number of printed CNT layers, we observe a temperature gradient on the surface of CNT/SMP composites. Therefore, by using SDM technique, it is convenient and flexible to manufacture CNT/SMP composites that used for heating areas need uniform temperature distribution or area needs different temperature at different locations.

To further investigate the electric heating performance of CNT/SMP composites, printed specimens are evaluated at different power densities. CNT/SMP composites are fabricated with different numbers of printed CNT layers from 10 to 50 layers with an interval of 10 layers. The CNT/SMP composites with 10 printed CNT layers is named as CS10, with 20 printed CNT layers as CS20, and so on. As Figure 3.8a shows, the surface temperature increases sharply in a short period of time after the voltage is applied, but as time goes, the surface temperature gradually reaches to its maximum point, which is called as the steady-state temperature. At the steady state, the input electrical power is balanced by the dissipated power, thus the surface temperature can keep stable at equilibrium state. The steady-state surface temperature of CS50 specimens increases when the power density increases. Steady-state temperature dependence on the input electrical power is plotted in Figure 3.8b. It is observed that the relationship between steady-state temperature and input power for these specimens is almost identical, with a slope of 286.39 °C/(W/cm$^2$), indicating that the final steady-state temperature of heating elements is mainly determined by the given input electrical power rather than the electrical resistance of specimens. However, the operation voltage required to reach the steady-state surface temperature is reduced by increasing the number of printed layers. For instance, to reach temperature around 90 °C, CS50 composite requires an input voltage of 40 V, whereas CS30 composite requires a higher input voltage of 50 V.

## 3.6 Electrical Actuation of the CNT/SMP Nanocomposites

Shape memory behavior is investigated according to bending tests, in which the specimens are firstly bent into “U” shapes at 100 °C and the “U” shapes should be fixed by cooling the specimens to room temperature. For bending tests, the angle between the two sides of the bent specimen at the time of 0 s is measured as $\theta_0$. When the specimens are electrically heated, the specimens start to recover to their permanent shape, and in this process the angle between the two sides of the bent specimen $\theta_t$ is measured at the variable time. The shape recovery ratio (SR) is calculated using the equation: $SR = (\theta_t - \theta_0)/(180 - \theta_0)$ [36, 37]. To investigate the localized actuation of CNT/SMP composites with different shape recovery ratios, the shape recovery process of the “L” shaped CNT/SMP composite (Figure 3.9a) is

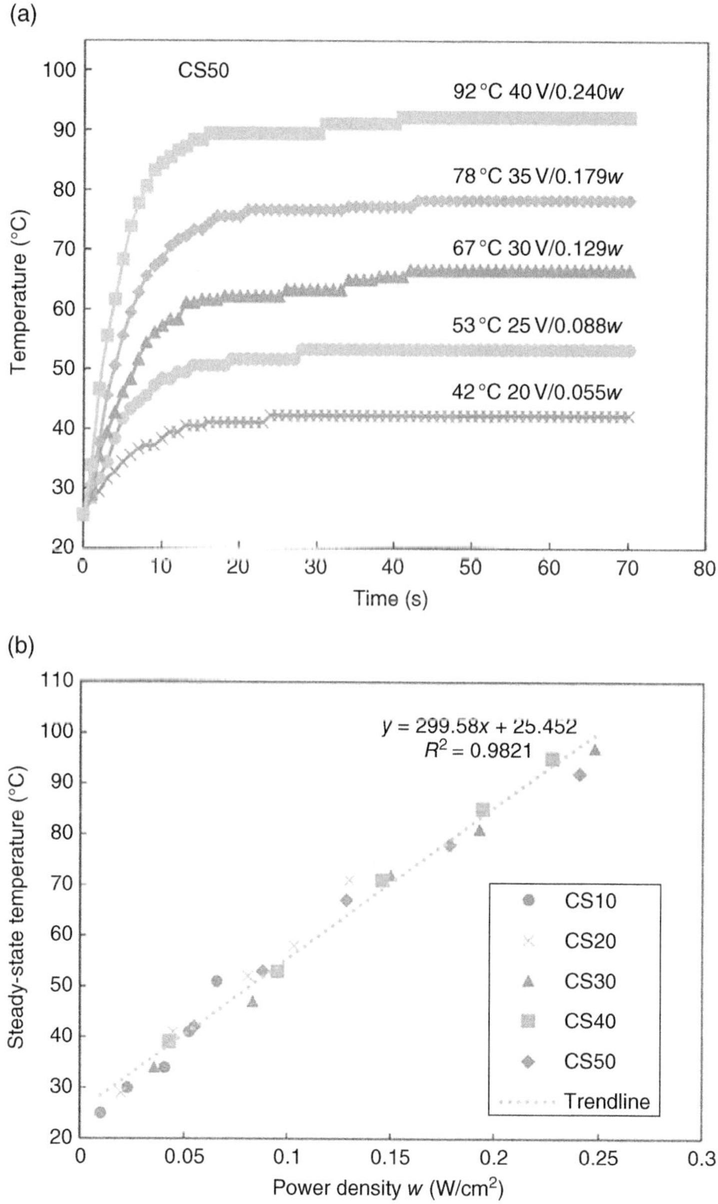


**Figure 3.8** (a) Time-dependent temperature of CS50 composites. (b) Steady-state temperature as a function of input power density.

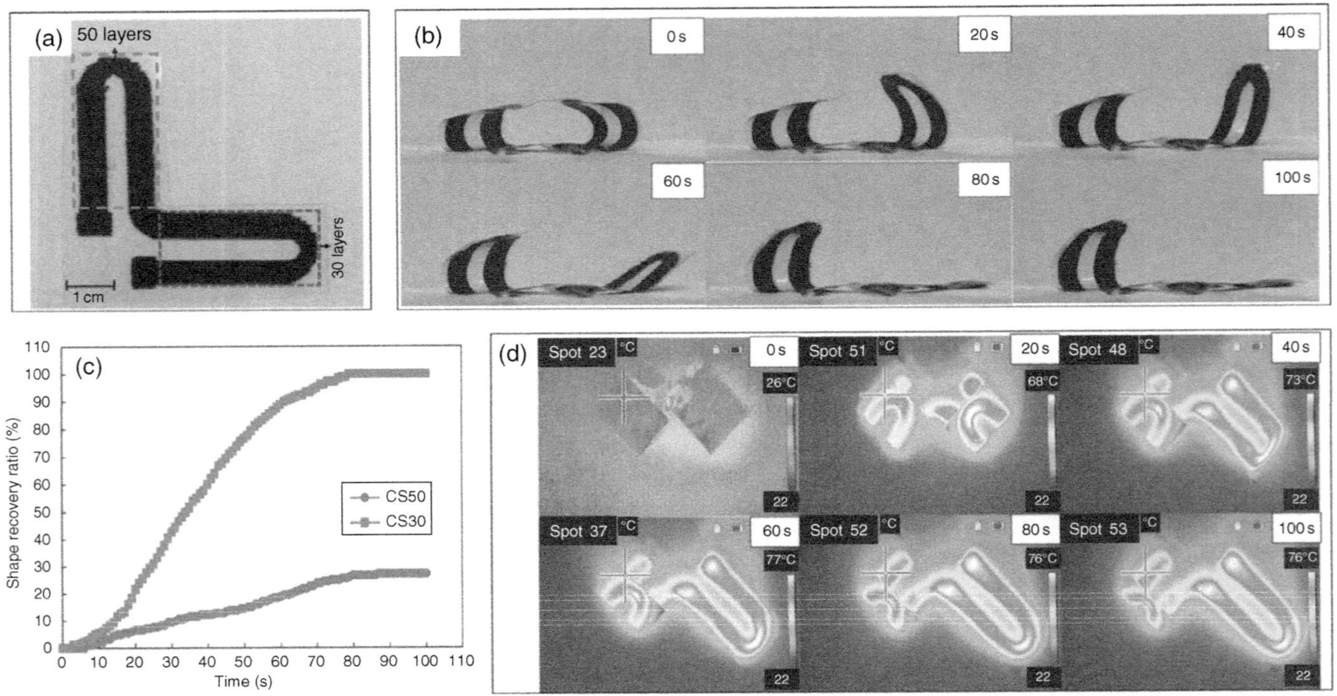


**Figure 3.9** (a) Picture of "L"-shaped CNT/SMP composites. (b) Snapshots of shape recovery process. (c) Shape recovery ratio over time. (d) Snapshots of temperature distribution.

studied. The vertical part of the "L" shaped specimen is fabricated with 50 CNT layers, and the horizontal part is fabricated with 30 CNT layers.

Both the vertical and horizontal parts are bent to "U" shape for the actuation test, which is done under a constant voltage of 60 V. Figure 3.9b shows the screenshots of the shape recovery process of the "L" shaped specimen. In this specimen, the part with 50 CNT layers and the part with 30 CNT layers can be treated as two series resistors. Since the electrical current is the same in these two parts, higher heating power can be achieved at the part with 30 CNT layers where higher electrical resistance is achieved. Therefore, the surface temperature of the part with 30 CNT layers id higher than the part with 50 CNT layers. As Figure 3.9d shows, the part with 30 CNT layers is heated to 75 °C that is about 20 °C above its glass transition temperature, whereas the temperature of CS50 is just above its glass transition temperature. Consequently, the part with 30 CNT layers is fully recovered to its original shape, but the part with 50 CNT could not have enough recovery torque to fully go back to its original shape and shows a recovery ratio of 27%. Thus, it can be seen that through designing the number of printed CNT layers at different locations, localized actuation of SMP composites with a desired shape recovery ratio could be performed, which gives great design flexibility for smart skin or tooling applications that need different deformations at different locations.

The actuation of the CNT/SMP honeycomb structure is also investigated. A basic honeycomb structure design with a dimension of 80 mm × 30 mm × 1 mm is used in this study. The printed CNT/SMP is stretched to 15% strain by a tensile testing machine. Figure 3.10 shows that the SMP nanocomposite is successfully actuated under 50 V. The shape recovery ratio reaches 100% in less than 90 s, and thermal images also show the temperature increase to 119 °C, which is much higher than its glass transition temperature of 55 °C. The excessive heat induces the shape recovery process. Thus, the successful demonstration of the honeycomb structure actuation brings a huge possibility for the fabrication of morphing wings with reduced weight.

## 3.7 Conclusions

This chapter presents a simple and accurate route to fabricate arbitrarily shaped carbon nanotube patterns onto different substrates based on a single-step spray deposition modeling (SDM) technique. The presented method significantly simplifies the fabrication process and improves its efficiency and flexibility. As a demonstration of applications, CNT-based strain sensors shows consistent resistance change under tensile strains. Up to a maximum strain amplitude of 45%, the strain-dependent resistance shows a good linearity. Strain sensing behavior under cyclic loading indicates that carbon nanotube strain sensors have high sensitivity at the high strain level. The SDM approach is able to control the

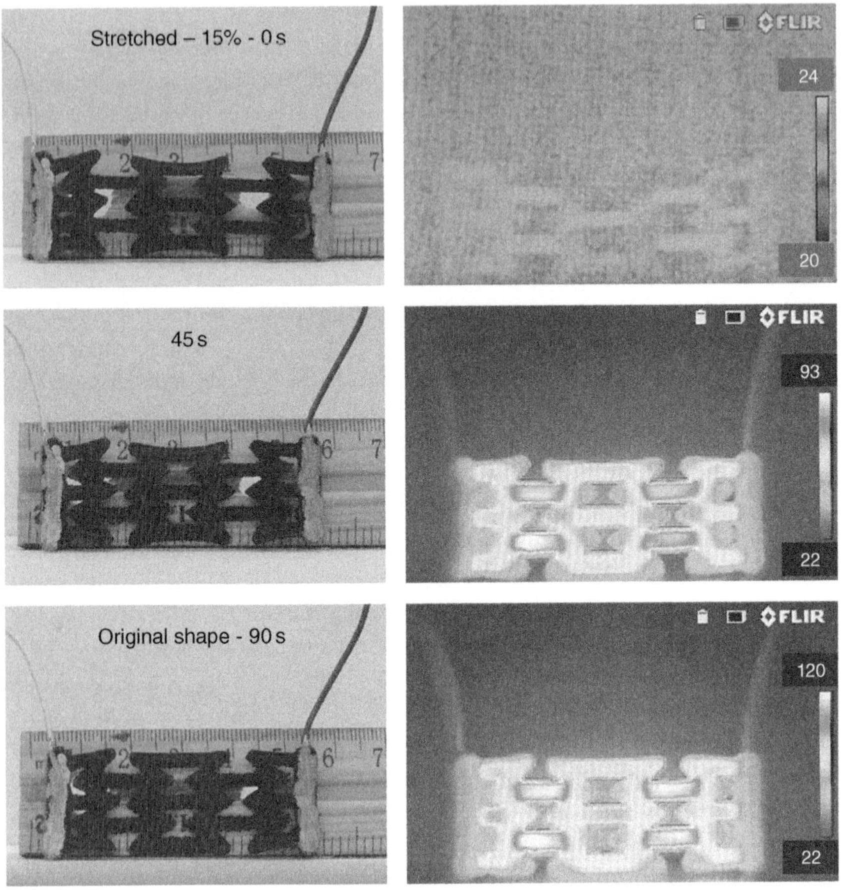


**Figure 3.10** Snapshots of shape recovery process and temperature distribution of printed nanocomposite with honeycomb structure.

sensing properties of the printed strain sensor by simply changing layers of printings. Multidirectional strain sensors are also demonstrated to monitor complicated deformation of human motion. The facial printing of this high-performance layered CNT/PDMS strain sensors, combining their high sensitivity, stretchability, and accuracy, paves a new way for the fabrication of wearable strain sensors for monitoring human motions. The self-heating performance of printed CNT-based films is also investigated. The self-temperature-control heating properties of printed CNT films are strongly dependent upon the numbers of printed CNT layers. The printed CNT/SMP composites show a fast thermal response to the applied voltage, and a steady-state temperature of ~92 °C is achieved in less than

20 s. The maximum temperature of the films could be controlled by adjusting the number of printed CNT layers. The quick heating and wide adjustability of temperature suggest its good application to high efficient heat pattern. The shape recovery effect of CNT/SMP composites can be triggered by their resistive heating without an external heater and the composite can reach a full shape recovery within 100 s. Particularly, by incorporating different numbers of printed CNT layers into SMP at specified locations, localized actuation of SMP with desired shape recovery ratio could be achieved. Finally, honeycomb-structured CNT/SMP composites are also successfully actuated through electrical stimulus and a full recovery of their original shape is observed in less than 90 s.

## References

1 Hull, C. W. (1986). Apparatus for production of three-dimensional objects by stereolithography. *Google Patents*. Washington, DC: U.S. Patent and Trademark Office.

2 Levy, G. N., Schindel, R., & Kruth, J.-P. (2003). Rapid manufacturing and rapid tooling with layer manufacturing (LM) technologies, state of the art and future perspectives. *CIRP Annals-Manufacturing Technology*, 52(2), 589–609.

3 Huang, S. H., Liu, P., Mokasdar, A., & Hou, L. (2013). Additive manufacturing and its societal impact: A literature review. *The International Journal of Advanced Manufacturing Technology*, 67(5–8), 1191–1203.

4 Kroll, E., & Artzi, D. (2011). Enhancing aerospace engineering students' learning with 3D printing wind-tunnel models. *Rapid Prototyping Journal*, 17(5), 393–402.

5 Wong, K. V., & Hernandez, A. (2012). A review of additive manufacturing. *ISRN Mechanical Engineering*, 2012, Article ID 208760, 10 pages.

6 Short, D. B. (2015). Use of 3D printing by museums: Educational exhibits, artifact education, and artifact restoration. *3D Printing and Additive Manufacturing*, 2(4), 209–215.

7 Murphy, S. V., & Atala, A. (2014). 3D bioprinting of tissues and organs. *Nature Biotechnology*, 32(8), 773–785.

8 De Volder, M. F., Tawfick, S. H., Baughman, R. H., & Hart, A. J. (2013). Carbon nanotubes: Present and future commercial applications. *Science*, 339(6119), 535–539.

9 Wood, J. R., Zhao, Q., Frogley, M. D., Meurs, E. R., Prins, A. D., Peijs, T., … Wagner, H. D. (2000). Carbon nanotubes: From molecular to macroscopic sensors. *Physical Review B*, 62(11), 7571.

10 Qiu, W., Li, Q., Lei, Z.-K., Qin, Q.-H., Deng, W.-L., & Kang, Y.-L. (2013). The use of a carbon nanotube sensor for measuring strain by micro-Raman spectroscopy. *Carbon*, 53, 161–168.

**11** Amjadi, M., Yoon, Y. J., & Park, I. (2015). Ultra-stretchable and skin-mountable strain sensors using carbon nanotubes? Ecoflex nanocomposites. *Nanotechnology*, 26(37), 375501.

**12** Cao, X., Wei, X., Li, G., Hu, C., Dai, K., Guo, J., … Guo, Z. (2017). Strain sensing behaviors of epoxy nanocomposites with carbon nanotubes under cyclic deformation. *Polymer*, 112, 1–9.

**13** Lee, D., Hong, H. P., Lee, M. J., Park, C. W., & Min, N. K. (2012). A prototype high sensitivity load cell using single walled carbon nanotube strain gauges. *Sensors and Actuators A: Physical*, 180, 120–126.

**14** Lipomi, D. J., Vosgueritchian, M., Tee, B. C., Hellstrom, S. L., Lee, J. A., Fox, C. H., & Bao, Z. (2011). Skin-like pressure and strain sensors based on transparent elastic films of carbon nanotubes. *Nature Nanotechnology*, 6(12), 788–792.

**15** Liu, H., Gao, J., Huang, W., Dai, K., Zheng, G., Liu, C., … Guo, Z. (2016). Electrically conductive strain sensing polyurethane nanocomposites with synergistic carbon nanotubes and graphene bifillers. *Nanoscale*, 8(26), 12977–12989.

**16** Chizari, K., Daoud, M. A., Ravindran, A. R., & Therriault, D. (2016). 3D printing of highly conductive nanocomposites for the functional optimization of liquid sensors. *Small*, 12(44), 6076–6082.

**17** Kim, K., Park, J., Suh, J.-h., Kim, M., Jeong, Y., & Park, I. (2017). 3D printing of multiaxial force sensors using carbon nanotube (Cnt)/thermoplastic polyurethane (Tpu) filaments. *Sensors and Actuators, A: Physical*, 263, 493–500.

**18** Wang, X., Jiang, M., Zhou, Z., Gou, J., & Hui, D. (2017). 3D printing of polymer matrix composites: A review and prospective. *Composites Part B Engineering*, 110, 442–458.

**19** Chu, K., Kim, D., Sohn, Y., Lee, S., Moon, C., & Park, S. (2013). Electrical and thermal properties of carbon-nanotube composite for flexible electric heating-unit applications. *Electron Device Letters, IEEE*, 34(5), 668–670.

**20** An, J.-E., & Jeong, Y. G. (2013). Structure and electric heating performance of graphene/epoxy composite films. *European Polymer Journal*, 49(6), 1322–1330.

**21** Tang, P., Zhang, R., Shi, R., & Bin, Y. (2015). Synergetic effects of carbon nanotubes and carbon fibers on electrical and self-heating properties of high-density polyethylene composites. *Journal of Materials Science*, 50(4), 1565–1574.

**22** Isaji, S., Bin, Y., & Matsuo, M. (2009). Electrical conductivity and self-temperature-control heating properties of carbon nanotubes filled polyethylene films. *Polymer*, 50(4), 1046–1053.

**23** Meng, H., & Li, G. (2013). A review of stimuli-responsive shape memory polymer composites. *Polymer*, 54(9), 2199–2221.

**24** Shabahang, S., Tao, G., Kaufman, J. J., Qiao, Y., Wei, L., Bouchenot, T., … Hoy, R. S. (2016). Controlled fragmentation of multimaterial fibres and films via polymer cold-drawing. *Nature*, 534, 529–533.

25 Lu, H., Wang, X., Yao, Y., Gou, J., Hui, D., Xu, B., & Fu, Y. (2015). Synergistic effect of siloxane modified aluminum nanopowders and carbon fiber on electrothermal efficiency of polymeric shape memory nanocomposite. *Composites Part B: Engineering*, 80, 1–6.

26 Everhart, M. C., Stahl, J. B., Traxler, E. W., & Havens, E. (2004). Shape memory polymer configurative tooling. *Smart Structures and Materials*, 5388, 87–94.

27 Sokolowski, W., Metcalfe, A., Hayashi, S., Yahia, L. H., & Raymond, J. (2007). Medical applications of shape memory polymers. *Biomedical Materials*, 2(1), S23.

28 Sun, J., Guan, Q., Liu, Y., & Leng, J. (2016). Morphing aircraft based on smart materials and structures: A state-of-the-art review. *Journal of Intelligent Material Systems and Structures*, 27(17), 2289–2312.

29 Mahapatra, S. S., Yadav, S. K., Yoo, H. J., Ramasamy, M. S., & Cho, J. W. (2014). Tailored and strong electro-responsive shape memory actuation in carbon nanotube-reinforced hyperbranched polyurethane composites. *Sensors and Actuators B: Chemical*, 193, 384–390.

30 Lu, H., Liu, Y., Gou, J. J., Leng, J., & Du, S. (2011). Surface coating of multi-walled carbon nanotube nanopaper on shape-memory polymer for multifunctionalization. *Composites Science and Technology*, 71(11), 1427–1434.

31 Zhao, H., & Bai, J. (2015). Highly sensitive piezo-resistive graphite nanoplatelet–carbon nanotube hybrids/polydimethylsilicone composites with improved conductive network construction. *ACS Applied Materials & Interfaces*, 7(18), 9652–9659.

32 Shimoni, A., Azoubel, S., & Magdassi, S. (2014). Inkjet printing of flexible high-performance carbon nanotube transparent conductive films by "coffee ring effect". *Nanoscale*, 6(19), 11084–11089.

33 Obitayo, W., & Liu, T. (2012). A review: Carbon nanotube-based piezoresistive strain sensors. *Journal of Sensors*, 2012, Article ID 652438, 15 pages.

34 Lu, N., Lu, C., Yang, S., & Rogers, J. (2012). Highly sensitive skin-mountable strain gauges based entirely on elastomers. *Advanced Functional Materials*, 22(19), 4044–4050.

35 Tang, Y., Zhao, Z., Hu, H., Liu, Y., Wang, X., Zhou, S., & Qiu, J. (2015). Highly stretchable and ultrasensitive strain sensor based on reduced graphene oxide microtubes–elastomer composite. *ACS Applied Materials & Interfaces*, 7(49), 27432–27439.

36 Tang, Z., Sun, D., Yang, D., Guo, B., Zhang, L., & Jia, D. (2013). Vapor grown carbon nanofiber reinforced bio-based polyester for electroactive shape memory performance. *Composites Science and Technology*, 75, 15–21.

37 Feng, X., Zhang, G., Zhuo, S., Jiang, H., Shi, J., Li, F., & Li, H. (2016). Dual responsive shape memory polymer/clay nanocomposites. *Composites Science and Technology*, 129, 53–60.

# Part II

# 3D printing challenges for production of complex objects

# 4

# Computational Design of Complex 3D Printed Objects

*Emiel van de Ven*[1], *Can Ayas*[2], *and Matthijs Langelaar*[2]

[1] *II Optomechatronics Group, Netherlands Organisation for Applied Scientific Research TNO, Delft, The Netherlands*
[2] *I Structural Optimization and Mechanics Group, Delft University of Technology, Delft, The Netherlands*

## 4.1 Introduction

Paradoxically, with the tremendous increase in design freedom offered by 3D printing compared with conventional manufacturing, in many ways the task of designers has become harder instead of easier. While there are fewer restrictions to consider due to manufacturing limitations, the number of design possibilities and the associated complexity have increased to a level that exceeds the capabilities of human designers to fully exploit. This has accelerated the demand for computational design techniques, algorithms that assist designers in finding optimized geometries that best meet application demands. A prominent computational design method often associated with 3D printing is *topology optimization* (TO). This method allows the generation of designs without requiring the designer to provide a specific initial concept: instead, TO methods have the capability to modify and optimize the topology and shape of a part simultaneously. In a TO-based computational design process, the task of the designer shifts to the specification of the appropriate design objectives and constraints that describe the part's desired functional characteristics.

A variety of TO methods exists, detailed overviews are provided by [1, 2]. However, in both the commercial and research domains, *density-based* TO method is most prominent. The focus of this chapter is also on this method. It is based on defining a part in terms of a material distribution in a given design volume, rather than using geometrical primitives to describe its geometry. A 3D part can thus be defined by a (normalized) density distribution in space, discretized in voxels or other

*3D Printing for Energy Applications*, First Edition. Edited by Albert Tarancón and Vincenzo Esposito.

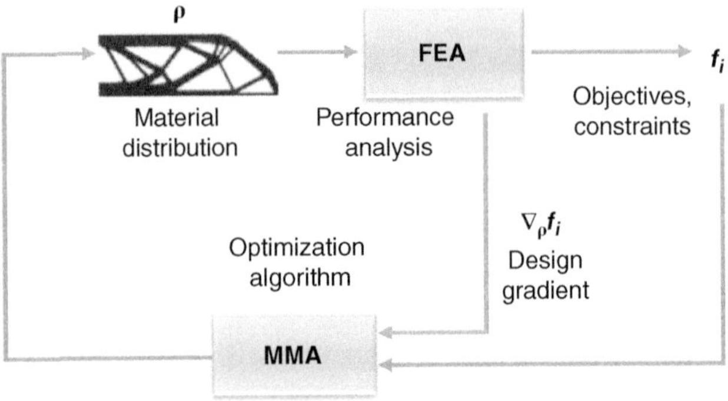


**Figure 4.1** Topology optimization flowchart. FEA: Finite element analysis, MMA: Method of moving asymptotes.

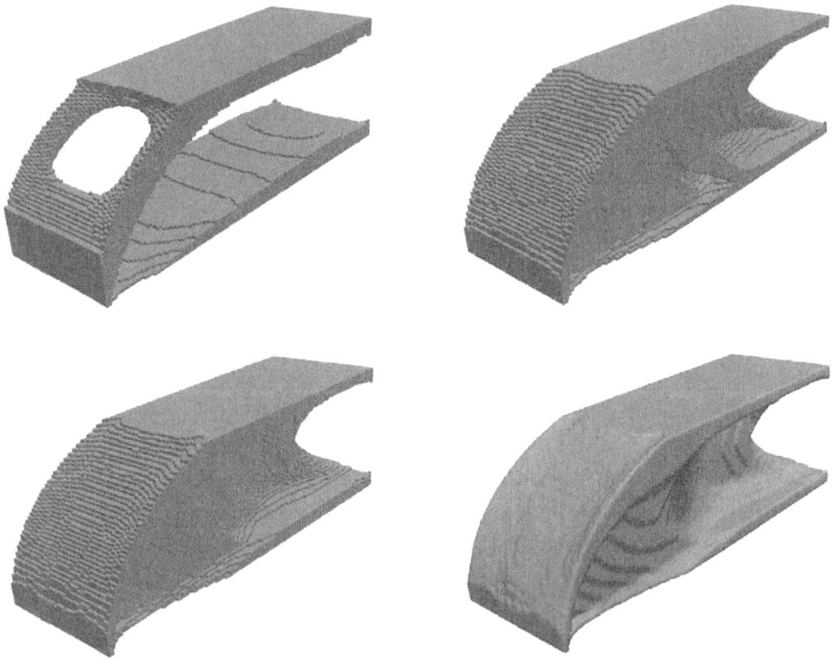

**Figure 4.2** Sequence of design evolution of a bracket (selected snapshots). The final geometry is obtained through post-processing the TO result.

subvolumes [3]. The chosen discretization defines the ultimate resolution and complexity of the generated part. Note the conceptual similarity with the way a 3D printer places matter in the build volume in order to create a component. Allowing continuous density variations (0–100%), comparable to porosity levels in printed parts, enables the use of gradient-based iterative mathematical optimization algorithms to efficiently optimize the density distribution, in order to extremize the chosen objective(s) while satisfying the constraints. In typical implementations, finite element analysis (FEA) is used to evaluate a part's performance (e.g., stiffness), while density design variables $\rho$ are associated to the FE mesh. Figure 4.1 depicts a flowchart of a typical TO process, where also the role of the design gradient is indicated. Efficient evaluation of this gradient information is what enables this form of computational design. An example of the evolution of a bracket optimized for maximum stiffness subject to a weight limit is shown in Figure 4.2. The final 3D density distribution that is obtained is typically converted to a boundary representation in CAD or STL format in a post-processing operation.

## 4.2 Dedicated Computational Design for 3D Printing

While 3D printing processes can realize parts of much higher geometrical complexity compared with conventional manufacturing, invariably current 3D printing technologies also involve specific limitations that must be accounted for. Typical examples are the achievable minimum feature or hole sizes, maximum part dimensions, or surface roughness. It is important to take such limitations into account during the design process, to ensure that designs can actually be realized. This philosophy is also known as *Design for Additive Manufacturing* (DfAM) [4, 5]. In order to allow designers to generate viable, manufacturable designs, it is imperative to extend computational design techniques such as TO with printing-specific design requirements.

Printability design rules such as minimum feature or hole dimensions can be incorporated into TO relatively easily using, for example, density or morphology filters which have already been developed for conventional TO procedures [6]. Another important geometric restriction that is common to many 3D printing processes is the *overhang angle* limitation. This refers to a limit of the angle between downfacing surfaces and the base plate (Figure 4.3): too shallow angles lead to unacceptable material or surface quality, depending on the process. For the popular

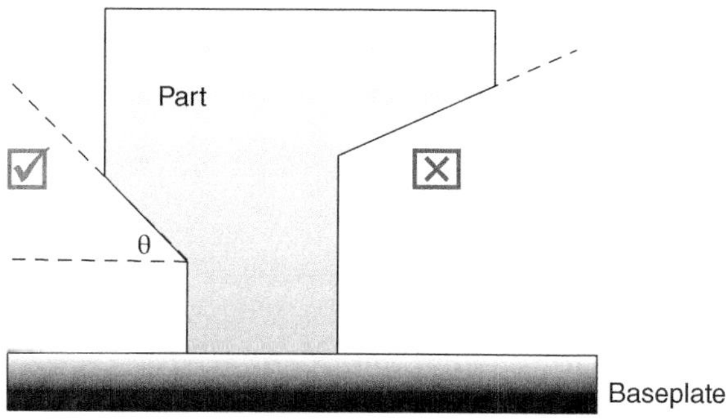


**Figure 4.3** Overhang angle definition.

selective laser melting (SLM) process used for metal printing [7], for example, typically a critical overhang angle of 45° is considered [8, 9]. When for a given part and build orientation this limitation is not respected, additional support material must be printed to realize the part. However, next to the added printing and material cost, this also leads to significant post-processing costs due to support removal, which is mostly a manual process. For these reasons, handling overhang constraints in a rational way is highly desired in computational design methods for 3D printing.

### 4.2.1 Overhang Angle Control Approaches

The seemingly simple geometric requirement of restricting the angle of downfacing surfaces can be included in TO in a number of ways. Instead of a detailed technical discussion, we present here three main directions: the full formulation and further details of the related approaches can be found in the respective references. The three main overhang angle control approaches that have been proposed are described.

#### 4.2.1.1 Local Angle Control

By detecting the part boundary and evaluating its angle with respect to the base plate (or build direction), local constraints can be defined to avoid unprintable surfaces. This approach has been attempted in density-based TO [10] and also in level-set-based TO [11], where boundary and angle information is more readily available. However, it turns out that only limiting boundary angles is insufficient to render printable geometries. Parts must also be sufficiently supported from the base plate during each stage of the printing process. The local angle control approaches do not take this into consideration. A typical inadequate result is seen in Figure 4.4a. With additional measures, such as a length scale constraint, this can be corrected [12].

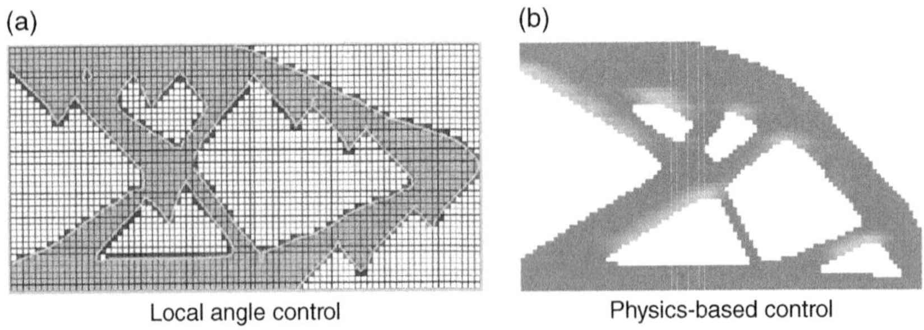


**Figure 4.4** Various overhang angle control approaches. (a) Limits the angle of downfacing surfaces, but introduces hanging points, (b) limits the maximum temperature obtained through in a thermal process simulation.

#### 4.2.1.2 Physics-Based Constraints

Instead of using geometrical (material distribution) criteria directly, this category of methods is based on evaluating the physical response of the geometry. Popular versions involve the evaluation of stiffness under self-weight [e.g. [11]], or the thermal response to a heat load at the top surface of a certain print stage [13]. By evaluating constraining this physical response at various intermediate stages of the printing process, the computational design process is stimulated to create geometries that have better stiffness against vertical loading, or improved effective conduction to the base plate. This discourages overhanging features, however in an indirect manner. Precise overhang angle control is not possible, and the computational cost of these methods is high due to the need for additional simulations in each design iteration. An example is shown in Figure 4.4b.

However, physics-based approaches should not be dismissed too easily. Ultimately, the limitations of a printing process are based on the process physics. With the appropriate models, these approaches may give printability predictions that are more precise than the current empirical geometric design rules.

#### 4.2.1.3 Simplified Printing Process

This class of methods introduces a simplified virtual printing process simulation to identify the regions of the part that are printable without additional supports. The performance of the printable remainder can then be used to drive the optimization, as illustrated in Figure 4.5. Since the printing simulation is evaluated in each design iteration, it is important that it has low computational cost. Existing methods make use of either local printability rules on structured or unstructured grids [14, 15] or employ front propagation to distinguish printable and unsupported regions [16]. The latter is discussed in more detail in Section 4.3.2. The virtual fabrication concept can be generalized and extended to include additional

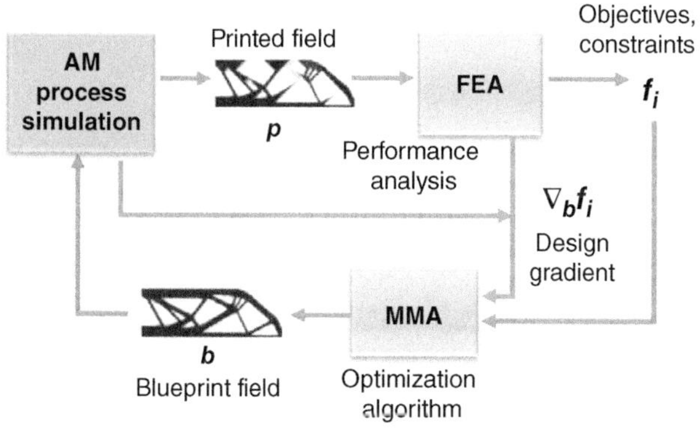


**Figure 4.5** Overhang angle control approach based on inclusion of a simplified AM process simulation in the TO loop.

process-induced restrictions. A modified version of [15] is currently available in commercial software [17].

Note that in addition the part orientation in the build chamber has a strong influence on the amount and extent of downfacing surfaces violating the critical overhang angle. Simultaneously controlling part orientation and part geometry generation is a challenging problem, and typically current overhang control approaches are applied for a fixed part orientation. An approach to solve the simultaneous problem has been demonstrated in 2D [18].

### 4.2.2 Design Scenarios

A computational method to determine the printable regions of a given part geometry enables different design scenarios for 3D printing. First, a designer can use conventional computational design approaches, for example, TO, to create an optimized part for a certain desired performance. Once this is available, the most favorable part orientation for printing can be determined, and commercial software packages exist that can assist in this task. Any remaining downfacing surfaces that violate the overhang restriction must be supported by sacrificial supports, which must be added to the print job and removed afterward. This scenario maintains the highest part performance, but also has the highest cost in terms of support material and removal. An example is shown in Figure 4.6a.

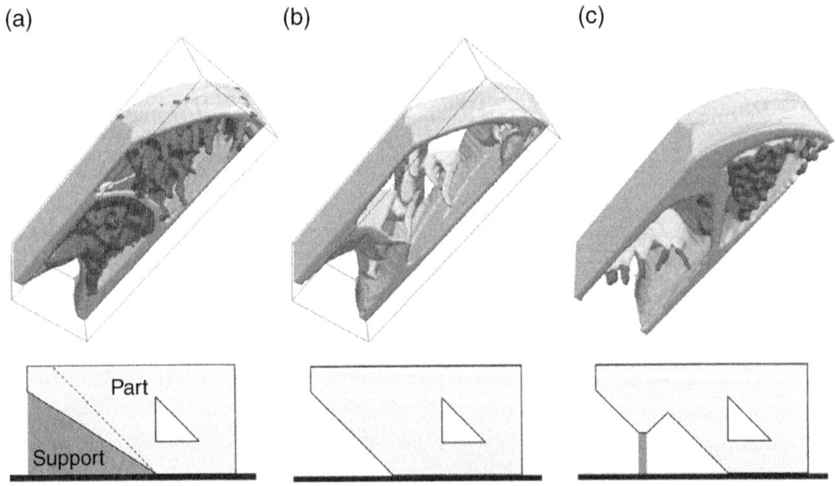


**Figure 4.6** Conceptual sketches and TO results in case of three design scenarios: (a) no overhang restriction, (b) overhang restriction (self-supporting part), and (c) trade-off solution allowing a moderate amount of support material. Designs (b) and (c) are 8% and 4% less stiff compared with case (a), respectively.

Second, one can design parts that are entirely self-supporting. This corresponds to the scheme previously depicted in Figure 4.5. By fixing the build orientation and excluding any use of support material, a relatively strict constraint is placed on the part's shape. Depending on the application, this can result in considerable reductions in performance, compared with designs generated without any manufacturing constraints. However, as supports are not needed in this case, added costs are minimal. This scenario is most often considered in the current literature on computational design for 3D printing. Figure 4.6b shows an example. In this case, a performance reduction of 8% was observed compared with case (a), but in contrast to the previous scenario, a fully self-supporting part is obtained.

Between these extremes of what is essentially a multi-objective design problem, an infinite number of trade-off solutions exist: cases where a certain amount of support material is allowed, which relaxes the overhang-based design constraints. Here, computational design methods can either use a strategy where regions that require support are detected and an appropriate support volume is estimated, or a secondary support density field is included in the TO process and the support layout is optimized simultaneously with the part itself [18, 19]. In both cases, it is up to the designer to set priorities the relative importance of performance versus support cost. For parts where the emphasis is on performance probably support costs are not the leading consideration, but for cost-sensitive printed parts in volume production, avoiding the use of support material is more important. For this scenario, an example is given in Figure 4.6c. The obtained design requires significantly less support material compared with case (a), and its performance is only reduced by 4%, instead of 8% for the self-supporting design. These values are of course case-dependent, and in other design problems much larger differences may occur.

## 4.3 Case Study: Computational Design of a 3D-Printed Flow Manifold

To illustrate the use of TO-based computational design for 3D printing, with specific attention to overhang constraints, a flow manifold is presented. This manifold facilitates fluid flow from an inlet to an outlet at predefined positions. Such components are found in numerous energy applications, for example, oil and gas, turbines in hydropower, process and aerospace industry, etc. Conventional manifolds often consist of straight cylindrical channels drilled in a solid block of metal, meeting at right angles. This results in significant drag forces in the fluid due to the suboptimal flow configuration, which results in considerable pressure loss which in turn increases power requirements for pumps. We here explore the potential to minimize this power dissipation over a manifold, by exploiting the design freedom that 3D printing offers. Moreover, we do so by employing a TO-based computational design approach, combined with overhang constraint control using a front propagation-based printing process simulation [16]. Generating self-supporting designs is critical in this application, because support material cannot be removed from inaccessible regions such as the inside of channels.

### 4.3.1 Fluid Flow TO

TO of fluid problems started with the work of Borrvall and Petersson [20], in which they optimized Stokes flow problems by modeling the design domain as a porous medium. The density field from TO is coupled to the flow problem by scaling the permeability of the medium: high-density regions are considered impermeable, while regions with a low density are considered porous. Since then, optimization of Stokes flow became an active research topic [21–23], and has been extended to laminar Navier–Stokes [24–28], and later to the turbulent flow regime [29–32]. Here, we do not focus on the fluid flow aspects of the problem but instead demonstrate the combination of fluid TO with 3D printing constraints to arrive at a manufacturable design. Therefore, although some publications show complete differentiation of the Navier–Stokes equations and turbulence model [29, 30, 32], we will follow the simplified "frozen turbulence" approach outlined in [33], where the turbulence model is not derived w.r.t. the design variables. For this, we used the OpenFOAM framework [34]. Before including the 3D-printing constraint, we will first look at plain fluid optimization.

Assuming steady-state incompressible flow and adopting the porous medium approach to TO, the Reynolds-averaged Navier–Stokes equations which describe the motion of the fluid are given by

$$\begin{aligned} \rho_0\left(\mathbf{v}\cdot\nabla\right)\mathbf{v} &= -\nabla p + \mu\Delta\mathbf{v} - \alpha\mathbf{u}, \\ \nabla\cdot\mathbf{v} &= 0, \end{aligned} \tag{4.1}$$

where $\rho_0$ and $\mu$ are the fluid density and viscosity assumed to be constant, $\mathbf{v}$ and $p$ are the mean velocity and pressure fields. $\alpha\mathbf{u}$ is the Darcy term, where $\alpha$ is a function of the pseudo-density field $\rho \in [0, 1]$, and varies between 0 (void) and a maximum value $\alpha_{\text{max}}$ (impermeable):

$$\alpha = \alpha_{\text{max}}\rho. \tag{4.2}$$

The value of $\alpha_{\text{max}}$ is dependent on the fluid and domain properties, and can be determined using Darcy's law [27]. By prescribing the inflow velocity and outlet pressure gradient, the fluid flow can be solved for any given manifold layout using Equation 4.1.

In order to optimize a manifold for minimum power loss, the following minimization problem is solved:

$$
\begin{aligned}
&\min_{\rho\in[0,1]} \quad -\int\left(p+\frac{1}{2}v^2\right)\mathbf{v}\cdot n d\Gamma \\
\text{s.t.} \quad &\rho_0\left(\mathbf{v}\cdot\nabla\right)\mathbf{v} = -\nabla p + \mu\Delta\mathbf{v} - \alpha_{\max}\rho\mathbf{u}, \\
&\nabla\cdot\mathbf{v} = 0, \\
&\int_{\Omega} 1-\rho d\Omega < V_{\lim},
\end{aligned}
\tag{4.3}
$$

where the power loss is calculated as the net inward flux of energy [31], and $V_{\lim}$ is the maximum allowable volume of the channel. The optimization problem is solved using gradient-based optimization algorithms, which require sensitivity information of the objective function w.r.t. the design variables. The sensitivities are obtained in a continuous manner by solving the adjoint RANS equations. For the optimization algorithm, a so-called one-shot approach is used, where the optimization updates the design before the SIMPLE algorithm [35] that is used to solve the RANS equations is converged.

The fluid TO is first demonstrated on a pipe-bend example. In this example, the flow enters the domain from the inlet, and has to make a 90° turn to reach to the outlet, as is shown in Figure 4.7a. The optimal design is shown in Figure 4.7b, where a channel is formed from the inlet to outlet. From here on, we will visualize the fluid domain instead of the optimized manifold, as shown in Figure 4.7c for ease of interpretation when the 3D-printing constraint is introduced.

Although the optimized channel shown in Figure 4.7c has a simple geometry, it cannot be printed without the addition of support structures inside the channel, which are difficult to remove afterward. Therefore, in order to avoid using support

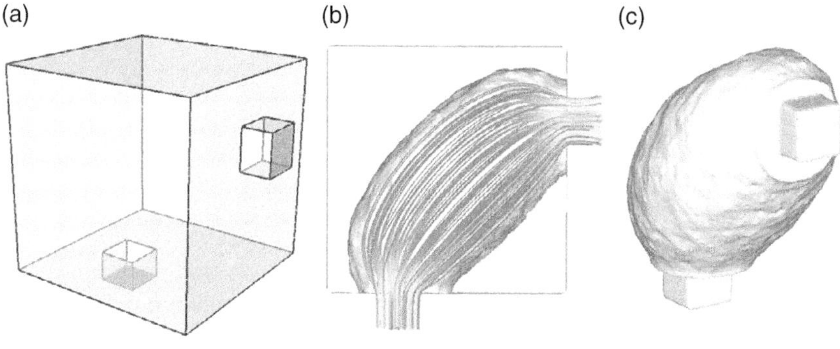


**Figure 4.7** Topology optimization of a pipe bend. (a) Design domain with inlet (red) and outlet (green). (b) Optimized manifold with flow. (c) Optimized channel.

structures, a post-processing step can be performed on the optimized domain to make the channel printable. For example, the shape of the channel can be changed from round to droplet-like, as shown in [36, 37]. For simple geometries, this is a feasible approach. However, for more complex geometries and systems with multiple adjacent channels, this might be impossible. Furthermore, the optimality will be compromised to an unknown degree. In order to obtain an optimal and printable design, the overhang limitation has to be included into the TO. In [16], it is shown how this can be done using a continuous approach, which we will demonstrate in the following.

### 4.3.2 Front Propagation-Based 3D Printing Constraint

Most overhang angle control schemes are implemented as a filter into TO, as shown in Figure 4.5. For a given density field $\rho$, the overhang filter returns a printable density field $\rho_p$, which excludes the regions that violate the overhang limitation. The numerical analysis performed to evaluate the part performance is then carried out on the printable densities $\rho_p$ instead of the original densities $\rho$. By providing sensitivities of this filter to the optimization algorithm, regions that are crucial for the performance can be properly supported, instead of being removed by the overhang filter.

The overhang filter presented in [16] is unique in detecting regions that violate the overhang limitation. While most algorithms are discrete in nature, the front propagation method of [16] is described by a partial differential equation, allowing discretization on unstructured grids, and continuously varying critical overhang angles. Front propagation algorithms calculate how a front propagates through space with time, for example, the wavefront produced by a rock thrown into the water. As output, it produces an arrival time field $T$, which for each location gives the time at which the front reaches that location. This is used as a crude approximation of the printing sequence, which can also be seen as a front starting at the base plate, propagating through space as the part is printed.

Front propagation is described by the Eikonal equation [38]:

$$\|\nabla T(\mathbf{x})\| = \frac{1}{f(\mathbf{x},\mathbf{n})}, \quad \mathbf{x} \in \Omega, \tag{4.4}$$

$$T(\mathbf{x}) = 0, \quad \mathbf{x} \in \Gamma, \tag{4.5}$$

where $T$ is the arrival time field, and $f(\mathbf{x}, \mathbf{n})$ is a function that determines the propagation speed at a position $\mathbf{x}$ at a given propagation direction $\mathbf{n} = \nabla T/\|\nabla T\|$. Consider the domain $\Omega$, which represents a manifold with an internal channel. It is to be printed in the given build direction $\mathbf{b}$, on the base plate $\Gamma$. In order to detect regions that are not printable, a front propagation is performed. The propagation speed is reduced when the propagation direction is below the allowable overhang

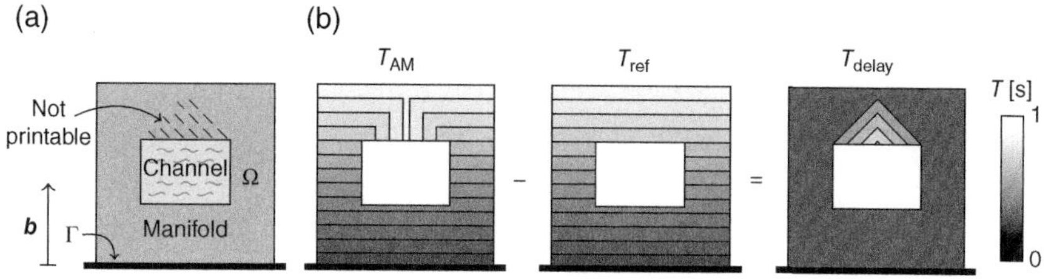


**Figure 4.8** Topology optimization of a pipe bend. (a) Initial geometry. (b) Procedure of OH detection. Regions where $T_{delay} = 0$ are printable.

angle. This, for instance, happens right above the channel, as can be seen for the $T_{AM}$ field in Figure 4.8b. The arrival time field $T_{AM}$ is compared with a reference field $T_{ref}$, which is simply the distance to the base plate Γ for each location. The difference between both fields reveals where the front propagation was delayed, and thus where the overhang limitation is violated. This is shown in $T_{delay} = T_{AM} - T_{ref}$. The field $T_{delay}$ is used to construct the printable densities $\rho_p$, by:

$$\rho_p(\mathbf{x}) = 2^{-\beta T_{delay}(\mathbf{x})}, \tag{4.6}$$

where $\beta$ is a parameter that influences the crispness of the design. With the relation given in Equation 4.6, regions that are not printable are removed: wherever $T_{delay} = 0$ (printable), $\rho_p = 1$ (material), and for $T_{delay} \gg 0$ (not printable), $\rho_p \approx 0$ (void).

### 4.3.3 Fluid TO with 3D Printing Constraint

The overhang filter described above is applied to the same pipe-bend case as previously optimized without overhang filter (Figure 4.7). As can be seen in Figure 4.9a, the shape of the channel is changed to a printable, droplet-like channel. If this would have been done as a post-processing step by removing the overhanging regions of the result shown in Figure 4.7c, one arrives at the design shown in Figure 4.9b. This is not a satisfactory result, as the volume of the channel increases, and the channel touches the boundary of the design domain at the top. These problems are avoided by using an overhang filter: as the overhang filter is active during the optimization, the volume constraint remains satisfied, and the solution is optimal.

Another example where the change in design is more prominent is given in Figure 4.10. Again the power dissipation is minimized, but now there are four outlets. Without overhang filter, the inlet is connected to all the outlets as can be seen in Figure 4.10b. However, with the introduction of the overhang filter, the outlet that is the furthest from the inlet is disconnected. With the limited volume that is available, due to the extra volume required for the droplet-like shaped channels, it is apparently more beneficial to only connect to the three closest

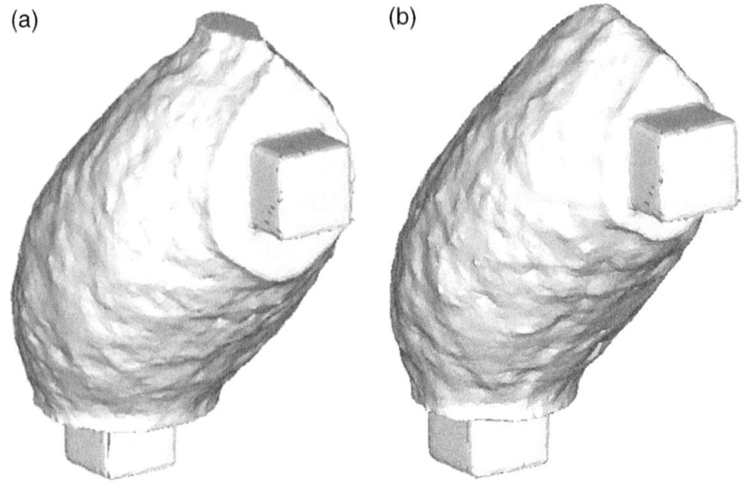


**Figure 4.9** Topology optimization of a pipe bend with an overhang filter. (a) Optimal overhang-free design. (b) Overhang-free design by post-processing.

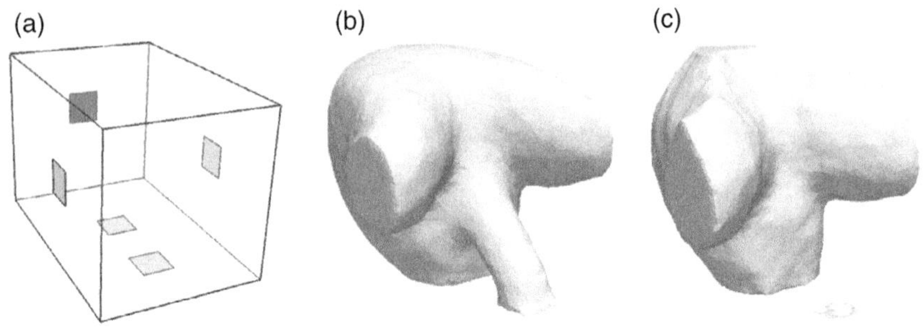


**Figure 4.10** Topology optimization of a simple manifold. (a) Design domain with inlet (at back) and outlets. (b) Optimized design using all outlets. (c) Optimized overhang-free design, using 3 out of 4 outlets.

outlets. Such a change in design would be extremely difficult to predict and perform in a post-processing setting, but happens naturally when the overhang limitation is included during the optimization phase.

## 4.4 Current State and Future Challenges

Additive manufacturing is in constant development, and computational design techniques for AM are constantly evolving as well. In the past years, the intensity of research on TO and AM has increased exponentially. Driven by customer demand, successful methods have rapidly been integrated in commercial software.

At the time of writing, several commercial TO packages offer their users some form of AM overhang control, mostly based on the simplified printing process approach discussed in Section 4.2.1. Other software packages are capable to perform suitably predictive thermal, mechanical, and thermomechanical AM process simulations that enable users to spot potential problems in advance or to find the root causes of printing problems that occurred during build.

At this moment, however, there is still a clear divide between these high-fidelity physics-based AM process simulations and computational design for AM. There are more aspects to consider at the design stage than only overhang control: temperatures, stresses, and distortions must remain within bounds during the printing process, the development of specific microstructures in the material could be targeted, process parameters, and, for example, hatching patterns can be tailored to desired effect, etc. Also in these aspects future designers will greatly benefit from computational design methods that integrate sophisticated yet efficient AM process models with generative design techniques such as TO. A major challenge to realize this is the computational effort that is required for both the AM process simulation itself, and the sensitivity analysis needed to perform gradient-based design optimization. Smart simplifications based on insight in the process physics, model reduction approaches, parallelization, and ways to reuse previously computed information during the optimization process are all ingredients that must be fused in an additive way to make this a reality.

## References


1 Deaton, J. D., & Grandhi, R. V. (2014). A survey of structural and multidisciplinary continuum topology optimization: Post 2000. *Structural and Multidisciplinary Optimization*, 49(1), 1–38.

2 Sigmund, O., & Maute, K. (2013). Topology optimization approaches. *Structural and Multidisciplinary Optimization*, 48(6), 1031–1055.

3 Bendsøe, M. P., & Sigmund, O. (2003). *Topology optimization – theory, methods and applications*. Berlin: Springer-Verlag.

4 Gibson, I., Rosen, D. W., & Stucker, B. (Eds.) (2010). Design for additive manufacturing. In *Additive manufacturing technologies* (pp. 299–332). Springer. https://link.springer.com/chapter/10.1007/978-1-4419-1120-9_11#citeas

5 Rosen, D. (2014). Design for additive manufacturing: Past, present, and future directions. *Journal of Mechanical Design*, 136(9), 090301.

6 Sigmund, O. (2007). Morphology-based black and white filters for topology optimization. *Structural and Multidisciplinary Optimization*, 33(4-5), 401–424.

7 Yap, C. Y., Chua, C. K., Dong, Z. L., Liu, Z. H., Zhang, D. Q., Loh, L. E., & Sing, S. L. (2015). Review of selective laser melting: Materials and applications. *Applied Physics Reviews*, 2(4), 041101.

**8** Mertens, R., Clijsters, S., Kempen, K., & Kruth, J.-P. (2014). Optimization of scan strategies in selective laser melting of aluminum parts with downfacing areas. *Journal of Manufacturing Science and Engineering*, 136(6), 061012.

**9** Wang, D., Yang, Y., Yi, Z., & Su, X. (2013). Research on the fabricating quality optimization of the overhanging surface in SLM process. *The International Journal of Advanced Manufacturing Technology*, 65(9–12), 1471–1484.

**10** Qian, X. (2017). Undercut and overhang angle control in topology optimization: A density gradient based integral approach. *International Journal for Numerical Methods in Engineering*. doi:10.1002/nme.5461

**11** Allaire, G., Dapogny, C., Estevez, R., Faure, A., & Michailidis, G. (2017). Structural optimization under overhang constraints imposed by additive manufacturing technologies. *Journal of Computational Physics*, 351, 295–328.

**12** Zhang, K., Cheng, G., & Xu, L. (2019). Topology optimization considering overhang constraint in additive manufacturing. *Computers & Structures*, 212, 86–100.

**13** Ranjan, R., Ayas, C., Langelaar, M., & Van Keulen, F. (2018, July 22–25). Towards design for precision additive manufacturing: A simplified approach for detecting heat accumulation. Proceedings of the 2018 ASPE and euspen Summer Topical Meeting: Advancing Precision in Additive Manufacturing of Berkeley, United States, 29–34. https://repository.tudelft.nl/islandora/object/uuid:8866dd52-247b-482f-bf56-238bca3321d1

**14** Gaynor, A. T., & Guest, J. K. (2016). Topology optimization considering overhang constraints: Eliminating sacrificial support material in additive manufacturing through design. *Structural and Multidisciplinary Optimization*, 54(5), 1157–1172.

**15** Langelaar, M. (2016). Topology optimization of 3D self-supporting structures for additive manufacturing. *Additive Manufacturing*, 12, 60–70.

**16** van de Ven, E., Maas, R., Ayas, C., Langelaar, M., & van Keulen, F. (2018). Continuous front propagation-based overhang control for topology optimization with additive manufacturing. *Structural and Multidisciplinary Optimization*, 57(5), 2075–2091.

**17** Hoffarth, M., Gerzen, N., & Pedersen, C. (2017, June 5–9). ALM overhang constraint in topology optimization for industrial applications. Proceedings of the 12th World Congress on Structural and Multidisciplinary Optimisation, Braunschweig, Germany.

**18** Langelaar, M. (2018). Combined optimization of part topology, support structure layout and build orientation for additive manufacturing. *Structural and Multidisciplinary Optimization*, 57(5), 1985–2004.

**19** Mirzendehdel, A. M., & Suresh, K. (2016). Support structure constrained topology optimization for additive manufacturing. *Computer-Aided Design*, 81, 1–13.

**20** Borrvall, T., & Petersson, J. (2003. ISSN 0271-2091, 1097-0363). Topology optimization of fluids in Stokes flow. *International Journal for Numerical Methods in Fluids*, 41(1), 77–107. doi:10.1002/fld.426

**21** Aage, N., Poulsen, T. H., Gersborg-Hansen, A., & Sigmund, O. (2008. ISSN 1615-147X, 1615-1488). Topology optimization of large scale Stokes flow problems. *Structural and Multidisciplinary Optimization*, 35(2), 175–180. doi:10.1007/s00158-007-0128-0

**22** Evgrafov, A. (2005. ISSN 0095-4616, 1432-0606). The limits of porous materials in the topology optimization of Stokes flows. *Applied Mathematics and Optimization*, 52(3), 263–277. doi:10.1007/s00245-005-0828-z

**23** Guest, J. K., & Prvost, J. H. (2006. ISSN 0029-5981, 1097-0207). Topology optimization of creeping fluid flows using a Darcy-Stokes finite element. *International Journal for Numerical Methods in Engineering*, 66(3), 461–484. doi:10.1002/nme.1560

**24** Deng, Y., Liu, Z., Zhang, P., Liu, Y., & Wu, Y. (2011. ISSN 00219991). Topology optimization of unsteady incompressible NavierStokes flows. *Journal of Computational Physics*, 230(17), 6688–6708. doi:10.1016/j.jcp.2011.05.004

**25** Gersborg-Hansen, A., Sigmund, O., & Haber, R. B. (2005. ISSN 1615-147X, 1615-1488). Topology optimization of channel flow problems. *Structural and Multidisciplinary Optimization*, 30(3), 181–192. doi:10.1007/s00158-004-0508-7

**26** Sebastian Kreissl, Georg Pingen, and Kurt Maute. Topology optimization for unsteady flow. *International Journal for Numerical Methods in Engineering*, pages n/a–n/a, 2011. ISSN 00295981. doi: 10.1002/nme.3151.

**27** Olesen, L. H., Okkels, F., & Bruus, H. (2006a. ISSN 0029-5981, 1097-0207). A high level programming-language implementation of topology optimization applied to steady-state Navier-Stokes flow. *International Journal for Numerical Methods in Engineering*, 65(7), 975–1001. doi:10.1002/nme.1468

**28** Pingen, G., Evgrafov, A., & Maute, K. (2007. ISSN 1615-147X, 1615-1488). Topology optimization of flow domains using the lattice Boltzmann method. *Structural and Multidisciplinary Optimization*, 34(6), 507–524. doi:10.1007/s00158-007-0105-7

**29** Dilgen, C. B., Dilgen, S. B., Fuhrman, D. R., Sigmund, O., & Lazarov, B. S. (2018. ISSN 00457825). Topology optimization of turbulent flows. *Computer Methods in Applied Mechanics and Engineering*, 331, 363–393. doi:10.1016/j.cma.2017.11.029

**30** Kontoleontos, E. A., Papoutsis-Kiachagias, E. M., Zymaris, A. S., Papadimitriou, D. I., & Giannakoglou, K. C. (2013. ISSN 0305-215X, 1029-0273). Adjoint-based constrained topology optimization for viscous flows, including heat transfer. *Engineering Optimization*, 45(8), 941–961. doi:10.1080/0305215X.2012.717074

**31** Othmer, C. (2008. ISSN 02712091, 10970363). A continuous adjoint formulation for the computation of topological and surface sensitivities of ducted flows. *International Journal for Numerical Methods in Fluids*, 58(8), 861–877. doi:10.1002/fld.1770

**32** Zymaris, A. S., Papadimitriou, D. I., Giannakoglou, K. C., & Othmer, C. (2009. ISSN 00457930). Continuous adjoint approach to the SpalartAllmaras turbulence model for incompressible flows. *Computers & Fluids*, 38(8), 1528–1538. doi:10.1016/j.compfluid.2008.12.006

**33** Othmer, C., de Villiers, E., & Weller, H. (2007). Implementation of a continuous adjoint for topology optimization of ducted flows. 18th AIAA Computational Fluid Dynamics Conference. Miami, FL: American Institute of Aeronautics and Astronautics, June 2007. ISBN 978-1-62410-129-8. doi:10.2514/6.2007-3947.

**34** OpenCFD (2019, September 25). OpenFOAM – Official home of The Open Source Computational Fluid Dynamics (CFD) Toolbox. Retrieved from https://www.openfoam.com/.

**35** Moukalled, F., Mangani, L., & Darwish, M. (2016). *The finite volume method in computational fluid dynamics: An advanced introduction with OpenFOAM and Matlab*. Fluid Mechanics and Its Applications. (Vol. 113). Cham, Heidelberg, New York, Dordrecht, London: Springer. ISBN 978-3-319-16873-9 978-3-319-16874-6. OCLC: 921234552

**36** Thomas, D. (2009). *The development of design rules for selective laser melting* (PhD thesis), University of Wales, Cardiff.

**37** Verboom, J. M. (2017). *Design and additive manufacturing of manifolds for Navier-Stokes flow: A topology optimisation approach* (Master's thesis), TU Delft, Delft.

**38** Sethian, J. A. (1999). *Level set methods and fast marching methods: Evolving interfaces in computational geometry, fluid mechanics, computer vision, and materials science*. Cambridge Monographs on Applied and Computational Mathematics. (9th ed., Vol. 3, pp. 2). Cambridge: Cambridge University Press.

# 5

# Multicomponent and Multimaterials Printing

## A Case Study of Embedded Ceramic Sensors in Metallic Pipes

*Cesar A. Terrazas, Mohammad S. Hossain, Yirong Lin, and Ryan B. Wicker*

*W.M. Keck Center for 3D Innovation, The University of Texas at El Paso, El Paso, TX, USA*

## 5.1 Multicomponent Printing: A Short Review

Additive manufacturing (AM) techniques, traditionally known as 3D printing, are processes that enable the manufacturing of components in a layer-by-layer fashion. The use of these AM techniques provides the ability of depositing material in a selective fashion with complete spatial control. This ability has enabled the production of components that have enhanced complexity at various scales, from the complexity of shape or geometry, to the complexity of materials. Different material systems, including polymeric, metallic, and ceramic have been processed to demonstrate multicomponent, multimaterial, and multifunctionality capability. Because of the ease of processing, polymer materials have seen many development in multimaterials, multicomponents. For example, the works by Medina et al. [1, 2] presented advances in AM stereolithography technology combined with direct write demonstrating multiprocess and multimaterial 3D printing technology. Later, the technology was used in the creation of functional 3D electronic components having embedded electronics and printed interconnecting traces of conductive ink, using direct print (DP), in the work by Lopes et al. [3]. The functionality of the hybrid approach employing SLA, AM, and DP was demonstrated in the fabrication of a module with low-power embedded electronics for the Trailblazer CubeSat that was launched into orbit in 2013 [4]. Similarly, Aguilera et al. [5] employed material extrusion technology for polymers, specifically the fuse

*3D Printing for Energy Applications*, First Edition. Edited by Albert Tarancón and Vincenzo Esposito.

deposition modeling (FDM), in a hybrid approach that included embedding of high-performance conductors within the thermoplastic component being printed, for the creation of a multicomponent assembly into a functional electromechanical device. Although the method used by Aguilera et al. [5] required the manual intervention in the printing process for insertion of components, it advanced the power requirements to up to 5 A for the fabrication of a working three-phase motor.

Other than polymers, the fabrication of multimaterial and multifunctional components using metals has also become an area of high research interest in the AM community in recent years. Three primary AM technologies can be employed to produce metal multimaterial and multicomponent parts comprising sheet lamination using ultrasonic additive manufacturing (UAM), powder bed fusion (PBF), and directed energy deposition (DED). For example, sheet lamination with metal foils has been researched to produce structures with embedded dielectric materials within a metal matrix to create multifunctional composite metal structures for various industrial applications [6]. The deposited layers of dielectric materials withstood the deposition process employing high-frequency ultrasound waves, and once deposited, provided a way to have insulation layers within an otherwise conductive metal matrix. Other works with the UAM technique permitted encapsulation of prepackaged circuitry along with printed electronic pathways employing the direct write (DW) process to create functional electronics within the metal matrix [7, 8]. With regard to multimaterial fabrication, the DED process has demonstrated to offer the greatest flexibility to achieve this outcome for AM components. The fact that various of the commercially available machines offer the ability to process multiple materials at once has been exploited to fabricate multimaterial components in which material has been designed. The review on the DED technology by Dass and Moridi [9] discussed at length the various combinations of materials including metal–metal, metal–ceramic, metal-oxide, amongst others, with applications that ranged from improving tribology properties, enhanced biocompatibility and osseointegration, ultra-high temperature resistance, the fabrication of components with embedded piezo sensors and transducers, and to produce thermal barrier protection systems.

Recently, the development of ceramic additive manufacturing has drawn increasing attention due to the excellent thermal, mechanical, and electrical properties of ceramics. There are several technologies developed for the printing of ceramics, mainly based on binder jetting, SLA, selective laser sintering, and direct write. Among all these ceramic AM technologies, direct write (DW, a.k.a. robocasting) holds the most promise to enable the printing of high-density ceramics [10, 11]. The use of multihead extruders for DW enables the fabrication of larger structures with more complex designs [12, 13]. Many researches have been focused on the development of sacrificial support structures that can be incorporated to DW. Li et al. [10] developed a calcium carbonate ($CaCO_3$) paste and used it as sacrificial support to fabricate Alumina ($Al_2O_3$) structures with high surface resolution. After presintering the ceramic parts, the residual material (CaO) can be dissolved in water. Afterwards, the sintering process can be completed [11]. Extrusion-based 3D printing techniques are promising since high-density ceramics could be fabricated (>95%) with improved geometry complexity. Yang et al. [12] developed a starch-based support paste to 3D print $Al_2O_3$ on which the supporting material was completely removed during sintering. Although they successfully fabricated different types of structures, the samples had rough surfaces due to clogging. The quality issues could be addressed by the optimization of the rheological properties and printing parameters. Xu, Quinn, Lebel, Therriault, and L'Espérance [13] 3D printed steel structures by DW using $Al_2O_3$ and Copper (Cu) as sacrificial support structure. The results showed no contamination in the samples after removal of sacrificial structure when printing with $Al_2O_3$ sacrificial support.

Finally, the PBF process has been demonstrated for the manufacture of multifunctional and multimaterial components, as well as smart components having embedded sensing capabilities. The works by Terrazas et al. [14], Hinojos et al. [15], and Segura et. al. [16] employed electron beam powder bed fusion (EB-PBF), commercialized as EBM by Arcam AB, to fabricate various structures using multiple materials. These works exhibited discrete interfaces for the materials being deposited which included dissimilar materials such as Ti-6Al-4V and copper, stainless steel with Inconel 718, and Inconel 690 cladded into 316L stainless steel. With regard to the use of PBF for the fabrication of smart components, the most relevant work is that from Mohammad et al. in which the EBM process was employed, using a custom developed "stop-and-go" process for the fabrication of a component with embedded piezoceramic material. The methodology presented in this work enabled the production of several prototype components that had the ability to sense external input in the form of external compressive forces. For the smart component produced in this work, the sensing capability was evaluated at various frequencies for a maximum compressive force of 40 MPa [17].

## 5.2 Multicomponent Printing: A Case Study on Piezoceramic Sensors in Smart Pipes

### 5.2.1 Brief Introduction to AM of Embedded Sensors for Smart Metering

Sensing technologies in energy system applications that can operate under harsh environments including those of high temperatures and pressures (regularly above 100 °C and 0.2 MPa) are in great demand, particularly in aerospace, automotive, and energy industries [18]. The inclusion of intelligent or smart components within a system can provide benefits to the system that include decreased maintenance costs, enhanced structural safety, optimized efficiencies, and improved functionality. For instance, for aerospace applications, propulsion systems would benefit from continuous monitoring under harsh environments to optimize control and operating parameters [19]. Similarly, the implementation of sensors within automotive electronic systems, usually exposed to environments with repeated thermal cycles and projected temperatures as high as 500 °C, can result in improved engine efficiency and reliability, as the monitored information data can be used to dynamically adjust operation parameters [20]. Further, the use of smart parts for *in situ* monitoring and inspection in next generation nuclear reactors, can improve operating safety, reduce life-cycle costs, and decrease exposure to radiation. Next generation nuclear reactors will have pressure vessel outlet temperatures as high as 1,000 °C, with the primary loops susceptible to corrosion, stress cracking, and irradiation embrittlement; thus, for these systems, *in situ* inspection and monitoring will improve safety, reduce life-cycle costs, and decrease exposure of workers to radiation [20, 21].

Previous research has been carried out in the design, fabrication, and testing of sensors to monitor temperature and pressure in harsh environments. These include surface thermocouples that even though are capable of withstanding high temperatures, can see their reliability affected during prolonged exposure to these conditions. Platinum-based thermocouples are very stable at high temperatures up to 2,000 °C, but low sensitivity and evaporation of the thermocouple material have been reported [19, 20]. Other devices such as fiber optic sensors have been developed and are available commercially for use as remote pyrometers, thermal extensometers, and thermometers; however, the temperature limitation (up to ~600 °C) of the optical fiber limits their use in high-temperature applications. By contrast, piezoelectric-based sensors can withstand high temperatures while providing certain advantages over other sensing approaches, such as reduced complexity, a faster response time, and ease of integration [22]. Piezoceramic materials using the piezoelectric or pyroelectric effects are potential candidates for use as sensing devices in environments of high-temperature and high-pressure conditions. The piezoelectric effect is the phenomenon that relates electric charge generation when a force is applied on the material and vice versa [23]. As a result, the piezoelectric effect can be used as a sensing mechanism with a wide range of applications for various stimuli that include strain, pressure, and force [24]. Similarly, the pyroelectric effect refers to the electrical charge generated by a change in temperature, which occurs in some piezoelectric materials, and that can be used for temperature sensing purposes [25]. Energy conversion components are frequently exposed to high temperature and high pressure; and piezoelectric and pyroelectric materials have become a promising approach to provide cost-effective, reliable, and passive sensing solutions.

In this work, we describe the AM fabrication of a smart coupling component, with embedded piezoelectric, for sensing of temperature and force. During experiments, the ability to sense temperature and force was conducted in separate experimental setups. Sensing of temperature and force was carried out in individual setups. Nevertheless, a coupled effect of temperature and force has been documented for piezoceramic sensor materials [17]. Decoupling can be accomplished by using a piezoelectric material (such as quartz, langasite, or $GaPO_4$) that has no pyroelectric property to isolate its force response and decouple it from the temperature responses [26]. The minimal impact of the high temperature during the AM printing process on the crystalline structure, and hence on the piezoelectricity and functionality of the sensor, was evaluated using X-ray diffraction (XRD). Future work will look at testing under simultaneous force and temperature stimuli to determine the potential effect of this coupling.

Currently, monitoring of components used in energy systems is mostly done through surface mounted sensors. This approach inherently exposes the sensors to high pressure, high temperature, and corrosive environments, thus decreasing their lifetime and accuracy. Surface mounting of the sensor can also be intrusive to the intended function as it alters the overall shape of the component, and perhaps its functionality. This might not be desirable in applications such as in a wind turbine, where modification in wind flow may alter the turbine's aerodynamic performance [27]. Additionally, a sensor attached to the surface may be unable to monitor the actual conditions experienced by a given component. For example, in the case of a pressure tube, a sensor attached to the external surface of the tube will be unable to monitor the actual internal pressure resulting from a hot and pressurized fluid flowing through the tube. Furthermore, if the sensors were to be attached to the internal surface, this could result in disturbance of the flow, and the potential degradation of the sensor.

A possible solution to these issues is the use of AM to build the specific component and embed a sensor within it during fabrication. With AM, the layer-by-layer fabrication process can be interrupted at any point to embed sensors without compromising the overall shape of the produced part. Interrupting the AM process and embedding additional functionality to the structure under fabrication has received considerable recent attention [5, 28–31]. However, these previous works were restricted to low-melting temperature materials, including thermoplastics, DuraForm™, and Zinc-Aluminum alloys, which made the fabricated components not suitable for high-temperature applications [5, 30, 31]. Recently, ultrasonic metal welding and consolidation AM technologies have been employed to embed sensors and electronics into metal structures. For these components, single added functionality, such as temperature sensing by thermocouples and optical fibers, or the enablement of actuation by using a NiTi shape memory alloy, have been successfully demonstrated [14, 32–35]. However, the widespread applicability of these components has been limited due to the low-melting point materials used to build them, or because of defects such as cracking and delamination that occur during their fabrication. Previous work by the authors demonstrated a methodology similar to that used here for fabricating a metallic smart component with embedded sensing capabilities, albeit with only a simple demonstration to sense compressive force [17]. While previous work by the authors demonstrated force sensing exclusively, the current work expands the concept of AM fabrication of smart parts by demonstrating multifunctionality capabilities of the manufactured smart coupling able to sense both temperature and force independently. Also, this work introduced a variation of the smart part fabrication technique presented before [34], to enable the fabrication of a higher shape complexity part resembling a tube-like component. The shape of this component can allow for its integration into a larger system to enable *in situ* monitoring.

In the following, a "stop and go" process was implemented using EB-PBF technology to fabricate a smart coupling with an embedded lead zirconate titanate (PZT) sensor. The smart coupling was tested in two setups to demonstrate force and temperature sensing independently. For example, during a compression–compression test, compressive forces in the range of 8–15 kN were obtained. Using a second experimental setup, in which heated air was flowed through a copper tube to which the smart coupling was attached, the temperature sensing capability of the device was also demonstrated from room temperature up to 130 °C. This work highlights the ability to control the AM process at any point during the layer-by-layer fabrication, to embed sensors within the main structure. Through this methodology multifunctionality can be imparted to a single component, in this case the ability to sense both temperature and pressure. Although there are overarching implications of this work for applications such as real-time monitoring in complex piping systems experiencing harsh environments, or for structural health monitoring, future research will seek to understand the potential pyroelectric–piezoelectric coupling effect that can develop from exposure of the smart coupling to simultaneous temperature and force stimuli.

### 5.2.2 Fabrication of the Embedded Piezoceramic Sensor in Metallic Pipes

#### 5.2.2.1 Smart Coupling Fabrication Process Using EPBF Technology

The EB-PBF process is a class of powder bed fusion (PBF) AM technology that is used for fabrication of metal parts in a layer-by-layer fashion by selectively melting precursor metal powder using an electron beam. The schematic view of an EBM (a tradename for the EB-PBF process) machine is shown in Figure 5.1. The regular fabrication process includes preheating and sintering powder using a defocused electron beam prior to selectively melting the metal powder for each layer using a focused and high-power electron beam. The build platform moves down after the melting step is complete, and the next layer of powder is deposited using a raking mechanism. Powder is grabbed from the containers situated at both sides of the build platform, from which powder flows assisted by gravity. The preheating and melting sequences are performed for each layer until the part is completed. For the

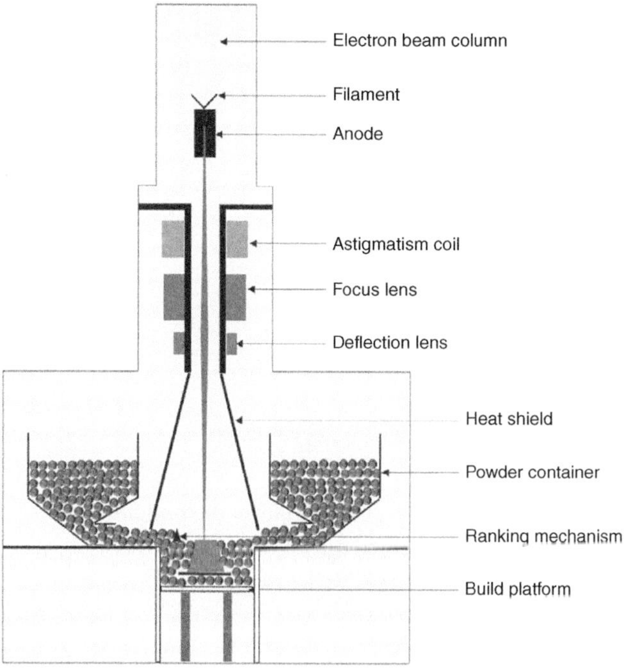


**Figure 5.1** Schematic diagram of EBM technology.

work presented here, an Arcam EBM S12 (Arcam AB, Mölndal, Sweden) was used to fabricate the structural component (i.e., smart coupling) into which a PZT was embedded. In this experiment, the smart coupling was fabricated using Ti-6Al-4V powder and standard 50 μm layer standard parameters.

A modified EB-PBF procedure, that had been previously described by Hossain et al. [17] as a "stop and go" process, was used for the fabrication of the smart coupling assembly presented in this work. The procedure consisted of the partial fabrication of the component, leaving a cavity available into which a sensor assembly (including the PZT) was inserted, following a process stop. After insertion of the sensor assembly, the EB-PBF fabrication process was resumed, and at the conclusion of the process, a smart part was obtained. For this work, two major modifications were introduced to the "stop and go" methodology presented by Hossain et al. These modifications addressed certain fabrication challenges that allowed manufacture of a functional smart coupling. The first modification was the use of a sacrificial insert part, fabricated in a previous EB-PBF run, to cover the hole created by the tube geometry. The location of this insert part (refer to Figure 5.2) prevented the electron beam from scanning in this region of loose powder and hence eliminated the risk of the detrimental powder spreading or powder smoke effect [14, 35]. The second innovation introduced was a change in

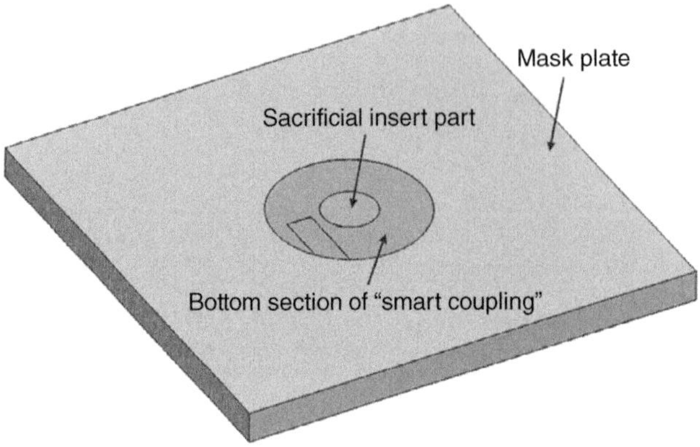


**Figure 5.2** Setup for fabrication of smart coupling using EPBF AM.

the orientation of the PZT sensor by 90° (i.e., vertical) from the build direction in EB-PBF. This embedding configuration enabled the force and temperature stimuli to be sensed in the principal direction of the PZT ($d_{33}$). Also, it allowed for embedding the sensor in a component with reduced thickness versus the previous smart part, corresponding to a reduction of the overall cavity size requirements of ~60%. As an added benefit, this embedded sensor configuration opens the prospect for using components of similar shape, for multistimuli, multidirectional sensing, such as temperature in the outside, and force or pressure at the inside.

#### 5.2.2.2 Materials

Figure 5.3a shows the exploded view of the computer aided design (CAD) assembly and the components used in the smart coupling fabrication. Figure 5.3b shows a view of the CAD assembly of the smart coupling as expected after the "stop and go" process is complete. The fabrication of the smart coupling was done using Ti-6Al-4V powder obtained from Arcam AB (Mölndal, Sweden). The sensor assembly was composed of the PZT piezoelectric, two tungsten electrodes, and an alumina housing that protected and insulated the piezoelectric material from electrical shorts. Commercially available PZT (from Piezo Systems, Inc, PZT 5A) was used as the piezoelectric ceramic material due to its high piezoelectric response; the piezoelectric coefficient (the property that dictates how much polarization is generated within the material in response to the applied stress) for PZT 5A is ~370 pC/N while those of other ceramic piezoelectric materials such as $PbTiO_3$ or $LiNbO_3$ are substantially lower at 50 and 6 pC/N, respectively [36]. Tungsten, having the highest melting point for all metals at 3,422 °C, was selected as the electrode material to survive the high heat conditions after resuming the EB-PBF process [37]. For the sensor housing, alumina was used given its

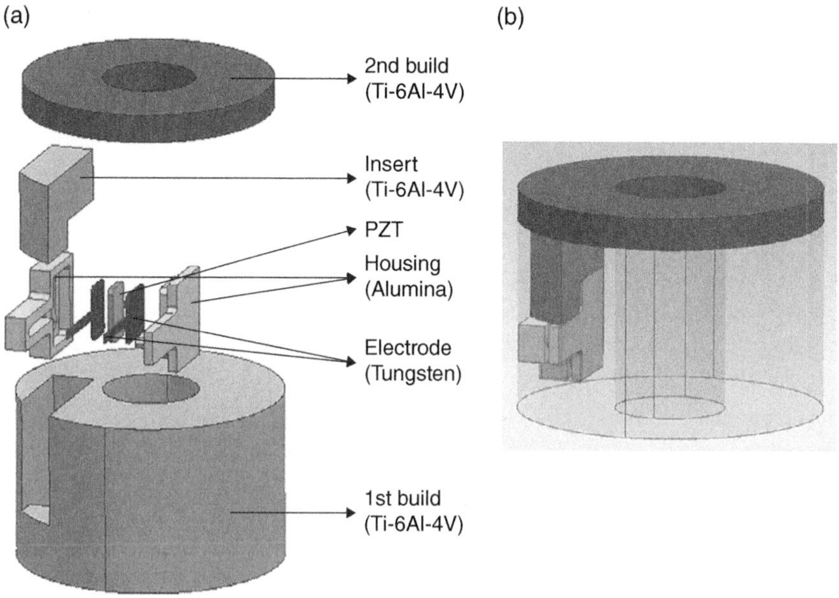


**Figure 5.3** Exploded (a) and assembled (b) schematic view of the "smart coupling" design.

high-temperature resistance, its ability to electrically insulate the remaining components of the smart coupling assembly [38], and for protection from metallization.

#### 5.2.2.3 Sensor Housing

Metallization during EB-PBF processing consists of the vaporization of light metal elements from the metal powder, such as aluminum and titanium, followed by condensation onto exposed surfaces. This phenomenon is exacerbated by the high (~$10^{-4}$Torr) vacuum used during the EB-PBF processing. In this work, the piezoelectric ceramic material and all other components of the sensor assembly were protected from metallization by enclosing them within an alumina housing (see Figure 5.3). The alumina housing protected the sensor from metal vapor during the fabrication process preventing electrical shorting of the sensor, and thus, the loss of sensing capability. However, the alumina housing also interfered with the force transferred to the sensor during the compressive load application, and with the heat transferred during the temperature sensing experiment, affecting the PZT sensitivity. A trade-off was made by the use of the ceramic housing, reducing the sensitivity of the embedded sensor while providing protection to assure sensing functionality.

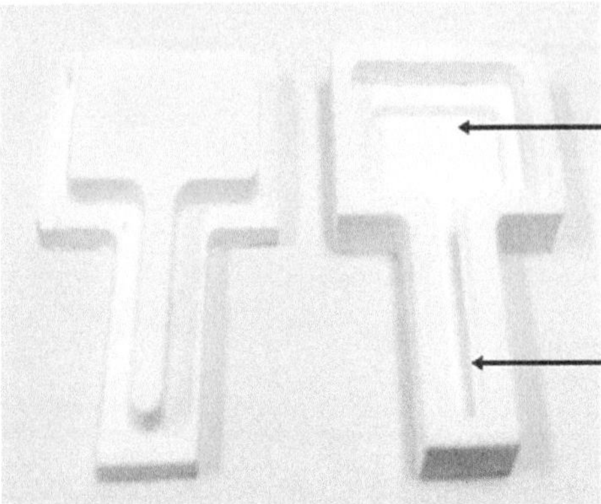


**Figure 5.4** Machined alumina sensor housing used to prevent metallization of piezoelectric ceramic during the EPBF fabrication process.

Figure 5.4 shows the machined alumina housing for the sensor assembly. The housing consists of two parts that were joined to create a cavity with a clearance fit (with a .04 mm gap) where the sensor was enclosed. The housing was machined from a block of alumina (Cotronics Corp., Brooklyn, NY [25]) using a CNC Mini Mill 2 (HAAS Automatic Inc., USA). The sensor assembly was placed in the cavity of the sensor housing, where the contact area of PZT was ~39 $mm^2$. Following a re-poling procedure, as described in the next section, the electrical current generated by the piezoelectric ceramic was read directly from the electrodes once they were exposed after removing the extended portion of the sensor housing. This procedure was done after the EB-PBF "stop and go" fabrication process was finished.

#### 5.2.2.4 Re-poling of PZT

Piezoelectric behavior might be lost when a material is heated above the Curie temperature. To regain the piezoelectric characteristics, the material needs to be subjected to a poling procedure. Re-poling can be obtained by application of both temperature and an electrical field stimulus to a piezoelectric material [39]. For the PZT employed in this research, a poling procedure was applied to the piezoelectric material since the fabrication temperature in EB-PBF exceeded 900 °C, which is higher than PZT's Curie temperature (350 °C). Re-poling of PZT was performed by applying a constant electrical field using a power supply, while maintaining it at a high temperature to reduce the overall poling time. During this poling process, the smart coupling was maintained at a temperature of 150 °C, while an electrical field of 1.4 kV/mm was applied for 1 hr. The re-poling of the piezoelectric material allowed for the voltage signal of the piezoelectric ceramic to be generated and transmitted appropriately through the electrodes.

#### 5.2.2.5 Impact in Sensing Properties Due to Heat-Treatment Induced By AM Process

The sensor embedded in one of the smart parts was recovered after fabrication to assess the impact of the manufacturing process in its sensing capabilities. The crystalline structure of the piezoelectric component of the sensor was characterized using XRD. A second piezoelectric sensor, which was not subjected to the heat from the AM fabrication process, was also analyzed and used for comparison.

#### 5.2.2.6 Smart Coupling Component

Although of small size, the prototype smart coupling produced highlights the possibility for integrating this type of component as part of larger and more complex energy systems for monitoring purposes. For instance, employing a similar approach, valves and flanges could be fabricated, and used to monitor flow conditions, by attaching them at discrete points along a duct or pipe. In addition, this methodology can be used for monitoring temperature conditions in air–fuel premixers, heat exchangers, or other piping components where the geometrical complexity prevents direct attachment of sensing devices, and for which the embedding process during AM manufacture provides the means of controlling the spatial integration of the sensors. The fabricated smart coupling prototype is shown in Figure 5.5. Figure 5.5a shows the first stage of the smart coupling with the open cavity for the sensor assembly visible and topped by a Ti-6Al-4V sacrificial insert. Figure 5.5b shows the smart coupling after the fabrication was completed thereby enclosing the sensor hardware within the metallic structure. The dimensions of the smart coupling were 50 mm for the outer diameter and 18 mm for the inner diameter with a height of 30 mm.

#### 5.2.2.7 Compressive Force Sensing

After "stop and go" fabrication, the force sensing capabilities of the smart coupling were tested using a compression–compression test. A MTS Landmark servo-hydraulic test system (Eden Prairie, MN) was used to perform this test. The setup included the use of two fixtures (machined out of aluminum) to ensure a distributed force transfer normal to the surface of the smart coupling main body. Before cyclical loading, the smart coupling was prestressed by compressing at ~0.3 mm. Then, the cyclical loading was performed by compressing at ±0.1 mm from the prestressed state at frequencies of 1, 10, and 15 Hz to simulate dynamic loading.

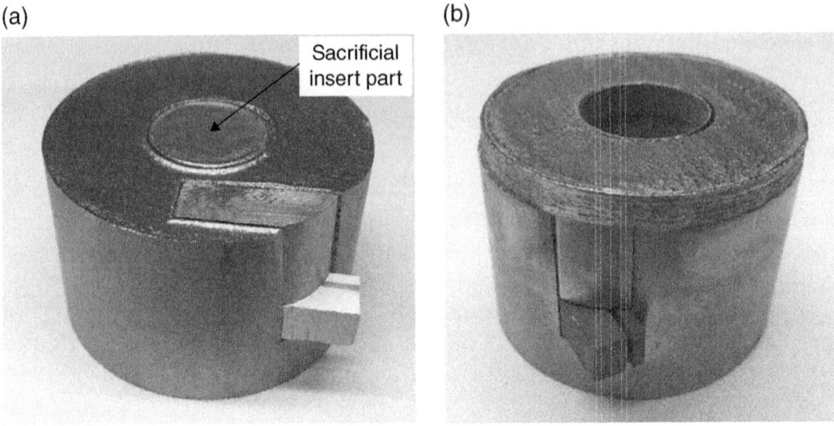


**Figure 5.5** Smart coupling after fabrication of the metallic structure with exposed cavity for sensor placement (a), and after continued fabrication to enclose the sensor within the metallic structure (b) metallization on the exposed faces of the ceramic sensor housing can be observed.

The frequencies selected allowed for testing the sensitivity of the smart coupling without the risk of compromising the structural integrity of the test setup since higher frequencies resulted in excessive vibration. The electrodes in the smart coupling were connected to a NI PCI-6221DAQ system (National Instruments, Austin, TX) to record the voltage response during this cyclic compression. A setup of the experiment is shown in Figure 5.6.

#### 5.2.2.8 Temperature Sensing

The temperature sensing capability of the smart coupling was demonstrated using the pyroelectric properties of the PZT. The pyroelectric effect consists of the generation of electrical current due to changes in the temperature over time [40]. Generated electrical current ($I$) through a pyroelectric material at temperature $T$ at any time ($t$) can be calculated using Equation 5.1:

$$I = \frac{dQ}{dt} = -pA\frac{dT}{dt} \tag{5.1}$$

In Equation 5.1, $Q$ is the charge generated due to temperature change, $p$ is the pyroelectric coefficient of the material, $A$ is the surface area of the electrode, and $dT/dt$ is the rate of temperature change of the material. To calculate the temperature of the pyroelectric material, Equation 5.1 can be written in integral form as follows:

**Figure 5.6** Setup of compression–compression test of the smart coupling. An aluminum fixture was machined and attached to the smart coupling to ensure safety and distributed force transfer. The voltage response was obtained by connecting the electrodes to a DAQ system.

$$T_f = -\frac{1}{pA}\int_{t_i}^{t_f} I dt + T_i \tag{5.2}$$

where, $T_i$ is the initial temperature of the pyroelectric ceramic material at time $t_i$, and $T_f$ is final temperature of the material at any time $t_f$. By considering $T_i$ as the measured room temperature, the pyroelectric coefficient ($p$), the electrode area ($A$), and the total amount of current ($I$) generated by the material within a certain period of time ($t_i$ to $t_f$), the final temperature ($T_f$) can be calculated at time ($t_f$).

To study the temperature sensing by the smart coupling, a simple experimental setup was employed. This setup consisted of attaching the smart coupling to a copper tube through which air, heated using a hot air gun, was flowed (shown in Figure 5.7). The variation of heat reaching the smart coupling was controlled by extending the travel distance of the hot air by changing the position of the smart coupling (distance $d$ in Figure 5.7) along the length of the copper tube to 7.6, 15.2, and 30.5 cm. As the sensor in the smart coupling was exposed to the heat, an electrical current was generated and measured. A Keithley 6485 picoammeter

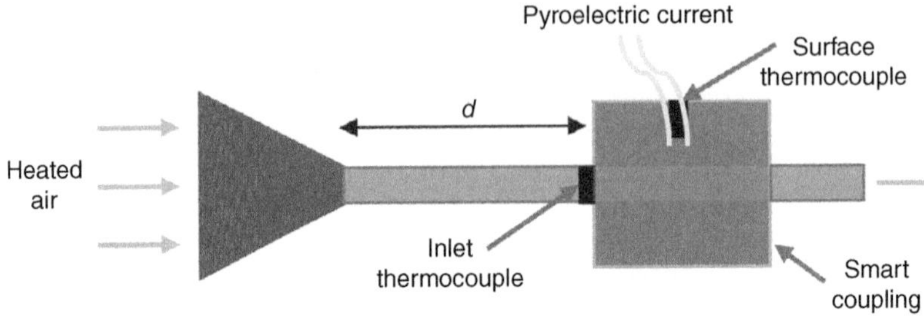


**Figure 5.7** Schematic of temperature variation experiment on the smart coupling. Two Type K thermocouples were attached to the setup, one fixed near the inlet and one on the surface of the coupling near the piezoelectric ceramic sensor. A picoammeter was attached to the electrodes to record the generated current. Parameter *d*, the length of the travel of the heated air inside the copper tube, was changed to vary the temperature across the smart coupling.

(Tektronix, Inc., Beaverton, OR) connected to the exposed electrodes of the smart coupling was used to measure the generated electrical current. The contact area of the PZT in the smart coupling was ~39 $mm^2$. The pyroelectric coefficient of the PZT reported from the manufacturer was 400 $\mu C/(m^2\,°C)$ [41]. The choice of contact area and pyroelectric coefficient was calibrated to obtain accurate temperature calculations.

### 5.2.3 Impact of the AM and Performance of the Multicomponent Printed Device

#### 5.2.3.1 Compressive Force Sensing

Three different frequencies (1, 10, and 15 Hz) were used to cyclically load the smart coupling and demonstrate its force sensing capabilities. The selection of these test frequencies allowed for the experiment to be performed before the onset of excessive vibration that could compromise the test setup. Figure 5.8a–c show the piezoelectric voltage responses obtained under the frequencies 1, 10, and 15 Hz, respectively, corresponding to the applied force. The average voltage–force ratios were .04, .011, and .0064 (V/kN) for the three frequencies employed. The ratios provided a comparison to characterize the integrity of the piezoelectric ceramic inside the smart coupling. The sensitivity of a cycle was defined as the ratio of the range of the voltage by the range of the force. The overall sensitivity was defined as the average sensitivity of all the cycles during the test. The following equation summarizes this approach:

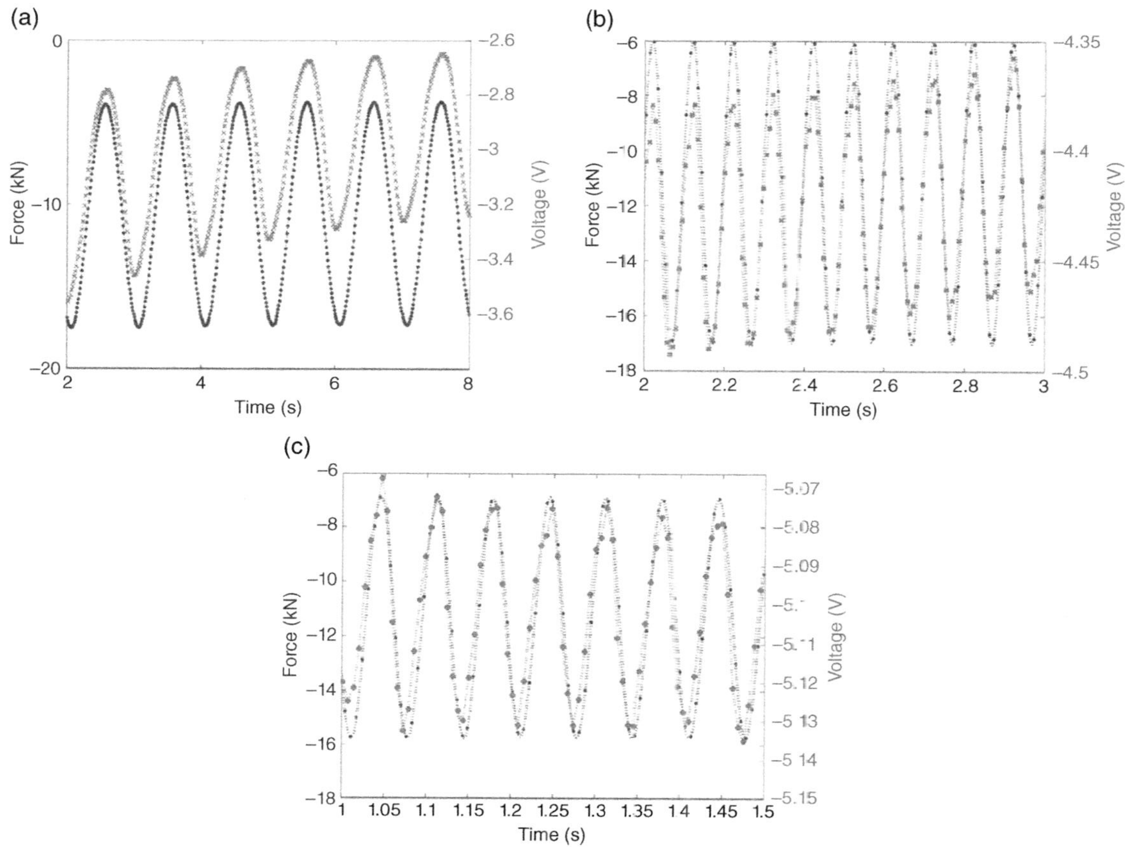


**Figure 5.8** Compression–compression cyclic testing results for three different frequencies tested: 1 Hz (a), 10 Hz (b), and 15 Hz (c). Each graph shows the compressive force in blue and the induced voltage in orange.

$$\text{Overall voltage} - \text{force ratio}\left(\text{V/kN}\right) = \frac{\sum_{i=1}^{N} \frac{\left(V_{\max,i} - V_{\min,i}\right)}{\left(F_{\max,i} - F_{\min,i}\right)}}{N} \tag{5.3}$$

where, $N$ is the total number of cycles applied during the compression–compression test, and $i$ is the cycle number.

The results obtained showed a decreasing trend of overall voltage–force ratio or sensitivity responses from the sensor material under cyclic loading with the increase in the loading frequency. Lower sensitivity values can be attributed to the clearance fit achieved during the sensor assembly including the alumina sensor housing. These clearances contributed to lowering the force transfer to the piezoelectric element. In addition, the fixture used during testing induced a triaxial stress state that created lower stress in the radial direction resulting in lower $d_{33}$ piezoelectric response from the sensor. Lastly, at a higher frequency, the dipole movement in the piezoelectric sensor fell behind the frequency of applied dynamic loading, which lowered the sensitivity.

#### 5.2.3.2 Temperature Sensing

Figure 5.9 shows the temperatures captured by the external thermocouple at the inlet (shown in Figure 5.7) and the electrical current generated from the embedded PZT sensor. In Figure 5.9a, the temperature captured by the sensor in the smart coupling varies based on the distance of travel of the hot air used in the experiment (distance $d$ in Figure 5.7). The maximum temperatures measured by the thermocouple fixed on the tube near the inlet of the smart coupling after six minutes into the heating stage were 198, 178, and 150 °C for corresponding values of $d$ of 7.6, 15.2, and 30.5 cm, respectively. For the measured temperatures, a corresponding variation occurred in the electrical current generation with maximum values of 281, 237, and 204 pA, respectively, as shown in Figure 5.9b.

The pyroelectric current generated during the temperature change presented in Figure 5.9b shows a shoulder in the electrical current signal at 350 s. A similar behavior did not occur in the chart of temperature ($dT/dt$) of the surface thermocouple. This behavior was observed during the verification of the PZT response by repeating the experiments under the same conditions. This change can be explained by the coexistence of different phases of the PZT that cause sudden slight variations of its piezoelectric and pyroelectric coefficients. For example, Sabat, Mukherjee, Ren, and Yang [42] showed that a change in piezoelectric coefficients of PZT occurred between the range from 125 to

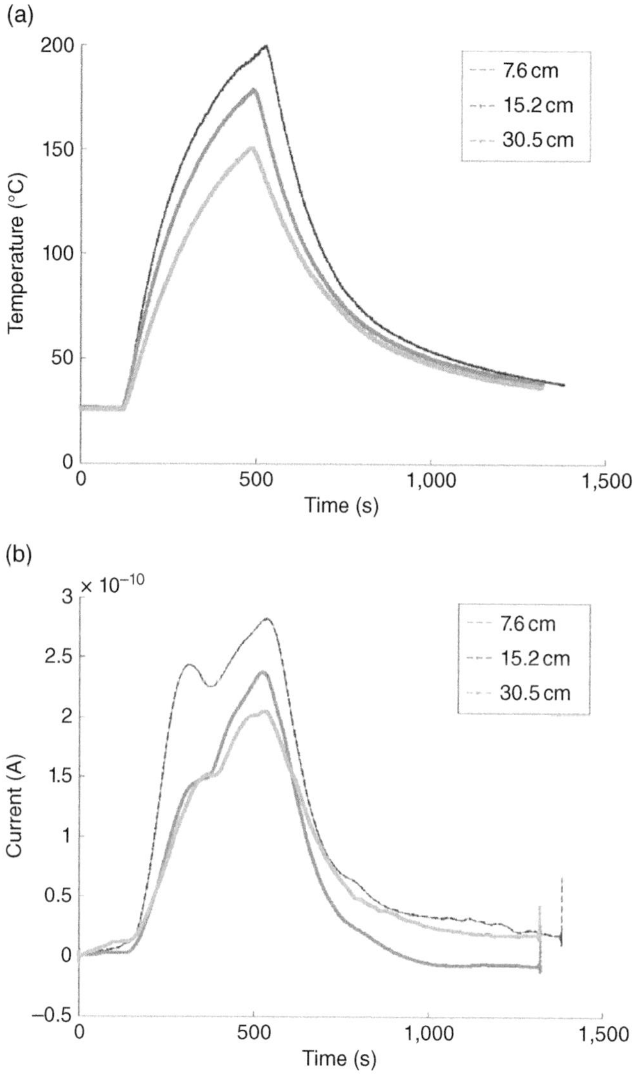


**Figure 5.9** Temperature values (a) occurring at the inlet of the smart coupling as measured by the K type thermocouple. Different lengths increased the heat loss and produced a variation in the temperature measured. Part (b) shows the current values generated from the embedded piezoceramic sensor material that were recorded using a picoammeter.

150 °C. Research has shown that although not clearly defined yet, tetragonal, rhombohedral, and monoclinic symmetries exist in the ferroelectric phase of PZT at different temperature regimes, resulting in changes in the piezoelectric constant of PZT [42].

To obtain more accurate temperature values, the generated electrical current was integrated and adjusted according to Equation 5.2. Using a calibrated coefficient for $p$, Equation 5.2 provided a sensing behavior for temperature using a piezoelectric component. The estimated coefficient $p$ was 750 pC/($m^2$ °C). This value shows that the PZT's contact area and the pyroelectric coefficient are lower than intended. This lower value was expected, since the complete contact area was likely not achieved during the sensor assembly. Figure 5.10 shows a comparison between the temperature sensed by the embedded pyroelectric sensor and that measured by the thermocouple mounted in the exterior surface of the smart coupling, in the same angular position as the cavity where the piezoelectric was embedded. Figure 5.10a–c corresponds to the experiment with the tube length of 7.6, 15.2, and 30.5 cm, respectively. It is shown that for all these three copper tube length settings, the embedded PZT sensor measured higher temperatures than those of the surface mounted thermocouple during the heating cycle lasting 6 min. This phenomenon can be explained due to the experimental setup used, and to the assembly of the sensor housing that resulted in a transient heat transfer effect. This transient effect resulted from the various thermal resistance layers in the sensor housing that insulated the PZT during experimentation. As a result, the PZT experienced lower temperatures than the surface thermocouple during the heating cycle. Similarly, during the cooling cycle, the thermal resistance from the sensor housing assembly reduced heat transfer, thus showing higher temperatures for the PZT versus the surface thermocouple. Moreover, during the cooling cycle the surface of the smart coupling was subjected to forced convection using a dedicated fan resulting in lower surface temperatures.

#### 5.2.3.3 Crystalline Structure Analysis

The results from the XRD studies are shown in Figure 5.11. The XRD plots in the figure indicate that the PZT sensor did not present any degradation or changes in its composition. All peaks index to those characteristic of PZT 52/48 in both the commercial and the embedded sample. However, the commercial sample did present more prominent "double peaks", such as the ones found at 2θ angles of 21.6° and 22°, 43.9° and 44.9°, and 54.9° and 55.5°. This "double peak" behavior is expected in piezoelectric materials as a result of the poling process [43]. From these results, we can conclude that the PZT embedded into the smart part maintains its sensing capability even after undergoing the high heat from the AM process, once the re-poling process has been performed.

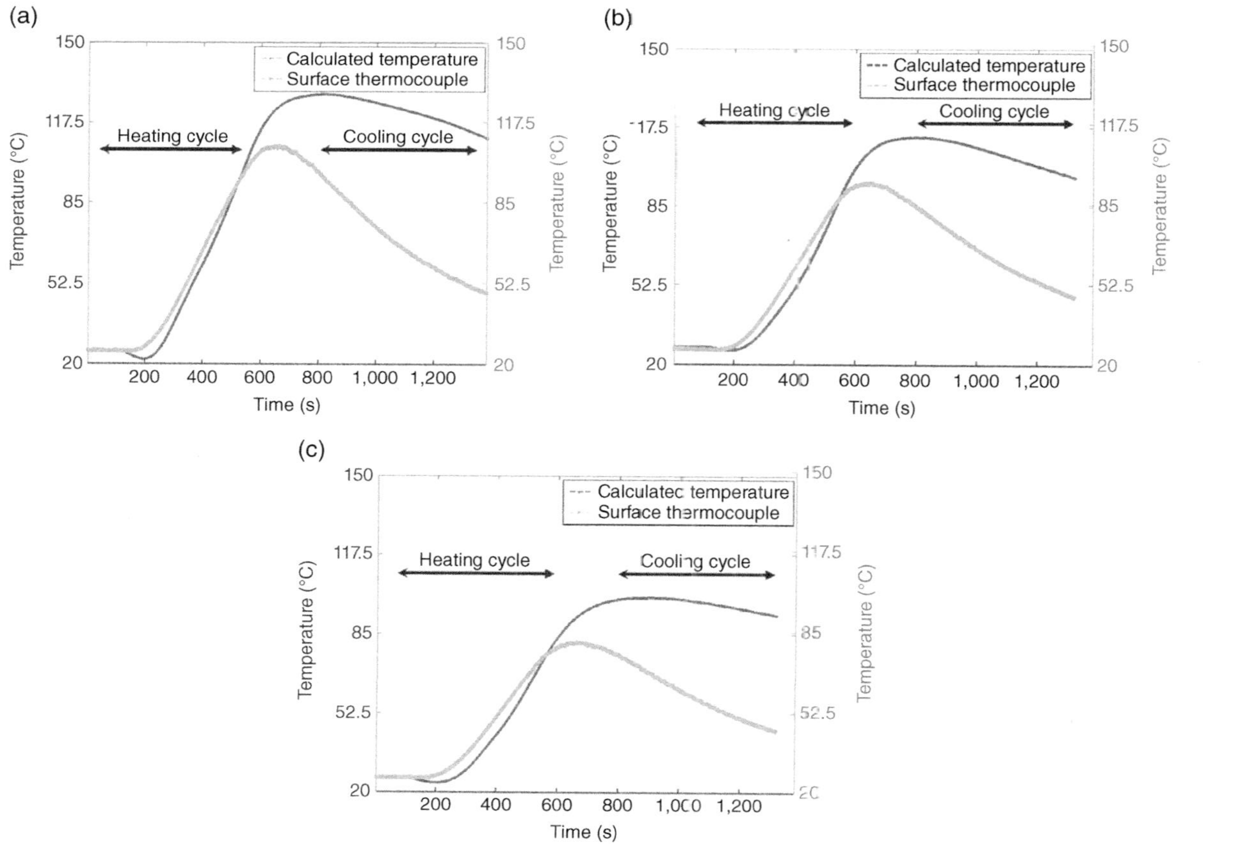


**Figure 5.10** Temperature comparison between the embedded sensor and the thermocouple located at the surface of the smart coupling. Heat tube length settings were: (a) 7.6 cm, (b) 15.2 cm, and (c) 30.5 cm. The heating and cooling cycles are indicated in the plots.

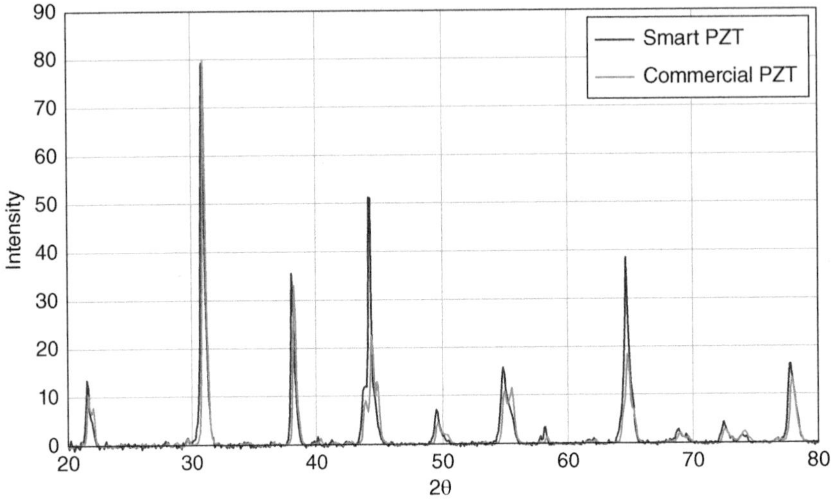


**Figure 5.11** XRD plots for the PZT sensor embedded in the smart-part (darker), and that of the unaffected commercial grade PZT material (lighter).

## 5.3 Summary and Outlook

A material system typically requires the use of multiple materials such as polymer, metal, and ceramic integrating together to achieve specific functions. Most of the AM technologies developed so far have been focusing on the fabrication of a single material only. However, with the maturation of AM technologies, the fabrication of multimaterials or multicomponents material system has become a reality. This opens doors to broader adoption of AM to enable products fabrication with system level multifunctional properties. Multimaterials AM can be achieved by integrating additive with subtractive manufacturing; modifying current AM technology to facilitate multiprocess fabrication; or using "stop and go" process to embedded secondary materials or material systems to a host structures. While there are many challenges that need to be addressed such as mismatch of material properties, stress concentration, alignment, and reliability, the potential of using multimaterial multicomponent AM will transform the future manufacturing industry. Distributed device fabrication with customized design will be enabled to on-site print functional parts and devices for biological, civil, and defense related applications.

An increasing demand of sensing technologies in energy system applications operating at high temperature and high pressure has motivated the development of monitoring devices that can operate under these harsh conditions. In this chapter, a case study on smart coupling prototype was produced by embedding a PZT sensor within a metal component that was fabricated through a "stop and go" procedure using the EB-PBF AM technique. This advanced fabrication process allowed for embedding the PZT within an alumina housing that protected the sensor from metallization during the EB-PBF process. After a re-poling process was performed on the smart coupling to restore piezoelectricity, the device was able to sense temperature and compressive forces when tested in two separate setups. The experiments performed showed that the smart coupling had force-sensing capability when subjected to cyclical compressive loadings at test frequencies of 1, 10, and 15 Hz, as demonstrated by an increase in the detected voltage–force ratios at each of these frequencies. Similarly, temperature-sensing capability was demonstrated for the same smart coupling, when exposed to temperatures in the range from 150 to 200 °C, as indicated by the measured currents detected by the picoammeter. This study has highlighted an advanced fabrication technique using the "stop and go" process in EB-PBF AM, that was improved from previous work to address fabrication challenges, and that enabled the fabrication of a higher shape complexity smart component. The technique eliminated the risk of powder spreading from occurring, it reduced the size requirements for the cavity by ~60%, and enabled the embedding of the PZT in a preferred orientation to sense force and temperature stimuli more efficiently. Overall, the technique enabled fabrication of a smart coupling with multifunctionality via the ability to sense both temperature and force, and it also opened the prospect for multidirectional sensing capabilities. This study has far reaching implications since the technique presented can enable the production of smart components, including valves, flanges, air–fuel pre-mixers, and heat exchangers among others. All these devices can be employed for *in situ* health monitoring in complex piping systems. Nevertheless, the development of dynamic effects leading to the potential coupling of both the pyroelectric and piezoelectric effects during exposure to both temperature and force stimuli were not accounted for in this study, and they will be the subject of future research.

## Acknowledgments

The fabrication of the smart coupling presented in this work was performed at The University of Texas at El Paso (UTEP), in the W.M. Keck Center for 3D Innovation, providing access to the state of the art facilities and equipment as a result of funding from the State of Texas Emerging Technology Fund. The research was supported by the US Department of Energy (DOE) under grant (DE-FE0012321) and the National Science Foundation under NSF-PREM Grant No. DMR-1205302. Overall project support was, in part, provided by the Mr. and

Mrs. MacIntosh Murchison Chair I Endowment at UTEP (RBW). The findings and conclusions contained herein are those of the authors and should not be interpreted as necessarily representing the official policies or endorsements, either expressed or implied, of the DOE or NSF. The authors greatly appreciate the insightful discussion with Ahsan Choudhuri, Norman love, and Ricardo Martinez.

## References

1 Medina, F., Lopes, A. J., Inamdar, A. V., Hennessey, R., Palmer, J. A., Chavez, B. D., & Wicker, R. B. (2005, May 10–12). Integrating multiple rapid manufacturing technologies for developing advanced customized functional devices. Rapid Prototyping & Manufacturing 2005 Conference Proceedings, Rapid Prototyping Association of the Society of Manufacturing Engineers. Hyatt Regency Dearborn, Dearborn, MI.

2 Medina, F., Lopes, A. J., Inamdar, A. V., Hennessey, R., Palmer, J. A., Chavez, B. D., … Wicker, R. B. (2005). Hybrid manufacturing: Integrating direct-write and stereolithography. Proceedings of the 2005 Solid Freeform Fabrication. Austin, TX: The University of Texas at Austin. Retrieved from https://repositories.lib.utexas.edu/handle/2152/80043

3 Lopes, A. J., Navarrete, M., Medina, F., Palmer, J. A., MacDonald, E., & Wicker, R. B. (2006). Expanding rapid prototyping for electronic systems integration of arbitrary form. Proceedings of the 2006 Solid Freeform Fabrication. Austin, TX: The University of Texas at Austin. Retrieved from https://www.sffsymposium.org/

4 Espalin, D., Muse, D. W., MacDonald, E., & Wicker, R. B. (2014). 3D printing multifunctionality: Structures with electronics. *International Journal of Advanced Manufacturing Technology*, 72, 963–978. doi:10.1007/s00170-014-5717-7

5 Aguilera, E., Ramos, J., Espalin, D., Cedillos, F., Muse, D., Wicker, R. & Macdonald, E. (2013). 3D printing of electro mechanical systems. Proceedings of the Solid Freeform Fabrication Symposium, 950–961.

6 Li, J., Monaghan, T., Masurtschak, S., Bournias-Varotsis, A., Friel, R. J., & Harris, R. A. (2015). Exploring the mechanical strength of additively manufactured metal structures with embedded electrical materials. *Materials Science and Engineering A*, 639, 474–481. doi:10.1016/j.msea.2015.05.019

7 Siggard, E. J., Madhusoodanan, A. S., Stucker, B., & Eames, B. (2006). Structurally embedded electrical systems using ultrasonic consolidation (UC). 17th Solid Freeform Fabrication Symposium. SFF 2006, 70–83. University of Texas, Austin (freeform).

8 Robinson, C. J., Stucker, B., Lopes, A. J., Wicker, R., & Palmer, J. A. (2006). Integration of direct-write (DW) and ultrasonic consolidation (UC) technologies to

create advanced structures with embedded electrical circuitry. 17th Solid Freeform Fabrication Symposium. SFF 2006. University of Texas, Austin (freeform).

9 Dass, A., & Moridi, A. (2019). State of the art in directed energy deposition: From additive manufacturing to materials design. *Coatings*, 9, 418. doi:10.3390/COATINGS9070418

10 Li, W., Ghazanfari, A., Mcmillen, D., Leu, M. C., Hilmas, G. E., & Watts, J. (2017). Fabricating ceramic components with water dissolvable support structures by the ceramic on-demand extrusion process. *CIRP Annals*, 66(1), 225–228.

11 Li, W., Ghazanfari, A., Mcmillen, D., Scherff, A., Leu, M. C., & Hilmas, G. E. (2017). Fabricating zirconia parts with organic support material by the ceramic on demand extrusion. Solid freeform fabrication Symposium. Austin, TX: The University of Texas at Austin.

12 Yang, L., Tang, S., Li, G., Qian, L., Mei, J., Jiang, W., & Fan, Z. (2019). Layered extrusion forming of complex ceramic structures using starch as removable support. *Ceramics International*, 45(17), 21843–21850.

13 Xu, C., Quinn, B., Lebel, L. L., Therriault, D., & L'Espérance, G. (2019). Multi material direct ink writing (DIW) for complex 3D metallic structures with removable supports. *ACS Applied Materials & Interfaces*, 11(8), 8499–8506.

14 Terrazas, C. A., Mireles, J., Gaytan, S. M., Morton, P. A., Hinojos, A., Frigola, P., & Wicker, R. B. (2015). Fabrication and characterization of high-purity niobium using electron beam melting additive manufacturing technology. *International Journal of Advanced Manufacturing Technology*, 84(5–8), 1115–1126. doi:10.1007/s00170-015-7767-x

15 Hinojos, A., Mireles, J., Reichardt, A., Frigola, P., Hosemann, P., Murr, L. E., & Wicker, R. B. (2016). Joining of Inconel 718 and 316 stainless steel using electron beam melting additive manufacturing technology. *Materials and Design*, 94, 17–27. doi:10.1016/J.MATDES.2016.01.041

16 Segura, I. A., Murr, L. E., Terrazas, C. A., Bermudez, D., Mireles, J., Injeti, V. S. V., Li, K., Yu, B., Misra, R. D. K., & Wicker, R. B. (2018). Grain boundary and microstructure engineering of Inconel 690 cladding on stainless-steel 316L using electron powder bed fusion additive manufacturing. *J. Mater. Sci. Technol.*

17 Hossain, M. S., Gonzalez, J. A., Hernandez, R. M., Shuvo, M. A. I., Mireles, J., Choudhuri, A., … Keck, W. M. (2016). Fabrication of smart parts using powder bed fusion additive manufacturing technology. *Additive Manufacturing*, 10, 58–66. doi:10.1016/j.addma.2016.01.001

18 Yoshida, Y., Kashiwai, Y., Murakami, E., Ishida, S., & Hashiguchi, N. (2002, June). Development of the monitoring system for slope deformations with fiber Bragg grating arrays. In *Smart structures and materials 2002: Smart sensor technology and measurement systems* (Vol. 4694, pp. 296–303). International Society for Optics and Photonics.

19 Yang, J. (2013). A harsh environment wireless pressure sensing solution utilizing high temperature electronics. *Sensors*, 13(3), 2719–2734. doi:10.3390/s130302719

20 Zhang, S., & Yu, F. (2011). Piezoelectric materials for high temperature sensors. *Journal of the American Ceramic Society*, 94(10), 3153–3170. doi:10.1111/j.1551-2916.2011.04792.x

21 Doctor, S. R. (2007). Nuclear power plant NDE challenges — Past, present, and future. AIP Conference Proceedings, 17–31. AIP. DOI: 10.1063/1.2717950.

22 Tressler, J. F., Alkoy, S., & Newnham, R. E. (1998). Piezoelectric sensors and sensor materials. *Journal of Electroceramics*, 2(4), 257–272. doi:10.1023/A:1009926623551

23 Vives, A. A. (Ed.) (2008). *Piezoelectric transducers and applications*. Berlin: Springer. doi:10.1007/978-3-540-77508-9.

24 Sirohi, J., & Chopra, I. (2000). Fundamental understanding of piezoelectric strain sensors. *Journal of Intelligent Material Systems and Structures*, 11(4), 246–257. doi:10.1106/8BFB-GC8P-XQ47-YCQ0

25 Webster, G. J. (2014). *The measurement, instrumentation and sensors handbook* (2nd ed.). CRC Press.

26 Rebillat, M., Guskov, M., Balmes, E., & Mechbal, N. (2016). Simultaneous influence of static load and temperature on the electromechanical signature of piezoelectric elements bonded to composite aeronautic structures. *Journal of Vibration and Acoustics*, 138(6), 064504. doi:10.1115/1.4034375

27 Cavalloni, C., Sommer, R., & Waser, M. (2011). D1. 1-High-temperature stability of new single crystal piezoelectric sensors. Proceedings of the SENSOR-TEST Conferences, 520–525.

28 Lescher, F., Camblong, H., Curea, O., & Briand, R. (2007). LPV control of wind turbines for fatigue loads reduction using intelligent micro sensors. Proceedings of American Control Conference, ACC'07, 6061–6066. IEEE.

29 MacDonald, E., & Wicker, R. (2016). Multiprocess 3D printing for increasing component functionality. *Science*, 353(6307), aaf2093. doi:10.1126/science.aaf2093

30 Pille, C. (2010). In-process-embedding of piezo sensors and RFID transponders into cast parts for autonomous manufacturing logistics. Smart Systems Integration, 1–10.

31 Rai, R., Campbell, M., & Wood, K. (2004). Extracting product performance by embedding sensors in SFF prototypes. Proceedings of International Solid Freeform Fabrication Symposium, 366–376. Austin, TX.

32 Li, Y., Liu, W., Feng, Y., & Zhang, H. (2012). Ultrasonic embedding of nickel-coated fiber Bragg grating in aluminum and associated sensing characteristics. *Optical Fiber Technology*, 18(1), 7–13. doi:10.1016/J.YOFTE.2011.09.004

33 Hahnlen, R., & Dapino, M. J. (2010). Active metal-matrix composites with embedded smart materials by ultrasonic additive manufacturing. In M. B. McMickell & K. M. Farinholt (Eds.), *Industrial and commercial applications*

*of smart structures technologies 2010* (Vol. 7645, pp. 764500). International Society for Optics and Photonics. doi:10.1117/12.848853

**34** Cheng, X., Datta, A., Choi, H., & Li, X. (2006, October). Metal embedded micro sensors for manufcaturing applications. 2006 5th IEEE Conference on Sensors, 1506–1510. IEEE. doi:10.1109/ICSENS.2007.355920.

**35** Murr, L. E. (2015). *Handbook of materials structures, properties, processing and performance* (1st ed.). Springer.

**36** Lee, H. J., Zhang, S., Bar-Cohen, Y., & Sherrit, S. (2014). High temperature, high power piezoelectric composite transducers. *Sensors*, 14(8), 14526–14552. doi:10.3390/s140814526

**37** Langmuir, I. (1913). The vapor pressure of metallic tungsten. *Physical Review*, 2(5), 329–342. doi:10.1103/PhysRev.2.329.

**38** Cotronics Corp. (2008). 960 Machinable Alumina. Retrieved from http://www.cotronics.com/vo/cotr/cm_machinable.htm. Viewed January 2018.

**39** Prewitt, A. D., & Jones, J. L. (2011). Effects of the poling process on piezoelectric properties in lead zirconate titanate ceramics. *Ferroelectrics*, 419(1), 39–45. doi:10.1080/00150193.2011.594725.

**40** Tichý, J., Erhart, J., Kittinger, E., & Privratska, J. (2010). *Fundamentals of piezoelectric sensorics: mechanical, dielectric, and thermodynamical properties of piezoelectric materials* (1st ed.). Springer Science & Business Media.

**41** Piezo Systems. PSI-5A4E Piezoceramic Sheets. Retrieved from http://www.piezo.com/prodsheet2sq5H.html. Viewed February 2018.

**42** Georges Sabat, R., Mukherjee, B. K., Ren, W., & Yang, G. (2007). Temperature dependence of the complete material coefficients matrix of soft and hard doped piezoelectric lead zirconate titanate ceramics. *Journal of Applied Physics*, 101(6), 064111. doi:10.1063/1.2560441

**43** Fan, H., & Kim, H. E. (2002). Perovskite stabilization and electromechanical properties of polycrystalline lead zinc niobate–lead zirconate titanate. *Journal of Applied Physics*, 91(1), 317–322.

# 6

# Tailoring of AM Component Properties via Laser Powder Bed Fusion

*Simon Ewald[1], Maximilian Voshage[1], Steffen Hermsen[1], Max Schaukellis[1], Patrick Köhnen[2], Christian Haase[2], and Johannes Henrich Schleifenbaum[1,3]*

[1] *Digital Additive Production, RWTH Aachen University, Aachen, Germany*
[2] *Steel Institute, RWTH Aachen University, Aachen, Germany*
[3] *Fraunhofer Institute for Laser Technology ILT, Aachen, Germany*

## 6.1 Introduction

The ongoing globalization and industrialization requires a more and more sustainable usage of the available resources and processes. Increasing the efficiency of machines and industrial processes as developing application-specific product solutions, for example, topology optimized components and adapted processing approaches, can be one solution. For example, the usage of customized lightweight components (e.g., in Automotive and Aerospace) reduces the fuel consumption and emission. Therefore, an increasing interest in new manufacturing methods, such as Laser Powder Bed Fusion (LPBF), for flexible production of high complex components can be observed. Another way to achieve these goals is the application of new materials with properties which are outperforming the current state of the art materials. The combination of customized components, due to the high degree of freedom in geometry, and advanced materials could accomplish even better results in order to achieve common goals, such as tackling the climate change [1–3].

LPBF is one of many Additive Manufacturing (AM) techniques which allow the production of high complex components directly from 3D-CAD models. After slicing the 3D-CAD model into several layers with a defined layer thickness, the components can be produced by selectively melting and fusion of metal powder

*3D Printing for Energy Applications*, First Edition. Edited by Albert Tarancón and Vincenzo Esposito.

layer-by-layer. This procedure allows the production of complex components (e.g., internal cooling channels and lattice structures), which cannot be manufactured by conventional methods [3, 4]. In comparison to components manufactured by conventional processing routes, AM and especially LPBF is much more flexible and have higher degrees of freedom regarding process control, materials, and geometrical design. Compared to conventional manufacturing processes, such as milling or casting, it is possible to influence the process parameters and thus the microstructure anywhere in the component due to the layer- and vector-based buildup. In addition, the powder-based process allows a rapid variation of the material chemistry through the use of powder blends. Since the properties of components depend essentially on the material (microstructure and chemical composition) and the component geometry, LPBF opens up completely new potentials to design components adapted to specific needs and applications.

Due to the vector- and layer-based manufacturing process of LPBF, the potential of adjusting the microstructure and therefore on the mechanical properties is large. Tucho, Lysne, Austbø, Sjolyst-Kverneland, and Hansen [5] already showed the influence of scanning speed, hatch distance, and energy density on the microstructure and because of this on the hardness. Additionally, Hovig, Holm, and Sørby [6] indicated that the relative densities of samples produced by LPBF increased with increasing preheating temperature. Both publications demonstrated that the microstructure and therefore the mechanical properties can be influenced by varying the energy input and cooling rate.

The high spectrum of LPBF process parameters offers a large variety of processable materials. The high energy intensity and the small laser focus leads to high cooling rates which results in less segregation which enables the production of hardly conventional processable materials [7, 8]. Since the feedstock for the LPBF process is metal powder, the approach of using elemental powder blends to design alloys and to control material properties offers additional flexibility. The possibility to use powder blends as input material for LPBF was successful shown in various previous studies [9, 10]. For example, the mechanical properties of high-manganese steels were adjusted by adding elementary Al powder to prealloyed steel powder [9, 11, 12]. The processability and usability of powder blends consisting of various elemental powders with different morphologies were additionally validated by Ewald et al. [13].

Due to the newly gained geometrical design freedom offered by AM technologies, it is now possible to design new shapes and geometries adapted to specific needs and applications. In particular, lattice structures are to be mentioned which are not possible to manufacture via conventional techniques. Due to the fine resolution of these structures and the potential to adapt the geometry on cell level, these structures are suitable to tailor component properties. Additionally, periodic lattice structures have an excellent stiffness-to-weight ratio and are characterized by high-energy absorption capability which can hardly be achieved in conventional geometries [1, 14, 15]. The potential lattice structures offer is well known, but only the emerging AM technologies made an economic manufacturing and integration in components of those structures possible. Therefore, fine and complex geometries, such as lattice structures, are promising for future light weight applications [1, 16].

Figure 6.1 shows the general approach applied in this contribution. Component properties can be tailored by means of the three factors LPBF process parameters, material chemistry, and geometrical design.

The aim of the present work is to examine and evaluate the potentials of the LPBF process regarding the production of tailored components experimentally. First, a process validation for the exemplary selected material system X30Mn22 was performed and then transferred to different preheating temperatures and material compositions. The preheating temperature was changed to 200 and 400 °C and 1 wt% Al was added to the selected base alloy in order to examine the influence on the microstructure and mechanical properties. The potentials, especially in absorption capability, of fine and complex structures were exemplary tested using lattice structures. Therefore, lattice structures with two different material compositions were produced. To investigate the microstructure light optical microscopy (LOM), scanning electron microscopy (SEM), energy

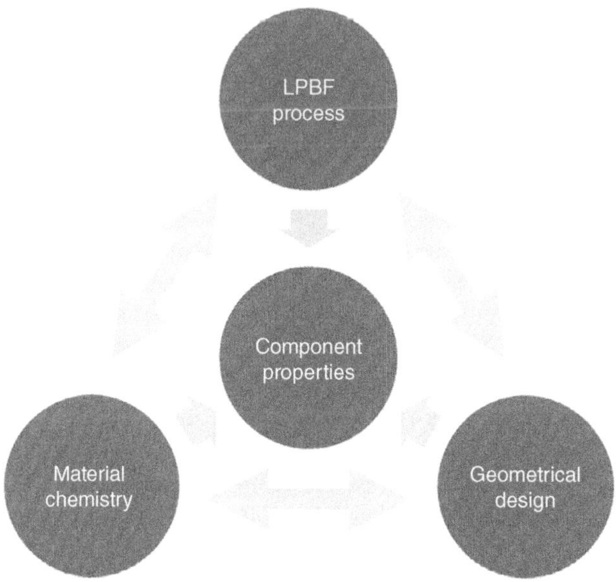


**Figure 6.1** Schematic overview of the correlations between the used factors to tailor AM component properties.

dispersive spectroscopy (EDS), and electron backscatter diffraction (EBSD) were used. To test the mechanical properties of bulk material tensile and hardness tests were performed while the lattice structures were tested via compression tests.

## 6.2 Machines, Materials, and Sample Preparation

For all LPBF experiments an AconityMINI system (Aconity 3D Herzogenrath, Germany) was used. This system has been specially designed for laboratory use and is characterized by a small building space with a diameter of 140 mm and a maximum height of 200 mm for reduced powder consumption. The energy source was a single-mode fiber laser with the wavelength 1,064 nm and up to 400 W power output and a beam diameter of 78 μm.

The material system used in this work was high-manganese steel (HMnS). These steels are iron based alloys with a Mn content of 15–30 wt%. HMnS have excellent mechanical properties with yield strength above 800 MPa and elongation at break up to 80% due to special deformation mechanisms, such as Twinning Induced Plasticity (TWIP) and Transformation Induced Plasticity (TRIP). Additionally HMnS are difficult to process by conventional manufacturing technologies because of segregation occurrence when processed via strip-casting [17]. The deformation mechanisms are mainly affected by the stacking fault energy (SFE). The SFE is formed from the individual energies of stacking errors in the austenite crystal and is a deviation of the stacking order of the atoms. Therefore, the SFE determines which deformation mechanism occurs during deformation [18, 19]. The SFE depends mainly on the temperature and chemical composition and can be controlled, when temperature kept constant, by the Al content in HMnS (Figure 6.2, Table 6.2). The BASE alloy is an HMnS called X30Mn22.

The pre-alloyed X30Mn22 powder was manufactured by gas atomization in an argon atmosphere with spherical particles of 15–50 μm in diameter. To achieve the different material compositions the pre-alloyed X30Mn22 powder is blended with elementary Al powder which is as well manufactured by gas atomization under argon atmosphere with spherical shape and a particle size distribution up to 45 μm. The blending was performed in a Turbular 2F tumbler mixer for 45 min to achieve homogenous mixed powder blends.

The cuboid ($5 \times 5 \times 10\,mm^3$) samples for microstructure characterization and process parameter development were synthesized on a C45 substrate plate using a bidirectional scanning strategy with 90° rotation between consecutive layers to keep a constant vector length. The samples used for mechanical testing were as well produced on a C45 substrate plate with a bidirectional scanning strategy, but with a scanning vector rotation of 33° between consecutive layers. For this

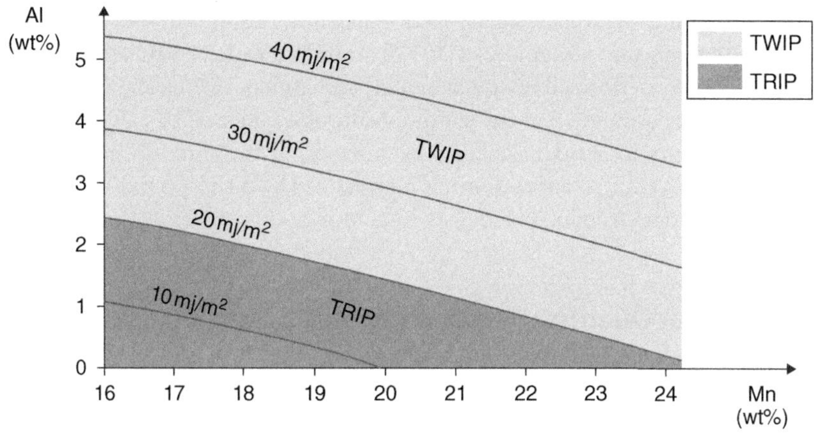


**Figure 6.2** Deformation mechanism and SFE map of steel with 0.3 wt%C and at 20 °C ambient temperature in dependence of the Al and Mn content. *Source:* Based on Twardowski [20].

cylinders with a diameter of 6 and 60 mm in length were manufactured and machined to round dog-bone tensile samples with B4X20 dimensions based on DIN 50125. The manufactured samples to examine fine and complex structures were a $f_2cc,z$ lattice structure by Rehme [1] and Merkt, Hinke, Bültmann, Brandt, and Xie [21] with 10 × 10 × 14 cells, a cell width of 3 mm, strut diameter of 500 μm, and overall dimensions of 30 × 30 × 42 $mm^3$.

## 6.3 Sample Preparation and Characterization Techniques

The produced specimens were mechanically removed from the substrate by wire eroding. Specimens for microstructure analysis were ground with SiC grit paper from 500 up to 1,200 and polished with 6 and 1 μm diamond suspension parallel to the build direction. The samples were etched with 3 and 10% aqueous natal solution.

Further microstructural examinations were performed via SEM, EDS, and EBSD on a Zeiss Sigmafield emission gun (FEG) with EBSD and SEM detectors by Oxford. EDS and EBSD were recorded at a voltage of 15 kV with high-current mode enabled in a working distance of 9 mm. For analysis and noise reduction of the collected data the MTEX toolbox within MATLAB® was used [22, 23].

Tensile tests were determined by quasi-static uniaxial tensile tests on a Z4204 Zwick/Roell device at room temperature and a strain rate of $2.5e^{-4}$/s. The lattice structures were compression tested at room temperature on a Schenk servo-hydraulic universal mechanical testing machine, equipped with a 400 kN load cell and a constant strain rate of $10^{-3}$/s according to DIN 50134:2008. The interpretation was performed after Ashby [24] and Tancogne-Dejean, Spierings, and Mohr [25] for evaluation of the energy absorption potential ($E_{s40\%}$). The specific energy absorption was calculated by integrating the force–displacement curve up until 40% ($E_{40\%}$) compression and normalized by the respective weight of the particular lattice structure [11, 26].

## 6.4 Material Qualification and Process Development

In order to evaluate only the process parameter and chemistry, the first step was to determine a standard set of process parameters which can be used for all following experiments. Therefore, experiments with various different sets of process parameters were performed to find standard hatch parameters as well as standard contour parameters. Cuboid ($5 \times 5 \times 10\,\text{mm}^3$) samples were manufactured with various process parameter combinations as shown in Figure 6.3. Following, the density and error distribution were measured to define the standard hatch parameter set. The definition of a standard contour parameter was determined by an additional experiment where the surface roughness, error distribution, and process stability were used for evaluation. After determination of a suitable process parameter set, it was transferred to different preheating temperatures and

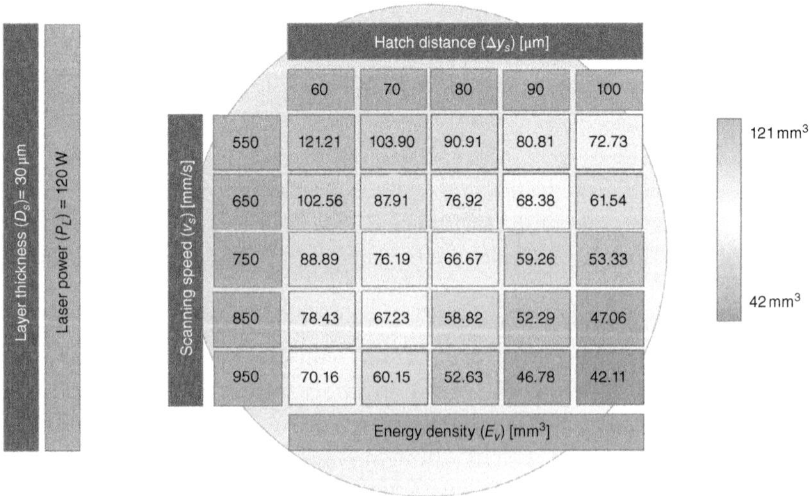


**Figure 6.3** Heat map of the investigated process parameter combinations with corresponding calculated energy densities ($E_V$).

material compositions in order to ensure the required relative density (>99.5%) for all experiments carried out in this contribution.

To achieve a minimal geometrical deviation from the CAD-model of the produced lattice structures, the track width compensation of the contour path was qualified. Therefore, lattice structure samples in 45° and 90° build direction with different track compensations were produced and analyzed regarding the resulting strut diameter of 500 μm.

For process development, the process parameters scanning speed and hatch distance were varied to define a set of process parameters which result in dense and homogenous samples. The layer thickness ($D_S$) was kept constant at 30 μm and the laser power r ($P_L$) at 120 W. The relative densities corresponding to the $E_v$ (Figure 6.3) are shown in Figure 6.4. For the LPBF process relative densities >99.5% can be considered as dense and qualified for the LPBF process [4]. Overall, most of the samples are within the required range. The required density was only not achieved for two samples with a scanning speed of $v_s$ = 950 mm/s and a hatch distance of 90 and 100 μm.

The blue marked set of process parameters in Figure 6.4 was chosen for the standard process parameter set ($P_L$ = 120 W, $D_S$ = 30 μm, $v_s$ = 750 mm/s, $\Delta y_s$ = 70 μm) which resulted in an $E_v$ of 76.19 J/mm$^3$. This standard process parameter set was following transferred to different preheating temperatures and material compositions. For all investigated preheating temperatures and material compositions, the required relative density of >99.5% could be achieved (Table 6.1).

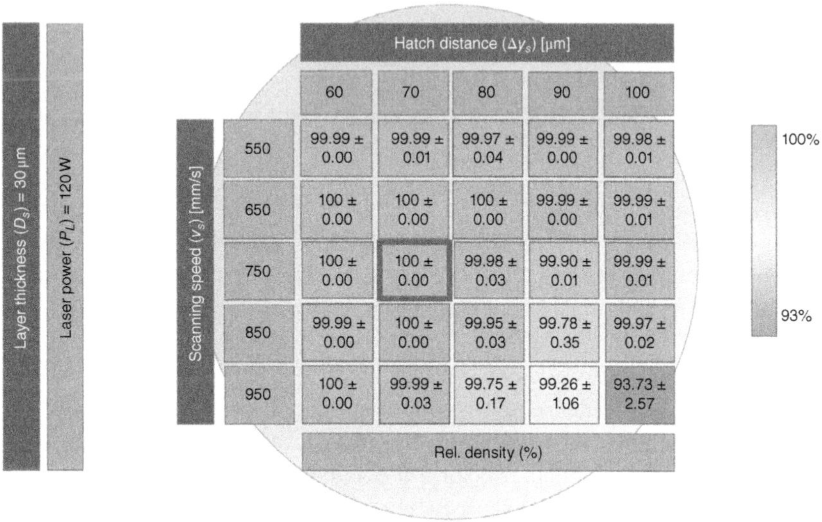


**Figure 6.4** Color-coded relative densities dependent on the different process parameters corresponding to the $E_v$ shown in Figure 6.3. The selected set of process parameters for further investigation is marked in bordered.

**Table 6.1** Measured relative densities of the chosen set of process parameters for the different preheating temperatures and material compositions

| Preheating temperature (°C) | Alloy | Density (%) |
|---|---|---|
| RT | BASE | 99.95 ± .04 |
| RT | BASE+1Al | 99.91 ± .01 |
| 200 | BASE | 100 ± .00 |
| 400 | BASE | 100 ± .00 |

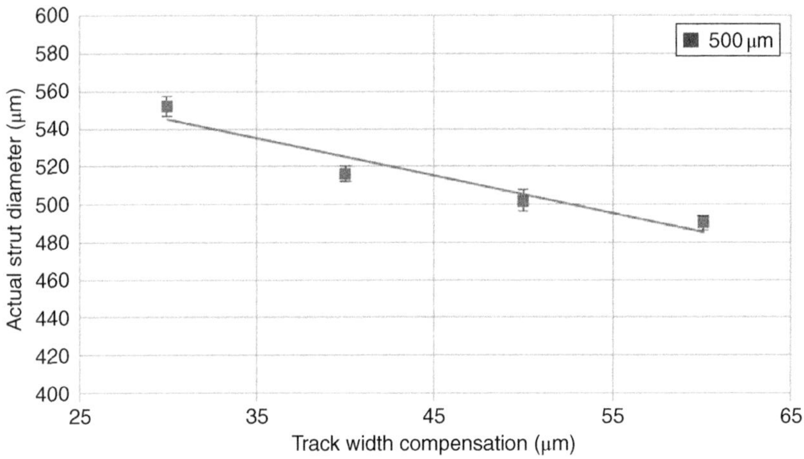


**Figure 6.5** Actual resulting strut diameter depending on the track width compensation for a targeted strut diameter of 500 μm.

To ensure geometric accuracy of the lattice structures, the track compensation was varied and analyzed. The results are shown in Figure 6.5. With increasing track width compensation the actual strut diameter decreased. The requested strut diameter of 500 μm could be reached with a track compensation of 50 μm (Figure 6.5).

The process parameter validation showed a broad spectrum of process parameters which resulted in dense and homogenous samples and was confirmed by density measurements (Figure 6.4 and Table 6.1). These experiments were performed to ensure the manufacturing of dense and chemical homogenous components. The transfer of the found dense process parameter set could be successfully transferred to different preheating temperatures and material compositions. For all states, relative densities of >99.9% achieved (Table 6.1). Thus it can be excluded that influences due to a change of the process parameters (Equation 6.1) on the microstructure and thus on the mechanical properties take place. But it can clearly be seen that the change in material chemistry and preheating temperature shifts the process parameter field and led to other densities (Table 6.1). In particular, by adding Al it can be seen that the density becomes worse even when kept the process parameters constant. An explanation could be the change in absorption due the added elementary Al powder and also the oxide layer of the Al powder could be reason for the process parameter shift [13]. However, since the density is still in the required range of more than 99.5% and the deviation is only 0.04% between BASE and BASE+1Al, it can be assumed that it has no significant influence on the mechanical properties. With the increase of the preheating temperature no change of the relative density could be observed and thus the chosen standard process parameter set are appropriate for the further experiments.

To conclude, varying LPBF process parameters (e.g., preheating), material chemistry (e.g., powder blending), and using fine and complex geometries (e.g., lattice structures) are excellent opportunities to tailor component properties to specific needs and applications. This could be successfully shown by varying preheating temperature, material chemistry, and by exemplary tested lattice structures.

## 6.5 Tailoring Grain Size via Adaptive Processing Strategies

In general the energy input during the manufacturing process in LPBF affects the heat flow and cooling rates of the produced component. The energy input can be controlled and described by the main process parameters shown in Equation 6.1 (laser power ($P_L$), layer thickness ($D_S$), scanning speed ($v_S$), and hatch distance ($\Delta y_S$) as well as by changing the preheating temperature of the powder and substrate. All of them can be used to control the cooling rate which directly affects the grain size [4, 27]. In this contribution exemplary, the preheating temperature is chosen to show the possibilities of tailoring the mechanical properties by control of the LPBF process parameters.

$$E_V = \frac{P_L}{D_S \bullet v_s \bullet \Delta y_s} \tag{6.1}$$

For evaluation samples with preheating temperatures of RT, 200 and 400 °C were produced. All other process parameters (Equation 6.1) were kept constant to eliminate multiple changing factors.

To evaluate the difference in grain size depending on the preheating temperature, EBSD measurements were performed. Figure 6.6 shows the EBSD results for samples manufactured at three different preheating temperatures. Each increase

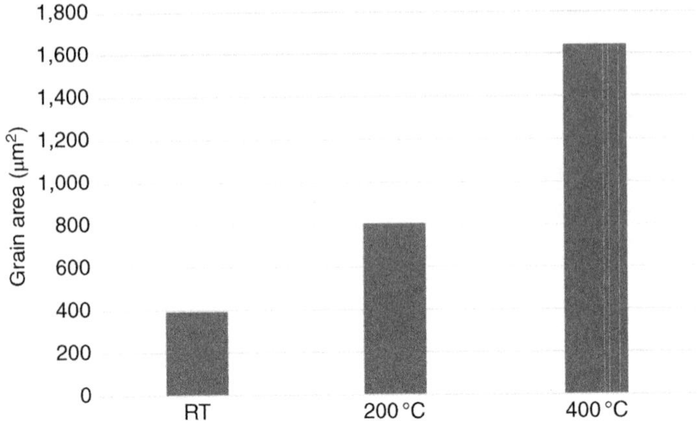


**Figure 6.6** Block diagram showing the mean value of the grain area depending on the preheating temperature.

in the preheating temperature from RT to 200 °C and from 200 to 400 °C results in an increased grain size.

Additionally, hardness and tensile tests were carried out to investigate the resulting change in grain size on the mechanical properties. In Figure 6.7a the results of the hardness mappings are shown. A transition in hardness (300–350 HV0.1) without preheating during manufacturing to 250–300 HV0.1 at 400 °C preheating temperature can be observed. The results of the tensile tests show similar behavior. The stress–strain diagram, shown in Figure 6.7b, presents the results of the tensile tests. A decreasing yield strength, but an increasing elongation can be observed with increasing preheating temperature while all the curves show a steady increase in stress and a failure without necking.

The possibility to tailor the mechanical properties by changing the process parameters is described above. The increasing preheating temperature leads to lower cooling rates, because of the lower heat gradient, which directly affects the grain growth. Different grain sizes resulted in various mechanical properties [27]. Figure 6.6 shows an incline in grain size with increasing preheating temperature which can be correlated to the decreasing mechanical properties in terms of strength, elongation, and hardness (Figure 6.7). This can be explained by the grain refinement, which is one of the hardening mechanisms of steel. Due to grain refinement the number of grain boundaries and for that reason the obstacles, dislocations need to overcome during the deformation, increases. This leads to a higher strain hardening and therefore higher tensile strength [28]. Summarized the mechanical properties decrease with increasing preheating temperature while all other process parameters remain constant.

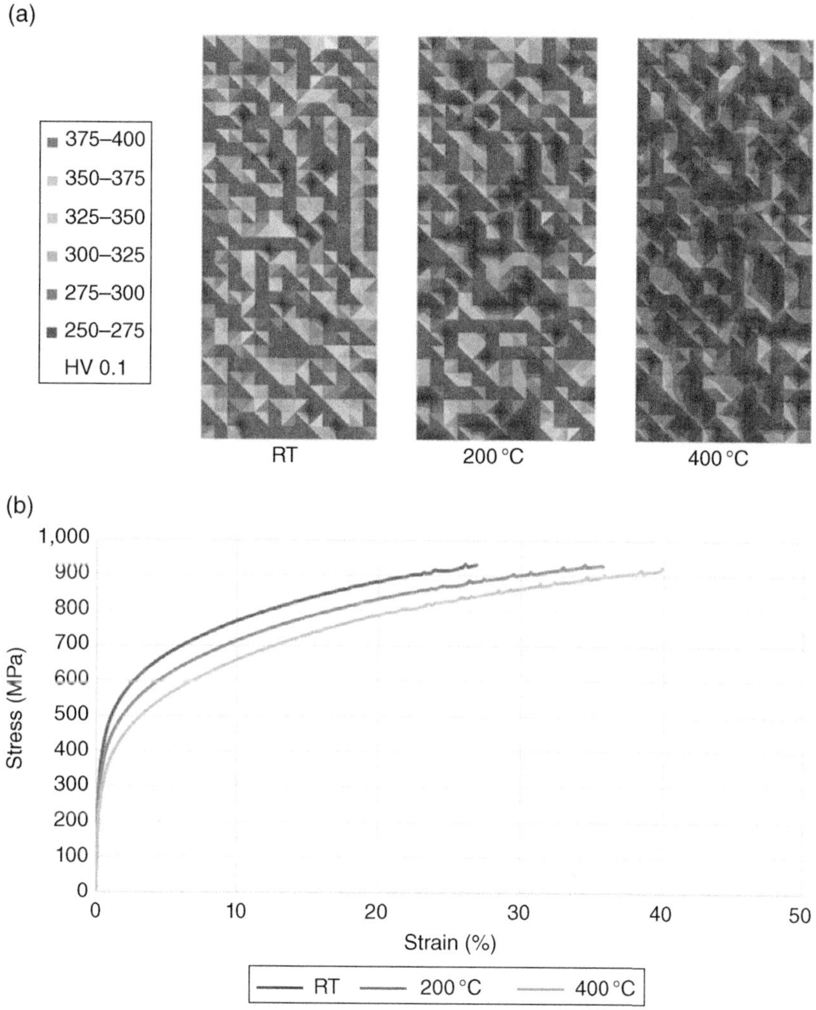


**Figure 6.7** (a) HV 0.1 hardness map depending on the preheating temperature (b) stress–strain diagram of tensile tests performed with different preheating temperatures and BASE composition.

In summary, increasing preheating temperature during the manufacturing process led to lower cooling rates and therefore larger grain sizes. This directly affected the mechanical properties and proofs the possibility to tailor component properties by varying LPBF process parameters.

## 6.6 Tailoring Material Properties By Using Powder Blends

To evaluate the possibility of tailoring mechanical properties by changing the material chemistry, two compositions with the respective SFEs (Table 6.2) were chosen. The two compositions were blended using pre-alloyed X30Mn22 and elementary Al powder. Since the SFE is decisively responsible for the resulting deformation mechanism, and mainly depending on the chemical composition, elementary Al can be used to control the SFE and therefore the deformation mechanisms and mechanical properties.

According to the SFE, BASE should mainly deform by TRIP and BASE+1Al by TWIP and TRIP simultaneously. Therefore, the mechanical properties of the different compositions will vary. TRIP is characterized by higher yield and tensile strength compared to TWIP while TWIP stands out by increased elongation at break and ductility.

The mechanical properties were tested by tensile tests. The results are shown along with the respective fracture surfaces in Figure 6.8a and b. BASE and BASE+1Al showed a constant increasing curve with constant strain hardening and without necking. BASE showed the highest tensile strength and BASE+1Al the highest elongation at break.

The next considered mechanism to affect the mechanical properties is the material chemistry. Within this contribution the alloying was performed by blending of metal powder using tumbler mixer which is a flexible and resource efficient approach [13]. The results of the tensile tests (Figure 6.8) showed an increasing yield strength and tensile strength with increasing Al content. By increasing the Al content in steel, the crystal lattice gets distorted which is known as solid solution hardening. The distorted crystal lattice as well works as obstacle for the moving dislocations. This increases the strain hardening and therefore affects the yield strength and tensile strength [19, 28]. Another explanation for changing mechanical properties is the different deformation mechanisms occurring. The deformation mechanisms are determined by the resulting SFE

**Table 6.2** Nominal chemical compositions and calculated SFE of the investigated alloys

| Alloy | Element | Fe | Mn | Al | C | SFE (mJ/m$^2$) |
|---|---|---|---|---|---|---|
| BASE | (at%) | 76.58 | 22.04 | — | 1.37 | 13 |
| | (wt%) | 77.70 | 22.00 | — | 0.30 | |
| BASE+1Al | (at%) | 75.06 | 21.60 | 2.02 | 1.32 | 21 |
| | (wt%) | 76.92 | 21.78 | 1.00 | 0.29 | |

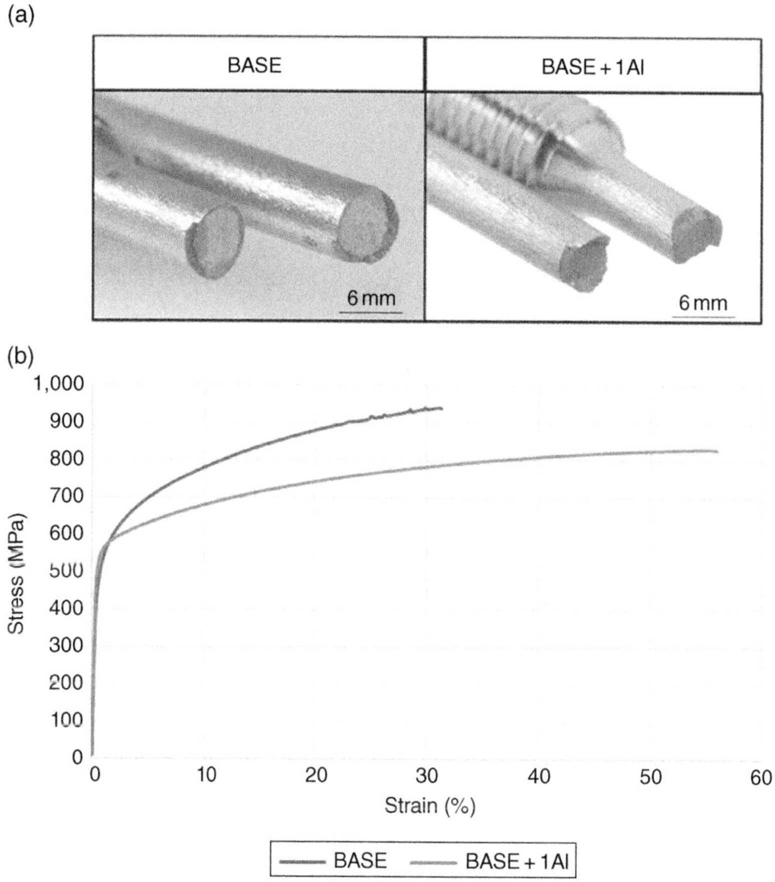


**Figure 6.8** (a) Fracture surface of the tensile test samples are fracture of BASE and BASE+1Al. (b) Stress–strain diagram of the tensile tests of BASE and BASE+1Al.

which can be controlled by the Al content as described in Section 6.2 [20]. For BASE a brittle fracture was observed during the tensile tests (Figure 6.8). With respect to Twardowski [20], this composition failed via TRIP due to martensite formation while deformation. BASE+1Al fails as well brittle, but with increased elongation at break. Regarding Twardowski [20], the BASE+1Al composition deforms via TRIP and TWIP because of the increased SFE. The brittle breaking edge is an indication for deformation by TRIP as well as for TWIP because constant strain hardening until break due to martensite formation and twinning both lead to brittle failure. The increased elongation at break and slightly decreasing tensile strength are indicators for the TWIP mechanism. Therefore, the assumption that BASE deforms via TRIP and BASE+1Al deforms via TRIP and TWIP can be confirmed. This was also proven by Haase et al. [17], Haase [29], and Kies et al. [11] who did further experiments on the same alloys used in this contribution with respect to the occurring deformation mechanisms.

To sum up, adding elementary Al powder to pre-alloyed X30Mn22 powder changed the SFE and therefore the deformation mechanisms, which directly affected the mechanical properties in bulk material as well as in lattice structures. By adapting the material chemistry by using powder blends, it was shown that the deformation behavior of lattice structures could be changed from brittle to ductile and the elongation at break for bulk material could be significantly increased.

## 6.7 Tailoring Properties By Using Special Geometries Such As Lattice Structures

To demonstrate the potential improvement of component capabilities due to use of special component geometries, lattice structures are examined exemplary. Lattice structures have excellent specific energy absorption capabilities and stiffness-to-weight-ratios. This is why lattice structures are promising in different applications and where tested in this contribution.

In Figure 6.9b the force–strain diagrams of BASE and BASE+1Al are shown. The BASE+1Al showed a plateau after the elastic deformation while the BASE alloy did not show such a plateau formation, but an increasing force until 18% strain followed by a rapid decrease and another increase of the force until a plateau formation starting at 30% strain. At 70% strain both curves started to increase again. In Figure 6.9a the lattice structure samples at 40% compression are shown. The BASE alloy showed a failure due to break up of struts along a 45° layer while BASE+1Al showed a constant elastic deformation of all struts.

Exemplary, to show the potential of fine and complex structures, such as lattice structures, which are only producible by the emerging AM technologies, lattice structures of two different material compositions were produced. Lattice structures have an exceptional energy absorption potential and stiffness to weight ratio which make them favorable for energy absorption applications, for example bumper systems, in the automotive industry [1, 21]. With a higher Al content the energy absorption capability increased. This can be explained by the different occurring deformation mechanisms as mentioned above. The BASE samples show brittle failure which leads to a decreased energy absorption capability compared to the BASE+1Al samples. The brittle failure is caused by the martensite occurrence due to TRIP as leading deformation mechanism (Figure 6.9). BASE+1Al with a combination of TRIP and TWIP showed higher absorption capability and a ductile and constant deformation of the lattice structure. Therefore, it can be successfully shown that these structures influence the component properties and the

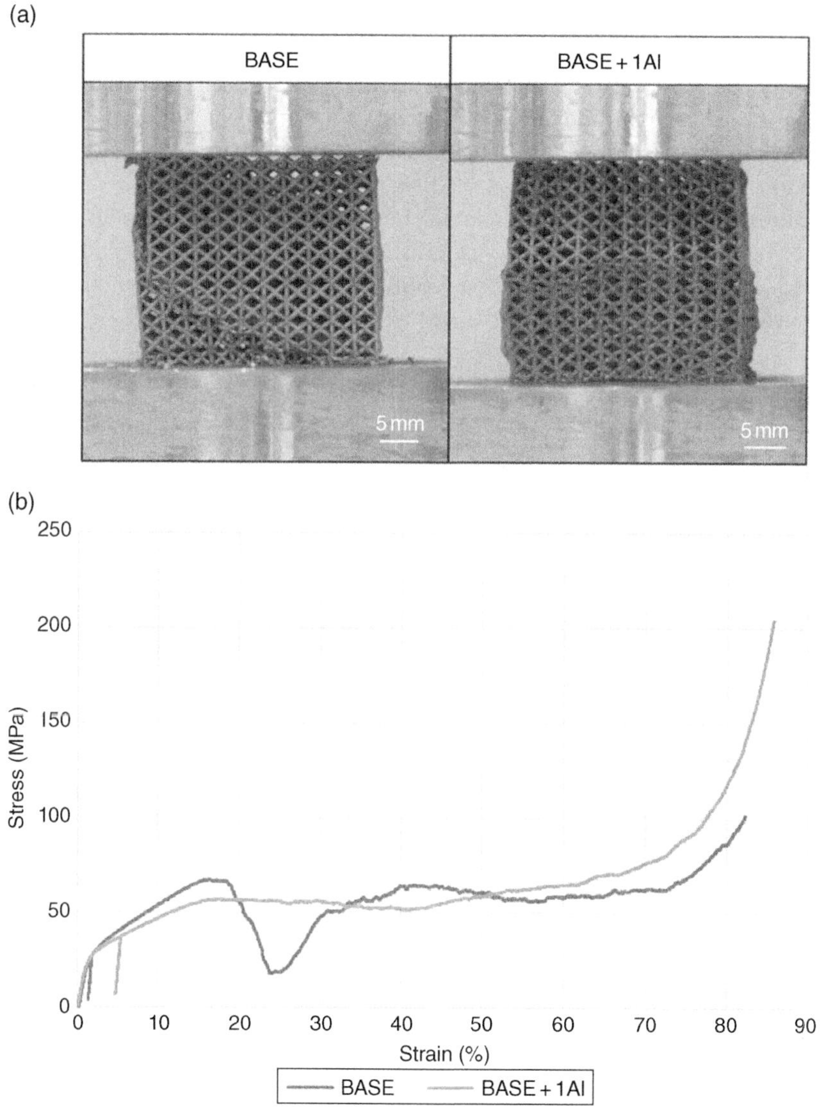


**Figure 6.9** (a) Photo of the BASE and BASE+1Al at 40% compression during a lattice compression test. (b) Force–strain diagram of lattice compression tests with the compositions BASE and BASE+1Al.

deformation behavior as well. While bulk material shows for BASE and BASE+1Al similar deformation behavior [constant strain hardening until brittle failure (Figure 6.8)], the behavior for lattice structures changed completely from brittle to ductile (Figure 6.9a, b). Therefore, the fine and complex structures are a good demonstrator to show the potential of tailoring the component properties by adjusting the component geometry. Lattice structures offer the possibility to tailor the geometry and therefore the component properties down to cell level which offer further control variables additionally to process control and material chemistry.

Concluding, the possibility of manufacturing complex structures (e.g., lattice structures), with LPBF using powder blends was successfully shown. Lattice structures could be reproducible manufactured with a strut diameter of 500 μm and a high geometrical accuracy for two different material compositions. By adding Al the energy absorption capabilities increased due to a changing deformation mechanism and enabled steady ductile deformation.

## Funding

Funded by the Deutsche Forschungsgemeinschaft (DFG, German Research Foundation) under Germany's Excellence Strategy—EXC 2023 Internet of Production—390621612.

## Conflicts of Interest

The authors confirm that there are no known conflicts of interest associated with this publication and there has been no significant financial support for this work that could have influenced its outcome.

## References

1 Rehme, O. (2010). *Cellular design for laser freeform fabrication* (1st ed.). Göttingen: Cuvillier Verlag. ISBN 9783736932739

2 Schleifenbaum, H., Meiners, W., Wissenbach, K., & Hinke, C. (2010). Individualized production by means of high power Selective laser melting. *CIRP Journal of Manufacturing Science and Technology*, 2, 161–169. doi:10.1016/j.cirpj.2010.03.005

3 Frazier, W. E. (2014). Metal additive manufacturing: A review. *Journal of Materials Engineering and Performance*, 23, 1917–1928. doi:10.1007/s11665-014-0958-z

4 Meiners, W. (1999). *Direktes selektives Laser Sintern einkomponentiger metallischer Werkstoffe* (Doctoral dissertation). Ing, Shaker Verlag: Aachen, Germany.

5 Tucho, W. M., Lysne, V. H., Austbø, H., Sjolyst-Kverneland, A., & Hansen, V. (2018). Investigation of effects of process parameters on microstructure and hardness of SLM manufactured SS316L. *Journal of Alloys and Compounds*, 740, 910–925. doi:10.1016/j.jallcom.2018.01.098

6 Hovig, E. W., Holm, H. D., & Sørby, K. (2019). Effect of processing parameters on the relative density of AlSi10Mg processed by laser powder bed fusion. In K. Wang, Y. Wang, J. O. Strandhagen, & T. Yu (Eds.), *Advanced Manufacturing and Automation VIII* (pp. 268–276). Singapore: Springer. ISBN 978-981-13-2374-4

7 Gu, D. (2015). *Laser additive manufacturing of high-performance materials.* Berlin: Springer. ISBN 978-3-662-46088-7

8 Marchese, G., Garmendia Colera, X., Calignano, F., Lorusso, M., Biamino, S., Minetola, P., & Manfredi, D. (2017). Characterization and comparison of Inconel 625 processed by selective laser melting and laser metal deposition. *Advanced Engineering Materials*, 19, 1600635. doi:10.1002/adem.201600635

9 Ewald, S., Schaukellis, M., Koehnen, P., & Schleifenbaum, J. H. (2019). Laser powder bed fusion of advanced high-strength steels: Modification of deformation mechanisms by increasing stacking fault energy. *Berg- und Huettenmaennische Monatshefte*, 10, 2. doi:10.1007/s00501-019-0830-4

10 Liu, Z. H., Zhang, D. Q., Sing, S. L., Chua, C. K., & Loh, L. E. (2014). Interfacial characterization of SLM parts in multi-material processing: Metallurgical diffusion between 316L stainless steel and C18400 copper alloy. *Materials Characterization*, 94, 116–125. doi:10.1016/j.matchar.2014.05.001

11 Kies, F., Köhnen, P., Wilms, M. B., Brasche, F., Pradeep, K. G., Schwedt, A.,…Haase, C. (2018). Design of high-manganese steels for additive manufacturing applications with energy-absorption functionality. *Materials & Design*, 160, 1250–1264. doi:10.1016/j.matdes.2018.10.051

12 Li, W., Chen, X., Yan, L., Zhang, J., Zhang, X., & Liou, F. (2018). Additive manufacturing of a new Fe-Cr-Ni alloy with gradually changing compositions with elemental powder mixes and thermodynamic calculation. *International Journal of Advanced Manufacturing Technology*, 95, 1013–1023. doi:10.1007/s00170-017-1302-1

13 Ewald, S., Kies, F., Hermsen, S., Voshage, M., Haase, C., & Schleifenbaum, J. H. (2019). Rapid alloy development of extremely high-alloyed metals using powder blends in laser powder bed fusion. *Materials (Basel)*, 12. doi:10.3390/ma12101706

14 Schaedler, T. A., Jacobsen, A. J., Torrents, A., Sorensen, A. E., Lian, J., Greer, J. R.,…Carter, W. B. (2011). Ultralight metallic microlattices. *Science*, 334, 962–965. doi:10.1126/science.1211649

15 Zheng, X., Lee, H., Weisgraber, T. H., Shusteff, M., DeOtte, J., Duoss, E. B.,…Spadaccini, C. M. (2014). Ultralight, ultrastiff mechanical metamaterials. *Science*, 344, 1373–1377. doi:10.1126/science.1252291

16 Yan, C., Hao, L., Hussein, A., & Raymont, D. (2012). Evaluations of cellular lattice structures manufactured using selective laser melting. *International Journal of Machine Tools and Manufacture*, 62, 32–38. doi:10.1016/j.ijmachtools.2012.06.002

**17** Haase, C., Bültmann, J., Hof, J., Ziegler, S., Bremen, S., Hinke, C., . . . Bleck, W. (2017). Exploiting process-related advantages of selective laser melting for the production of high-manganese steel. *Materials (Basel)*, 10(1), 56. doi:10.3390/ma10010056

**18** Grässel, O., Krüger, L., Frommeyer, G., & Meyer, L. W. (2000). High strength Fe–Mn–(Al, Si) TRIP/TWIP steels development – Properties – Application. *International Journal of Plasticity*, 16, 1391–1409. doi:10.1016/S0749-6419(00)00015-2

**19** Gottstein, G. (2014). *Materialwissenschaft und Werkstofftechnik*. Berlin, Heidelberg: Springer. ISBN 978-3-642-36602-4

**20** Twardowski, R. *Mikrostrukturelle Beschreibung von Verformung und Schädigung hochmanganhaltiger Stähle mit TRIP- und TWIP-Effekt*. Aachen: Publikationsserver der RWTH Aachen University.

**21** Merkt, S., Hinke, C., Bültmann, J., Brandt, M., & Xie, Y. M. (2015). Mechanical response of TiAl6V4 lattice structures manufactured by selective laser melting in quasistatic and dynamic compression tests. *Journal of Laser Applications*, 27, S17006. doi:10.2351/1.4898835

**22** Bachmann, F., Hielscher, R., & Schaeben, H. (2011). Grain detection from 2d and 3d EBSD data--specification of the MTEX algorithm. *Ultramicroscopy*, 111, 1720–1733. doi:10.1016/j.ultramic.2011.08.002

**23** Nolze, G., & Hielscher, R. (2016). Orientations: Perfectly colored. *Journal of Applied Crystallography*, 49, 1786–1802. doi:10.1107/s1600576716012942

**24** Ashby, M. F. (2006). The properties of foams and lattices. *Philosophical Transactions. Series A, Mathematical, Physical, and Engineering Sciences*, 364, 15–30. doi:10.1098/rsta.2005.1678

**25** Tancogne-Dejean, T., Spierings, A. B., & Mohr, D. (2016). Additively-manufactured metallic micro-lattice materials for high specific energy absorption under static and dynamic loading. *Acta Materialia*, 116, 14–28. doi:10.1016/j.actamat.2016.05.054

**26** Köhnen, P., Haase, C., Bültmann, J., Ziegler, S., Schleifenbaum, J. H., & Bleck, W. (2018). Mechanical properties and deformation behavior of additively manufactured lattice structures of stainless steel. *Materials & Design*, 145, 205–217. doi:10.1016/j.matdes.2018.02.062

**27** Saewe, J., Gayer, C., Vogelpoth, A., & Schleifenbaum, J. H. (2019). Feasability investigation for laser powder bed fusion of high-speed steel AISI M50 with base preheating system. *Berg- und Huettenmaennische Monatshefte*, 5, 359. doi:10.1007/s00501-019-0828-y

**28** Bleck, W. (Ed.) (2012). *Spezielle Werkstoffkunde der Stähle für Studium und Praxis* (2nd ed.). Mainz, Aachen: RWTH Aachen University. ISBN 3-8107-0060-6

**29** Haase, C. *Texture and microstructure evolution during deformation and annealing of high-manganese TWIP steels*. Aachen: RWTH Aachen.

# 7

# 3D Printing Challenges and New Concepts for Production of Complex Objects

*Hayden Taylor, Hossein Heidari, Chi Chung Li, Joseph Toombs, and Sui Man Luk*

*Department of Mechanical Engineering, University of California, Berkeley, CA, USA*

## 7.1 Introduction

One of the many exciting aspects of additive manufacturing today is that promising new technologies seem equally likely to emerge from small startup companies, open-source practitioners, academic research labs, and well-established corporations. This is a field in which a large impact can be made with a little ingenuity—and in some cases with very limited resources. Moreover, some of the most versatile additive processes are applicable all the way from the experimentalist's benchtop to the factory floor. To give just one example, fused deposition modeling (FDM)—the extrusion of molten polymer from a nozzle moving along a controlled path, also known as fused filament fabrication (FFF)—can be carried out to varying levels of sophistication by equipment ranging in cost from a few hundred dollars to a few hundred thousand. Such immense scalability promises to decentralize and democratize manufacturing over the coming decades. Nevertheless, additive manufacturing faces many unsolved challenges, and ways are still being sought to:

- achieve higher production rates,
- eliminate unwanted porosity, defects and distortions, and achieve materials properties comparable with those of traditionally processed bulk materials and composites,
- deliver the final desired surface finish without secondary processing,
- create complex geometries without wasteful temporary supporting structures, and
- quantify and limit the embodied energy of additively manufactured objects.

*3D Printing for Energy Applications*, First Edition. Edited by Albert Tarancón and Vincenzo Esposito.

Digging through the daily avalanche of new publications on additive processes could easily be a full-time job. Fortunately, there is an active specialist press which keeps tabs on progress and produces informative e-mail digests of the latest developments [1, 2]. There are also already many excellent review articles describing the state of the art in the more established additive process technologies [3], including selective laser sintering (SLS), selective laser melting (SLM), and electron beam melting (EBM) of powdered materials [4, 5], FDM/FFF [6], single-material stereolithography (SLA) [7, 8], inkjet-based manufacturing [9], and direct ink writing [8]. The purpose of this chapter, in contrast, is to highlight some highly promising but less conventional material transformation principles that are beginning to be applied to additive fabrication, and that have received especially strong attention in the research literature over the last 1–2 years. We focus on four particular themes that have emerged:

- Promising new metal deposition approaches that overcome some of the speed, safety, and energy consumption limitations of metal powder processing: these methods include the fusion of filaments and laminae, and electrodeposition;
- Novel machine designs and material formulations for stereolithography that allow multiple materials to be integrated into a printed component, or enable the properties of the solidified material to be varied spatially;
- Rapid progress on bioprinting: increasing the size and complexity of printed tissue models, and ordering cellular structures by combining top-down patterning with emergent growth—in particular of microvascular networks;
- Emerging *volumetric* additive processes, in which all points in an object are formed simultaneously, as opposed to being defined relatively slowly, one point, line or layer at a time.

These topics are not intended to be an exhaustive representation of current research, but to provide a reasonable cross-section of technologies that can be expected to progress rapidly over the coming years. The entrepreneurial nature of the field means that some of the most exciting developments need to be pieced together from patents, company presentations and websites, as well as from archival publications.

## 7.2 Geometrical Complexity

A common thread in these new methods is their potential to create structures with greater *complexity* than is possible with established processes. What do we mean by "complexity"? Most usually, complexity is taken to mean that an object contains a large number and variety of *geometrical* features—varying in shape, scale, or both. Large geometrical complexity may be beneficial in several ways, including to improve the mechanical or thermal performance of a structure. Patterning a material to form triangulated lattice structures, for example, can offer exceptionally high specific bending stiffness and strength compared to the bulk material [10]. These mechanical benefits can be further accentuated by integrating multiple length-scales to print hierarchical lattices [11]. Such "mechanical metamaterials" can be rendered more defect-tolerant by introducing discontinuities into manufactured lattices that mimic the hardening mechanisms of natural crystal structures [12]. Through these means, additive processing can vastly reduce the mass needed for a given structural function, by creating geometries that are inaccessible with other methods. Alternatively, additive manufacturing can organize material into geometries that readily bend or buckle, resulting in highly compliant, resilient or energy-absorptive structures. Meanwhile, complex, hierarchical fluid channel geometries reaching down to sub-millimeter sizes allow higher surface-to-volume ratios than are possible with conventional manufacturing. These larger heat transfer areas enable, e.g., optimized heat exchanger performance [13] and enhance cooling of turbine blades [14].

The growing geometrical complexity of manufactured objects requires innovation in the computational methods used to design them and in the data structures used to record them [15]. Computer-driven topological optimization is developing rapidly, and multiscale gigavoxel optimization (e.g. for aircraft wings) [16] has been achieved. While most examples of additively manufactured complexity are created entirely "top-down" with deterministic geometries—whether human-designed or computer-generated—it is also possible to incorporate less spatially organized, emergent, or "bottom-up" structures at the smaller length-scales. For example, $MoS_2$–graphene aerogels have been formed by inkjet printing of an ink laden with precursor materials, followed by freeze-drying and annealing [17]. In this case, the aerogel nanostructure was random, but the larger structure composed by the aerogel was determined by the material deposition process. Other nanocomposite printed materials such as graphene-laden resins can have a complex internal structure that is not directly controlled by the process [18].

## 7.3 Material Complexity

Geometry is not the only aspect of additively manufactured objects that can be complex. Additive processes offer enormous scope to combine multiple *materials* and also to vary the properties of a given material spatially. Interspersing voxels of different materials in varying ratios—in particular through inkjet-based printing—is now a fairly mature way of creating spatial heterogeneity of elastic modulus and color, via both step-changes and gradients [19–21]. The biggest challenges in fact now lie in the combination of completely different *classes* of materials, when forming gradients of property of a single type of material (e.g., a polymer) will not suffice. If, for example, a metal and a polymer need to be combined in an object, a single material deposition method may not be appropriate.

The need for multimaterial structures is especially acute in the production of energy systems, in which, for example, heat conduction paths need to be tailored and electromagnetic fields need to be optimally directed. Laminated structures for modulating conduction properties are of course already prevalent in electrical and electromechanical systems such as motors and transformer cores, and have the potential to be further optimized as reliable multimaterial additive processes are developed. Bonds between dissimilar metals and alloys are also of great interest for co-optimizing the mechanical and thermal performance of heat exchangers, and such bonds can be achieved through additive processes such as the deposition and ultrasonic welding of thin laminae. Achieving sharp spatial transitions in material properties is also increasingly important for the integration of thermoelectric devices and phase-change materials into systems for energy conservation and harvesting. A further example of an emerging need for precision multimaterial deposition is the production of metasurfaces integrating electrically conductive and dielectric geometries. Current additive processes for achieving this aim are embryonic (e.g., dipping an SLA-printed polymeric pattern in a conductive paste to produce mm-wave "metamaterial-embedded geometrical optics" [22]). The sophistication of such processes will doubtless grow to meet the potential needs.

So, a "complex" design can be one that is multiscale, multi-geometry, multimaterial, multi-property, multicomponent, or that combines these sources of richness in some way. A popular related term is "4D printing," generally meaning the production of objects that change their form upon demand in use [23]. 4D-printed structures are often multimaterial, so that some differential response to an external stimulus (e.g., a temperature change leading to differential thermal expansion and hence shape change [24]) can be accentuated. The concept of 4D printing also encompasses materials that will initiate certain temporal behavior in use, such as cell-adhesive, remodelable tissue scaffolds that promote blood vessel growth [25].

## 7.4 Energy Requirements

Underpinning all of this process ingenuity is increasing debate and analysis surrounding the embodied energy of additively manufactured objects, compared to those produced by more traditional techniques. This is a formidably complex question to answer precisely; the energy invested in the production of the source material (e.g., powder, filament, or billet) can be particularly hard to determine accurately. There is certainly cause for concern, with some analysis suggesting that the energy consumption of additive processes may be ~100 times higher than *bulk forming* processes, such as injection molding, for a comparable geometry [26].

It remains very unclear, on the other hand, whether additive processes are more or less energy-consuming than traditional *subtractive* machining. If the power consumption of the material *shaping* process is considered alone, then subtractive processes based on cutting tend to appear far less energy-intensive than additive ones based on molten polymer extrusion or fusion/sintering of particles. Not only is it necessary in many additive processes to remelt material that has already been melted at least once during the preparation of the feedstock, but a large fraction of the heat delivered during material addition does not in fact contribute to melting, being transported instead into the surrounding material or environment and wasted.

If, however, the embodied energy of the feedstock material is also considered, the picture is more complex. In powder-bed processes, for example, a large fraction of the unused material can generally be reused without recycling, whereas in subtractive processes the scrapped material, which is usually in the form of chips, needs to be recycled before being reused, incurring an additional energy input. Depending on the geometry of the component, the scrap fraction, and the energy cost of recycling the material, an embodied energy analysis could favor either traditional or additive processes.

These embodied energy considerations nevertheless pale into insignificance for applications in which the mass reduction enabled by the geometric freedom of additive manufacturing translates to major energy savings in service. This is why additive processes have seen their most vigorous adoption in aerospace applications, where mass savings during design and manufacture correspond to enormous fuel savings during the lifetime of an aircraft or spacecraft. The energy implications of maintenance and decommissioning also need to be considered. While additive manufacturing has enabled part counts and assembly times to be reduced, larger, more complex components may be harder to separate into their constituent materials for recycling, and there may be a tendency to scrap a larger mass of material when just one element of a complex component fails.

There are some positive signs that resource usage is being increasingly considered in the development of additive processes. For example, in FFF, work has been done to optimize extrusion rates and nozzle temperatures for minimal energy consumption [27]. Printers are increasingly becoming available with extruders that are fed directly with thermoplastic pellets as opposed to filament—eliminating one remelting cycle from the preparation of the feedstock. More sustainable feedstock materials are also being explored, such as plant-derived cellulose acetate [28]. Such thoughtful approaches to machine, process, and materials design will no doubt become more prevalent.

## 7.5 Promising Metal Deposition Approaches

The vast majority of metal additive processes today rely on processing metal powders in some way, whether directly sintering or fusing them with a scanning laser or electron beam, or bonding the powders together with a polymeric binder that may later be burned off. These methods are increasingly capable, and satisfy many application areas. Metal powder-based processing, however, has significant challenges, many of which stem from the nature of the powders themselves and their shape and size variability [29]. Many powders of engineering relevance, including aluminum alloys, can be explosive, and there are occupational health concerns surrounding inhalation. The porosity of printed components has only recently begun to be understood and controlled [30], and there is a severe trade-off between printing speed and the ability to produce fully dense parts. Equally concerning are the thermal considerations. Not only is an enormous fraction of the heat delivered to the component lost to the surroundings without contributing to fusion, but the transport of this heat creates temperature gradients that result in residual stresses and component distortions [31]. Given these challenges, it is not surprising that alternative metal additive processes are being sought.

One of the most exciting alternatives recently to emerge is "Joule printing," in which a solid metal or bulk metallic glass filament is fused onto the target surface by electrical resistance heating [32, 33]. The use of a solid filament of the structural material avoids the safety concerns of powders, and removes the need to handle any form of polymeric binder. As with polymeric FFF, one might anticipate challenges from thermal gradients or from voids between deposited filaments, especially at the highest deposition rates. Solid metal can also be deposited in *laminar* form by ultrasonically welding a thin shim of material onto the object, machining the layer to the desired shape, and repeating many times to construct an object [34]. This technique has been shown to form strong bonds between dissimilar materials (e.g., Cu and Al alloys), and permits the formation of internal channels and voids at the resolution of the machining process. These capabilities lend the process particularly well to the production of, for example, advanced heat exchangers in which thermal conductance and mechanical strength both need to be considered and fluid–solid interactions need to be optimized.

Returning to the particle scale, electrophoretic deposition, directed either by electrodes [35] or light [36], offers the ability to construct multimaterial objects from colloids without delivering heat to the particles. Finally, selective ion-by-ion metal deposition is being extensively explored. One of the most ambitious efforts in this area is analogous to a hybrid of SLA and electrodeposition [37, 38]. A square array of independently switchable electrodes drives layer-by-layer plating of a metal onto a 3D component. A particular challenge with this process is localizing deposition to the vicinity of the desired electrodes; fringing fields may lead to parasitic plating further from the electrodes. Chemical passivation of the printed component's surface away from the electrode array—rather like in electrochemical mechanical polishing [39]—may offer one way to localize metal deposition. Point-by-point electrodeposition through nozzles has also been demonstrated, enabling greater spatial resolution than layer-based deposition, as well as allowing multiple metallic materials to be combined in an object [40–42]. Photosensitive materials have also been engineered to yield metallic structures, for example, via photoreduction of metal in direct laser writing [43], or by pyrolyzing a Ni-rich photoresist after exposure [44].

Taken together, the emerging metal-printing processes described in this section offer ways to deposit and assemble atoms, particles, filaments and laminae of metallic materials in controlled and complex ways. It seems likely that in the future, processes will emerge that blend two or more of these deposition principles to optimize material properties or fabrication time.

## 7.6 Multimaterial and Multi-property SLA

Stereolithography—which can be accomplished either point-by-point with a scanning laser beam or layer-by-layer through the projection of 2D light patterns—is sometimes referred to as digital light printing (DLP) or continuous liquid interface production (CLIP) [45]. In spite of SLA's high profile throughout industrial prototyping—and, increasingly, production—few options existed until recently for SLA printing of objects with multiple materials or with spatial variation of physical properties such as elastic modulus. There has, however, recently been an explosion of material and process innovation to this end.

## 7.7 Temporal Multiplexing

SLA has traditionally struggled to handle multiple materials because the object being printed is usually immersed in a tray of a single specific resin. One way to accomplish multimaterial printing is to change the object's resin surroundings repeatedly so that features composed of one material after another can be fabricated. Han, Yang, Fang, and Lee's [46] microstereolithographic system, for example, uses a flow cell that repeatedly exchanges, by pumping, the uncured material that is being illuminated around the target component, to build up composite structures layer-by-layer. Material exchange was accomplished within ~15 s, which is on the same order as the printing time of one layer of material, meaning that this process is likely to be feasible for small parts with microscopic feature dimensions. It is not yet clear how a flow cell scheme might scale to the production of macroscopic components. An alternative approach, as shown by Choi, is to move the component from one vat of resin to another on a rotating carousel as layers are built up [47]. Hohnholz et al. [48], meanwhile, combined aerosol spray deposition of multiple materials with laser scanning photopolymerization, allowing higher-viscosity materials (10–1,000 cP) to be processed than in the inkjet-based systems that are typically used for multimaterial or multi-property fabrication.

## 7.8 Resin Formulations with Multiple End-States

Given the practical mechanical difficulties of cycling repeatedly through different resins, ways of modulating the final properties of a single resin formulation are desired. One approach is to mix two or more materials whose polymerization chemistries are quite different (e.g., free radical and cationic polymerization), can be independently triggered by two separate wavelengths, and have different final properties (e.g., elastic modulus). Such a system of "orthogonal" chemistries was demonstrated by Schwartz and Boydston [49] and allows spatial control of elastic modulus. Dolinski's et al. [50] "solution mask liquid lithography," similarly, uses two orthogonal photopolymerization chemistries (radical and cationic, sensitive to different wavelengths of light) within a single mixture to vary elastic modulus from ~100 kPa to several GPa.

Kuang et al. [51], meanwhile, have introduced a two-stage "grayscale" DLP process in which spatially varying the intensity of illumination controls the extent of a free-radical photopolymerization reaction, forming a mesh with a spatially variable density. A subsequent thermal curing step connects crosslinker molecules between the unreacted acrylates to render the material unreactive while accentuating the stiffness variations. Young's modulus was varied in this way from ~1 MPa to ~1 GPa simply by modulating the intensity illumination. Yin, Ding, Zhai, Tan, and Yin [52] similarly achieved spatial control of a PEGDMA hydrogel's modulus (in the range 2–15 kPa) and geometry via grayscale illumination. Spatial control of cellular adhesion, and not just modulus or structure, is critical for bioprinting applications and synthetic polymers may offer greater control of these behaviors than naturally derived ones. For example, RGDS (biotin) peptide–polymer conjugates can readily be produced and incorporated into printable materials [53].

## 7.9 Associated Processing Considerations

One of the primary rate- and geometry-limiters of DLP processes is the hydrodynamic stress that arises in the film of liquid resin that exists between the component being printed and the window through which the illumination is directed [45]. These stresses may distort the partially cured component, and they grow with the viscosity of the resin, the lateral dimensions of the individual features being printed, and the linear speed of retraction of the object from the resin. Addressing viscosity limitations becomes increasingly important as DLP processes seek to be used in production.

The challenge is particularly relevant to the processing of particle-laden resins, in which an organic resin loaded with a high solid volume fraction is printed to produce a green (partially-cured) part [7]. Such materials, when the particles are ceramics such as zirconia or silica, are of interest for printing, (e.g., dental restoratives or optical components [54]). Printed nanocomposites can offer desirable engineering properties such as high toughness or electrical conductivity. Acoustic fields have been used to position and orient electrically [55] and thermally [56] conductive particles within resin. An electrical field has been used to align graphene nanoplatelets to produce laminar, nacre-like nanocomposites within SLA-printed material [57]. Magnetic manipulation of composite structures within SLA has also been demonstrated [58].

These particle-laden fluids have higher viscosities than standard photosensitive resins. For example, Song, Park, Lee, Lee, and Yun [59] found that a suitable $ZrO_2$ particle-laden resin had a viscosity of 20,000 cP—about seven times that of a typical commercial production SLA resin [60]—and yet could be successfully processed by DLP, albeit more slowly than in typical processes. Printable alumina slurries with solid particles up to 60 wt% show a comparable viscosity of 15,300 cP [61]. The high viscosity of the green component may have advantages for producing and maintaining unsupported spans until sintering [59].

One way of reducing hydrodynamic stresses would be to increase the thickness of the liquid film between the printed component and the transparent illumination window. In processes where this liquid film is maintained by oxygen inhibition of the free-radical crosslinking process and the requisite oxygen diffuses in through the window itself, the liquid film is referred to as a "dead zone" and printing rate may be limited by the diffusion of oxygen through this window. Increasing the thickness of the dead zone has been accomplished by Beer, who replaced the oxygen inhibition with radical quenching by a light-activated molecule [62] and eliminated oxygen diffusion as a rate-limiting factor.

Other, mechanically oriented, approaches to improving the performance of this fluid–solid interface have been shown. Systems are available that heat the liquid resin to lower its viscosity [63]. Meanwhile, Walker, Hedrick, and Mirkin [64] recently demonstrated linear print speeds up to 430 mm/hr by replacing the oxygen permeable window with a flowing layer of fluorinated oil which not only removes the speed limitation imposed in other systems by oxygen diffusion but also convects heat from the polymerization site. The most desirable temperature for the resin during printing is likely to require a balance between achieving favorably low viscosity and avoiding thermal curing or thermal stress-induced distortions.

As the performance of SLA/DLP systems has improved there has been increased attention on overcoming the so-called "stair-step" effect, whereby surfaces that are intended to be smoothly curved become roughened by the microscopic offsets between the sharp edges of adjacent printed layers. Recent work has shown that this effect can be countered by using computation to design grayscale edges to the illumination patterns to smoothen the layer-to-layer transitions [65], or by microscopic oscillation of the projection optics during printing [66].

A more radical way of avoiding hydrodynamic stresses in a layer-by-layer process is to deposit the material electrostatically in the form of a solid thermoplastic powder, and then fuse it thermally onto the component—analogously to a 3D laser printer. This process is referred to as electrography, or by its inventors as the "Selective Thermoplastic Electrophotographic Process" (STEP) [67–69]. The process inherently enables multimaterial processing and can in principle use a wide range of thermoplastics, since it does not require photosensitive molecules.

## 7.10 Bioprinting of Realistic and Vascularized Tissue

The rapid recent advancements in SLA/DLP printing have opened a new front in bioprinting, where organ-scale printing of vascularized tissue is an unsolved problem. Multiscale vascular constructs have been achieved, some incorporating patterned, pro-angiogenic factors [25], but the size and geometrical complexity of such models is still limited. One successful example of the use of light-based additive manufacturing for vascular structures is the work of Zhu et al. [70], in which vascular networks were defined in methacrylated gelatin–hyaluronic acid composite gels with elastic moduli in the range 2–6 kPa. Multiple cell types, including human umbilical vein endothelial cells, were encapsulated, and the constructs were successfully implanted into mice and naturally connected (anastomosed) with the animals' vascular networks. Recently, Grigoryan et al. [71] used a form of DLP to fabricate interpenetrating networks of oxygen channels and blood vessels with unprecedented size and complexity, to simulate a lung structure [72].

SLA/DLP joins several other established processes as candidates for tissue and organ printing. Coaxial extrusion of multilayered hydrogel lumens has been widely used to produce vessels, although with limited geometrical complexity. Extrusion and UV curing have been combined to enhance the ability to build up complex shapes in cell-laden methacrylated gelatin (GelMA) [73]. To address the need for mechanical support in the layer-based printing of soft hydrogels, a water-soluble sacrificial ink, Pluronic F-127, has been integrated with DLP of methacrylated hyaluronic acid and methacrylated alginate [74]. Additionally, extrusion of sacrificial and functional materials into a volume of supporting gel is now a widely adopted fabrication approach. For example, Lee used the previously developed "freeform reversible embedding of suspended hydrogels" (FRESH) method to print a collagen scaffold for a neonatal-scale human heart model [75]. Similar support gel-based methods have been demonstrated by, for example, Senior, Cooke, Grover, and Smith [76] and Wang et al. [77].

Obtaining a physiologically realistic volumetric density of cells within a printed construct remains a challenge since hydrodynamic shear stresses during extrusion or limited illumination penetration during DLP can constrain cell density. To address this challenge, recent work by Skylar-Scott et al. [78] has shown how extrusion printing of sacrificial materials can define channels within a granular volume of cell spheroids, minimizing the need to expose cells to substantial shear stresses during printing and yielding vascularized, perfusable cardiac tissue. Ultrasound acoustic fields have also been proposed to aggregate cells into spheroids [79] and to manipulate them into structures localized within a suspension [80]—analogously to the use of acoustic energy to assemble engineering composites as described in Section 7.9.

For initially non-cellularized scaffold components in particular—such as some heart valve implants—mechanical properties are critical because they can direct subsequent cell growth, migration, and differentiation. Mechanical stiffness and its directionality can be more powerfully controlled in polymer filament deposition methods if the filaments do not need to be confined to planes. For example, stress-line additive manufacturing [81] enables extruded filaments to be extruded in directions of principal stress. Silicone mesh structures with precisely designed mechanical compliances have been extruded by direct ink writing onto a mandrel which was itself printed by SLA from patient computed tomography (CT) data. The extrusion process was combined with subsequent spray deposition of a fine silicone membrane. This hybrid process, which uses several different additive deposition methods, is a sign of how specific applications can be served by customized process flows [82]. Meanwhile, shear stress in an extrusion nozzle, while generally considered detrimental when depositing cell suspensions, has been shown to be useful in controlling collagen fibril alignment for corneal tissue engineering [83].

## 7.11 Emerging Volumetric Additive Processes

Today's polymeric "3D printing" processes are overwhelmingly accomplished through the repetition of *lower*-dimensional unit processes, ranging from point-by-point stereolithographic curing of photosensitive resins and extrusion from nozzles and needles, to line-by-line printing from nozzle arrays, to layer-by-layer DLP exposure of photopolymers. Given the throughput limitations of layer-based processing that we described in Section 7.9, it would be desirable to have a printing process available that could rapidly create all points of a 3D polymeric object *simultaneously*. This goal has been pursued in recent years in a number of ways.

The first approach is sometimes referred to as *holographic* light patterning, in which a customized wavefront of light is created (e.g., with a phase mask or spatial light modulator) and diffracts as it propagates, creating a 3D intensity distribution essentially via constructive and destructive interference at different locations. Conceivably, this 3D intensity distribution could be created inside a photopolymer to solidify an object. Although many possible algorithms for generating 3D light fields have been described [84, 85], the geometrical versatility of such a method is inherently limited by the fixed physical relationship between the light fields at various propagation distances. One consequence of this relationship is that for reasonably sized print volumes (e.g., at least several mm), the numerical aperture of projection optics will be low enough that the spatial resolution achievable in the axial (propagation) direction is far inferior to that achievable within planes normal to the propagation direction. Moreover, because holographic methods rely on modulating the phase of light to control interference, the intensity pattern generated is highly sensitive to even small inhomogeneities in the refractive index of the propagation medium—as might indeed be expected during photopolymerization, when refractive index can rise by up to ~10%. Efforts to develop holography-based 3D printing commercially [86] do not seem to have grown. Recently, hope for a holographic approach has been rekindled by progress in *metasurfaces*, which offer enhanced control over phase and polarization and could enable higher axial resolution throughout large print volumes. In current applications of metasurfaces to stereolithography, however, printing depths per exposure step are limited to about 10 μm [87].

An alternative to holography was therefore proposed by Shusteff et al. [88], in which three collimated and patterned beams were intersected orthogonally in a photopolymer solution. Where the beams intersected, the intensity exceeded a critical value for polymerization to occur during the exposure period. This approach does not rely on any form of interference between the intersecting beams, but rather on the simple superposition of intensities. The method generated objects within a few seconds, but the range of geometries that could be produced was limited by the need for the beams to intersect.

As a result, some of us have recently introduced *computed axial lithography* (CAL) [89], an approach to synthesizing a 3D light dose distribution through tomographic reconstruction. The method is analogous to carrying out a CT scan in reverse but using light at the curing wavelength of the photopolymer solution, rather than X-rays. The system (Figure 7.1) consists of a rotating cylindrical volume of photosensitive material and a video projector that shines patterns of light sideways into and through the volume, propagating approximately perpendicularly to the axis of rotation. As the volume rotates, the projected pattern changes in synchrony, with typically >1,000 video images being used per revolution. Over time, the superposition of these many patterns controls the total

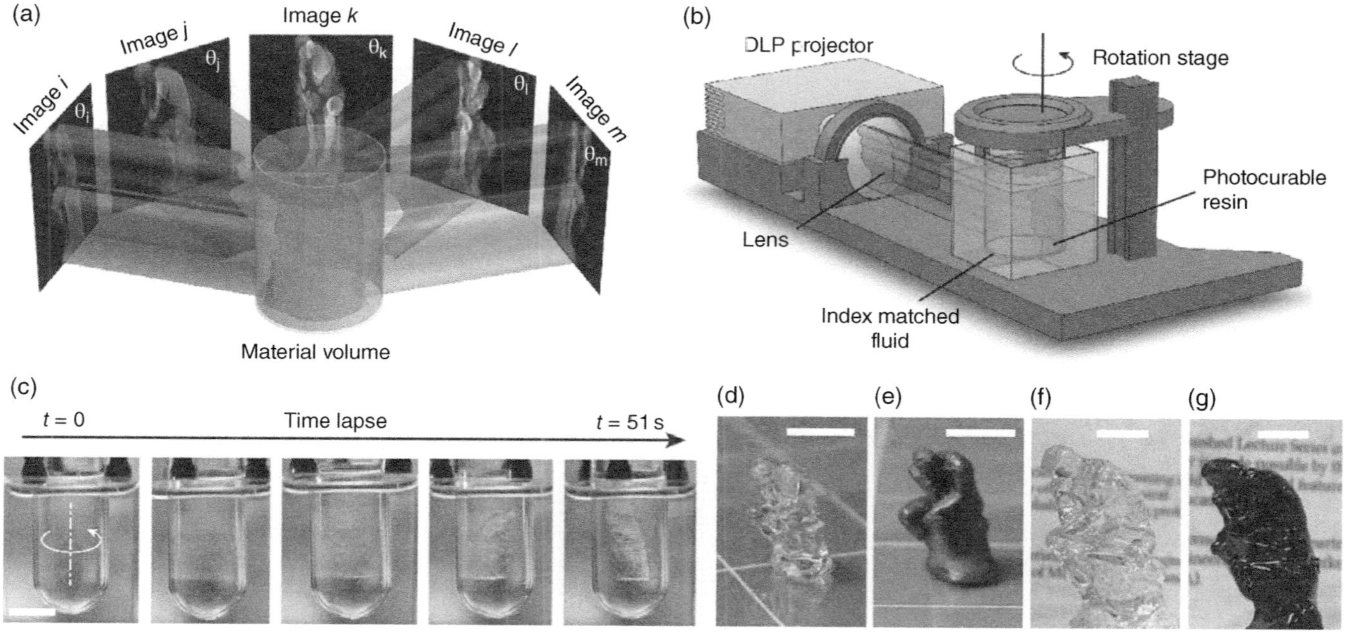


**Figure 7.1** CAL volumetric fabrication. (a) Underlying concept: patterned illumination from many directions delivers a computed 3D exposure dose to a photoresponsive material. (b) Schematic of CAL system used in this work. (c) Sequential view of the build volume during a CAL print. A 3D geometry is formed in the material in less than a minute. (d) The 3D part shown in (c) after rinsing away uncured material. (e) The part from (d) painted for clarity. (f) A larger (40 mm-tall) version of the same geometry. (g) Opaque version of the geometry in (f), using crystal violet dye in the resin. Scale bars: 10 mm [89]. *Source:* From Kelly, B.E., Bhattacharya, I., Heidari, H., Shusteff, M., Spadaccini, C.M., and Taylor, H.K. (2019) Volumetric additive manufacturing via tomographic reconstruction. *Science*, 363 (6431), 1075–1079. Courtesy of Hayden Taylor.

dose of photons absorbed at each location within the photopolymer. Where the absorbed dose exceeds a threshold, which is determined by the amount of oxygen dissolved in the resin at the start of the process, the polymer solidifies and the desired object is formed. Again, this method is not relying on controlling phase; just intensity. Inexpensive LED light sources can therefore be used and temporal multiplexing of the many angular projections via rotation is possible.

## 7.12 Computation for CAL

CAL employs the same core tomographic principles [90] that are widely used in X-ray CT for 3D imaging [91, 92]. There has also previously been exploration of optical tomography for 3D biological microscopy down to ~10 μm features [93] and, by introducing an interfering optical reference beam in addition to tomographic projection, micron-level reconstruction of 3D refractive index distributions (e.g., of red blood cells) in "tomographic phase microscopy" [94].

The process of designing a sequence of video images to be projected during CAL begins with the voxelization of a computer model of the desired object. This voxelized representation is used to generate a 3D matrix of target light dose for the printing process. Currently, our illumination dose objective is usually uniform within the object and zero outside of it. For each plane of voxels at a specific position along the rotation axis ($z$), we carry out the Radon transform of the resulting 2D matrix for a selected number of angular sampling points (1,440 equally spaced samples is typical). The Radon transform computes the line integral of the variable of interest—in this case the dose objective—along lines of constant $r$, where $r$ is the transverse position relative to the center of rotation and $\theta$ is the rotation angle. (Ultimately, the axis $r$ will correspond to rows of pixels in the projected video images, and different $\theta$ values will correspond to the different frames of the video.) The quality of the reconstruction can be expected to rise with the number of angular sampling points, and this tendency represents a core advantage of CAL: the achievable geometrical complexity can be increased simply by raising the number of sampling points, while the complexity of the apparatus does not need to increase.

The direct Radon transform cannot, however, simply be backprojected into the photosensitive material, because of the oversampling of lower spatial frequencies. Therefore, the well-known "Ram-Lak" filter is applied to the Radon-transformed values [95]: this is a ramp filter, whose magnitude is proportional to the absolute value of spatial frequency. Unfortunately, this filtering process introduces negative values in the output. Some conceivable ways of addressing this problem to create realizable projection images might include adding a constant offset to make the minimum value of the function equal to zero; however, this approach may render the contrast of the reconstructed dose impracticably low. The negative values could simply be truncated to zero, and for some simple target geometries we have found this approach to be reasonably effective. A more generally applicable approach, however, is to carry out an iterative optimization of the patterns to be projected, in which the acceptance of negative values is gradually reduced to zero over multiple iterations. This iteration process also allows us to capture explicitly the nonlinear dose response of real resin materials, which we do with a step function of dose. We attribute the nonlinear response to the presence of dissolved oxygen, which quenches free radicals generated by illumination up to a certain threshold dose; beyond this dose, the oxygen is largely depleted and polymerization can proceed. The optimization seeks to minimize deviations between the target dose and the thresholded dose delivered by the optimized projections. Matlab code for this procedure is available online for education, research and not-for-profit purposes [96].

The use of the plain Radon transform within this optimization algorithm is appropriate for cases where the absorption coefficient of the resin is low enough that intensity does not fall appreciably as the projected rays pass through the printing volume. Some resins, however, may absorb an appreciable fraction of the illumination during passage (Figure 7.2a), even though for the tomographic effect to work the intensity at the center of the volume still needs to be a substantial fraction of the incident intensity. This characteristic is quite unlike DLP printing, where dyes are introduced to absorb most light within ~100 μm of the resin's surface. If absorption in CAL is appreciable, the *exponential* Radon transform can be used in the projection computations [97]. We have analytically shown (Figure 7.2b) that the printing time can be expected to be minimized when the absorption coefficient is the reciprocal of the printing container's radius: that is, when about 60% of the incident intensity has been absorbed by the time the light reaches the center of the printing volume.

## 7.13 Material–Process Interactions in CAL

The practical advantages of CAL flow from the lack of any substantial relative motion between the uncured resin and the object being printed during the exposure process. As a result, hydrodynamic stresses that are problematic for SLA/DLP are not present in CAL. What limits CAL's printing speed is simply the rate of the polymerization reaction, which in turn depends on the brightness of the light source that can be delivered. The exposure time itself can also potentially be much shorter than DLP for a given illumination power: since we do not need to add passive dyes to CAL resins—as are required in DLP to limit penetration depth—CAL does not waste light energy by absorbing it unnecessarily. Overall, printing speeds could be several orders of magnitude higher than in SLA/DLP [89]. A model for

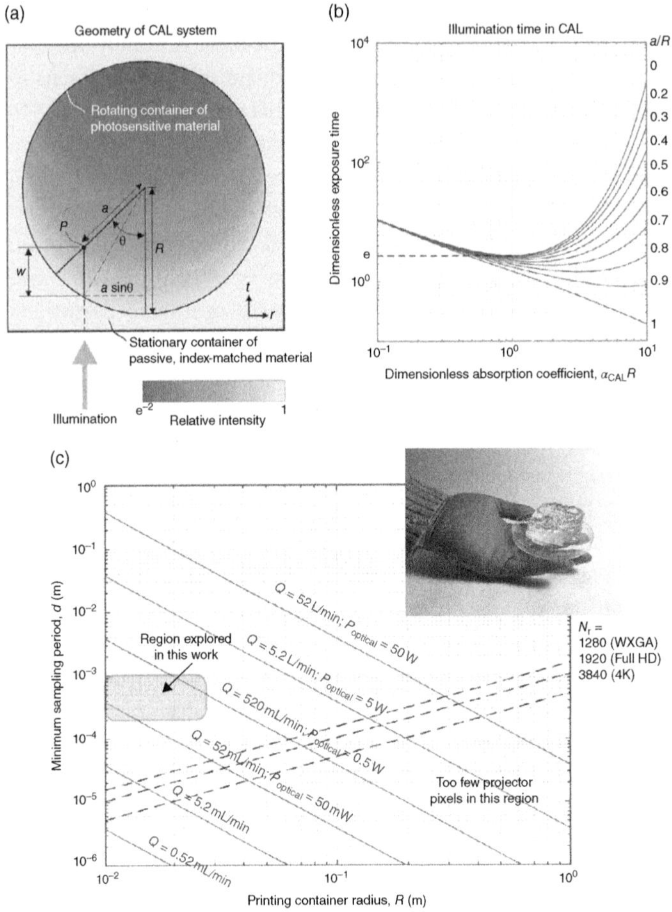


**Figure 7.2** Performance of CAL. (a) Beer–Lambert model of light propagation through printing volume, showing exponential decay of intensity with propagation distance $w$ for the minimal required exposure time to print a point at the center of the volume. (b) Highest print speed is associated with absorption coefficient equaling the reciprocal of the container radius: $\alpha_{CAL}R = 1$. (c) Relationship between minimum sampling period $d$ (representing spatial resolution), container radius $R$, peak optical output power $P_{optical}$ and volumetric print speed $Q$ in CAL. In this chart, a curing dose of $D_{c0} = 4.9 \times 10^{22}$ photons/m$^3$ (as found experimentally for our current resin), a projector refresh rate of $f_p = 25$ Hz, and illumination wavelength of 455 nm are assumed. These conditions are representative of the projector and resin used for many geometries printed so far with CAL; the region of geometries explored until now is indicated with a shaded box. Contours of illumination power $P_{optical}$ and volumetric solidification rate $Q$ are shown (solid lines), as are constraints imposed by the number of pixels $N_r$ across the width of projector hardware (dashed lines). Inset: dental model [89]. *Source:* Adapted from Kelly, B.E., Bhattacharya, I., Heidari, H., Shusteff, M., Spadaccini, C.M., and Taylor, H.K. (2019) Volumetric additive manufacturing via tomographic reconstruction. *Science*, 363 (6431), 1075–1079.

the relationship between printing speed, spatial resolution, and printing volume size (Figure 7.2c) captures two primary effects. The first is the limitation on spatial resolution imposed by a finite number of projector pixels being distributed across the printing volume. The second effect is that of motion blurring, which is determined by the angular velocity of the printing volume, the frame rate of the projector, and the radius of the printing volume. The largest component produced so far with CAL, a dental model (Figure 7.2c, inset), is between 5 and 10 cm in diameter and was produced in less than 10 min. The analysis predicts that with a commercially available 4K projector delivering 5 W optical power, we expect in the future to be able to fabricate 0.1 mm features in a 0.5 m diameter printing volume.

The lack of relative motion and hydrodynamic stresses also broadens the range of materials that can be printed. We have printed into exceptionally high-viscosity materials (>100,000 cP, cf. 3,000–4,000 cP for a typical DLP resin) and even into solid gels, which would be impossible in layer-based processes. High accessible viscosities mean that resins do not need to be diluted, as they do with many SLA/DLP processes, so the final material properties do not need to be compromised in CAL. Moreover, highly viscous materials provide comprehensive support for objects being printed, so there is no need to incorporate dedicated solid supporting structures, for example, large overhanging features, as there often is with layer-based processing. The self-supporting nature of the process enables delicate structures to be fabricated in soft, highly compliant materials such as hydrogels. It is encouraging to see that some first steps have been taken to exploit this capability for bioprinting [98]. Rapid production of complex, multiscale, perfusable vascular networks will be the next major hurdle to overcome if complete organs are to be produced.

The absence of discrete layers from the CAL process also means that the surfaces produced are especially smooth. CAL may therefore be able to compete with surface tension-smoothened inkjet printing [99] for the production of custom optical components (e.g., for eyeglasses), contact lenses, or even corneal regeneration [100]. Finally, CAL can solidify 3D polymeric geometries around preexisting solid objects that could be made by other methods or with different materials—a capability we term “overprinting.”

## 7.14 Current Challenges in CAL

The theoretical relationship between speed, printing size, and spatial resolution shown in Figure 7.2c does not consider several observed effects that may limit resolution, dimensional accuracy, or printing speed. The first of these effects is that when photopolymers solidify, they generally undergo shrinkage which results in an increase in mass density and refractive index. The optical changes are helpful for visually detecting the completion of a CAL printing process, but may also perturb the propagation of light through the material prior to the completion of a print, causing small geometrical distortions. Moreover, localized inhomogeneities in the polymerization process can arise because of a well-known effect whereby polymerization of a small region of material causes lensing of the light passing through it, increasing the intensity nearby and further accelerating the conversion [101]. This phenomenon may occasionally lead to visible "striations" in printed components.

A second challenge also stems from curing shrinkage: as components materialize, they become slightly denser than the surrounding resin and may begin to sink. While these sinking rates are typically limited to at most a few tens of microns per second, they may limit micron-scale definition of patterns if not counteracted. We have found that sinking can be rendered completely undetectable by increasing the viscosity to above about 100,000 cP or using a thermally gelled precursor material, but it would be desirable to address this effect in lower-viscosity materials as well.

A helpful factor is that most of the exposure time of a CAL print is occupied with consuming the inhibiting oxygen; only during the final stage of the exposure does appreciable polymerization and shrinkage take place. Although the target illumination dose is set to be equal in all points of the desired object, there is in practice some temporal distribution of polymerization time. Generally, larger features and points closer to the center of a printed object solidify slightly earlier. We attribute this effect primarily to the diffusion of oxygen from surrounding resin into more highly illuminated zones. Computational correction of the delivered dose to compensate for these transport effects could potentially tighten the temporal distribution of polymerization and thereby reduce the potential for light scattering or component sinking during exposure.

The photopolymerization process is exothermic, and it will need to be considered whether dissipation of the generated heat becomes a possible rate-limiter as printed objects become very large. Nevertheless, resin absorption coefficients are orders of magnitude lower in CAL than in SLA/DLP, so the *volumetric* rate of generation of heat is expected to be far lower in CAL for a given printing rate.

The overprinting capability of CAL has been demonstrated on convex, slender metallic objects, and further work could valuably explore the use of coupling agents between the preexisting surface and the photosensitive resin, to strengthen the intermaterial bond. The potential to polymerize resin around concave objects, where resin can be accessed from a limited range of projection angles, needs exploration.

## 7.15 Expanding the Capabilities of CAL

The very high anticipated volumetric printing rate of CAL suggests that the process could be scaled up to printing volumes that lie well beyond the capabilities of SLA/DLP printers. However, the CAL process as currently implemented relies on parallel beams of light being projected into the printing volume, which implies that the projection optics need to be at least as large as the printed object. Instead, a cone-beam implementation of CAL—in which rays would spread out in a cone from the projector—could potentially allow larger volumes to be addressed using standard projector optics. A spiral path of the projector around the printing volume's central axis could allow yet larger volumes to be printed.

On the materials front, an ability to pattern optically opaque media, such as dental composites, would open new applications for CAL. One approach that we have demonstrated is to mix into the resin a dye (crystal violet) that is transmissive to a curing wavelength (e.g., 405 nm) but opaque to most visible wavelengths [89]. This approach might be extended with other dyes to tune the spectral response of the printed material more specifically.

A more speculative possible research direction is to explore how tomographic fabrication could be carried out with other, nonoptical, sources of energy. The use of an array of electrodes to direct electropolymerization has been proposed by Bowyer [102] and could employ the electrocuring principle demonstrated in, for example, adhesive films [103]. The tomographic synthesis of 3D dose distributions from ultrasonic waves, electron beams, microwaves, or X-rays are all conceivable. An intriguing recent patent application publication from Bharti, Patra, and Rakshit [104] proposes creating a 3D acoustic hologram with phased arrays of ultrasound emitters and applying (e.g., spraying) a material to the surface of the hologram, which the acoustic energy would then solidify.

## 7.16 Concluding Remarks and Outlook

It has not been possible in this chapter to describe *all* current developments that could enhance the complexity and sophistication of 3D printing. One additional process technology that has potential for fabrication of layered multimaterial structures is plasma spray deposition [105, 106]. Its use in forming composite films and surface finishing of additively manufactured components can be expected to grow. Another important area of rapid recent development has been the induction of continuous reinforcement fibers into printed polymeric matrices [107]. Carbon fiber-reinforced composites of both thermoplastic and thermoset materials, deposited by FFF, have been shown to attain tensile strengths exceeding 160 MPa [108].

Developments in printing with continuous fibers could greatly benefit 3D printing for building construction. This is a field with enormous potential, but where breakthroughs in process technology are needed. Concrete extrusion printing currently dominates the landscape, and the challenges of formulating printable, robust, and affordable concretes for extrusion have been well documented [109]. However, automated methods for incorporating crucial reinforcement materials into printed concrete remain in their infancy. One might well ask whether concrete is in fact the best material to pursue for construction 3D printing: a material that is responsible for 8% of global $CO_2$ emissions [110] and has minimal tensile strength seems unlikely to offer the most sustainable route forward. Indeed, it may represent a step backward in sustainability if it displaces simple timber frames in residential construction. More promising might be to print complex building components from renewable composites (e.g., cellulose-based materials [28, 111]) and possibly to integrate these 3D-printed elements with glue-laminated timber frame elements, which are becoming increasingly popular even for high-rise construction.

This chapter has not touched either on the increasingly important role of machine learning in optimizing additive processes to reduce defects [112], nor on that of computation in optimizing geometries. It is likely that computational methods for geometrical design and for process optimization will become increasingly interdependent as process physics are better understood.

We can expect greater consolidation and *combination* of additive processes over the coming years as new applications develop and as some of the nontechnical barriers to adoption of additive manufacturing—such as the need for specialized education and training—begin to be addressed [113]. Additive manufacturing of complex objects remains an exciting field with enormous potential for researchers to make contributions on many fronts: from machine and materials development to computer-aided design and process modeling.

## Acknowledgments

We thank Maxim Shusteff, Chris Spadaccini and Laura Waller for helpful discussions. H.T. is an inventor on two US patent applications relating to CAL, and H.H. is an inventor on one of those patent applications.

## References

1 3DPrint. Retrieved from https://3DPrint.com
2 3dPrinting Industry. Retrieved from https://3dprintingindustry.com/
3 Ngo, T. D., Kashani, A., Imbalzano, G., Nguyen, K. T. Q., & Hui, D. (2018). Additive manufacturing (3D printing): A review of materials, methods, applications and challenges. *Composites. Part B, Engineering*, 143, 172–196.
4 Ameen, W., Ghaleb, A. M., Alatefi, M., Alkhalefah, H., & Alahmari, A. (2018). An overview of selective laser sintering and melting research using bibliometric indicators. *Virtual and Physical Prototyping*, 13(4), 282–291.

5 Galati, M., & Iuliano, L. (2018). A literature review of powder-based electron beam melting focusing on numerical simulations. *Additive Manufacturing*, 19, 1–20.
6 Brenken, B., Barocio, E., Favaloro, A., Kunc, V., & Pipes, R. B. (2018). Fused filament fabrication of fiber-reinforced polymers: A review. *Additive Manufacturing*, 21, 1–16.
7 Layani, M., Wang, X., & Magdassi, S. (2018). Novel materials for 3D printing by photopolymerization. *Advanced Materials*, 30(41), 1706344.
8 Truby, R. L., & Lewis, J. A. (2016). Printing soft matter in three dimensions. *Nature*, 540(7633), 371–378.
9 Murr, L. E. (2016). Frontiers of 3D printing/additive manufacturing: From human organs to aircraft fabrication. *Journal of Materials Science and Technology*, 32(10), 987–995.
10 Zheng, X., Lee, H., Weisgraber, T. H., Shusteff, M., DeOtte, J., Duoss, E. B., . . . Spadaccini, C. M. (2014). Ultralight, ultrastiff mechanical metamaterials. *Science*, 344(6190), 1373–1377.
11 Zheng, X. R., Smith, W. L., Jackson, J., Moran, B. D., Cui, H., Chen, D., . . . Spadaccini, C. M. (2016). Multiscale metallic metamaterials. *Nature Materials*, 15(10), 1100–1106.
12 Pham, M.-S., Liu, C., Todd, I., & Lertthanasarn, J. (2019). Damage-tolerant architected materials inspired by crystal microstructure. *Nature*, 565(7739), 305–311.
13 Jafari, D., & Wits, W. W. (2018). The utilization of selective laser melting technology on heat transfer devices for thermal energy conversion applications: A review. *Renewable and Sustainable Energy Reviews*, 91, 420–442.
14 Magerramova, L., Vasilyev, B., & Kinzburskiy, V. (2016, June). Novel designs of turbine blades for additive manufacturing. Proceedings of ASME Turbo Expo 2016: Turbine Technical Conference and Exposition of ASME, Seoul, South Korea.
15 Engineering Software for Advanced Manufacturing. Retrieved from https://ntopology.com/.
16 Aage, N., Andreassen, E., Lazarov, B. S., & Sigmund, O. (2017). Giga-voxel computational morphogenesis for structural design. *Nature*, 550(7674), 84–86.
17 Brown, E., Yan, P., Tekik, H., Elangovan, A., Wang, J., Lin, D., & Li, J. (2019). 3D printing of hybrid MoS2-graphene aerogels as highly porous electrode materials for sodium ion battery anodes. *Materials and Design*, 170, 107689.
18 Feng, Z., Li, Y., Xin, C., Tang, D., Xiong, W., & Zhang, H. (2019). Fabrication of graphene-reinforced nanocomposites with improved fracture toughness in net shape for complex 3D structures via digital light processing. *C — Journal of Carbon Research*, 5(2), 25.
19 Loh, G. H., Pei, E., Harrison, D., & Monzón, M. D. (2018). An overview of functionally graded additive manufacturing. *Additive Manufacturing*, 23, 34–44.

**20** Hosny, A., Keating, S. J., Dilley, J. D., Ripley, B., Kelil, T., Pieper, S., . . . Weaver, J. C. (2018). From improved diagnostics to presurgical planning: High-resolution functionally graded multimaterial 3D printing of biomedical tomographic data sets. *3D Printing and Additive Manufacturing*, 5(2), 103–113.

**21** Bader, C., Kolb, D., Weaver, J. C., Sharma, S., Hosny, A., Costa, J., & Oxman, N. (2018). Making data matter: Voxel printing for the digital fabrication of data across scales and domains. *Science Advances*, 4(5), eaas8652.

**22** Sadeqi, A., Nejad, H. R., Owyeung, R. E., & Sonkusale, S. (2019). Three dimensional printing of metamaterial embedded geometrical optics (MEGO). *Microsystems & Nanoengineering*, 5(1), 1–10.

**23** Zhang, Z., Demir, K. G., & Gu, G. X. (2019). Developments in 4D-printing: A review on current smart materials, technologies, and applications. *International Journal of Smart and Nano Materials*, 10(3), 205–224.

**24** Boley, J. W., van Rees, W. M., Lissandrello, C., Horenstein, M. N., Truby, R. L., Kotikian, A., . . . Mahadevan, L. (2019). Shape-shifting structured lattices via multimaterial 4D printing. *Proceedings of the National Academy of Sciences*, 116(42), 20856–20862.

**25** Miri, A. K., Khalilpour, A., Cecen, B., Maharjan, S., Shin, S. R., & Khademhosseini, A. (2019). Multiscale bioprinting of vascularized models. *Biomaterials*, 198, 204–216.

**26** Yoon, H.-S., Lee, J.-Y., Kim, H.-S., Kim, M.-S., Kim, E.-S., Shin, Y.-J., . . . Ahn, S.-H. (2014). A comparison of energy consumption in bulk forming, subtractive, and additive processes: Review and case study. *International Journal of Precision Engineering and Manufacturing-Green Technology*, 1(3), 261–279.

**27** Tian, W., Ma, J., & Alizadeh, M. (2019). Energy consumption optimization with geometric accuracy consideration for fused filament fabrication processes. *International Journal of Advanced Manufacturing Technology*, 103(5), 3223–3233.

**28** Pattinson, S. W., & Hart, A. J. (2017). Additive manufacturing of cellulosic materials with robust mechanics and antimicrobial functionality. *Advanced Materials Technologies*, 2(4), 1600084.

**29** Vock, S., Klöden, B., Kirchner, A., Weißgärber, T., & Kieback, B. (2019). Powders for powder bed fusion: A review. *Progress in Additive Manufacturing*, 4, 383–397.

**30** Martin, A. A., Calta, N. P., Khairallah, S. A., Wang, J., Depond, P. J., Fong, A. Y., . . . Matthews, M. J. (2019). Dynamics of pore formation during laser powder bed fusion additive manufacturing. *Nature Communications*, 10(1), 1–10.

**31** Hajializadeh, F., & Ince, A. (2019). Short review on modeling approaches for metal additive manufacturing process. *Material Design & Processing Communications*, 2(2), e56.

**32** Gibson, M. A., Mykulowycz, N. M., Shim, J., Fontana, R., Schmitt, P., Roberts, A., . . . Schroers, J. (2018). 3D printing metals like thermoplastics: Fused filament fabrication of metallic glasses. *Materials Today*, 21(7), 697–702.

**33** Digital Alloys. Retrieved from https://www.digitalalloys.com/

**34** Fabrisonic. Retrieved from https://fabrisonic.com/

**35** Sullivan, K. T., Zhu, C., Tanaka, D. J., Kuntz, J. D., Duoss, E. B., & Gash, A. E. (2013). Electrophoretic deposition of thermites onto micro-engineered electrodes prepared by direct-ink writing. *The Journal of Physical Chemistry B*, 117(6), 1686–1693.

**36** Pascall, A. J., Qian, F., Wang, G., Worsley, M. A., Li, Y., & Kuntz, J. D. (2014). Light-directed electrophoretic deposition: A new additive manufacturing technique for arbitrarily patterned 3D composites. *Advanced Materials*, 26(14), 2252–2256.

**37** Fabric8Labs. (2019). Next generation metal 3D printing. Retrieved from https://fabric8labs.com/.

**38** Wirth, D. M., Pain, D. F., & Herman, J. W. (2017). Three dimensional additive manufacturing of metal objects by stereo-electrochemical deposition. *United States US20170145584A1*, filed January 25, 2017 and issued May 25, 2017.

**39** Liu, F. Q., Du, T., Duboust, A., Tsai, S., & Hsu, W.-Y. (2006). Cu planarization in electrochemical mechanical planarization. *Journal of the Electrochemical Society*, 153(6), C377–C381.

**40** Daryadel, S., Behroozfar, A., & Minary-Jolandan, M. (2019). Toward control of microstructure in microscale additive manufacturing of copper using localized electrodeposition. *Advanced Engineering Materials*, 21(1), 1800946.

**41** Chen, X., Liu, X., Ouyang, M., Chen, J., Taiwo, O., Xia, Y., . . . Wu, B. (2019). Multi-metal 4D printing with a desktop electrochemical 3D printer. *Scientific Reports*, 9(1), 1–9.

**42** Reiser, A., Lindén, M., Rohner, P., Marchand, A., Galinski, H., Sologubenko, A. S., . . . Spolenak, R. (2019). Multi-metal electrohydrodynamic redox 3D printing at the submicron scale. *Nature Communications*, 10(1), 1–8.

**43** Waller, E. H., & von Freymann, G. (2018). From photoinduced electron transfer to 3D metal microstructures via direct laser writing. *Nanophotonics*, 7(7), 1259–1277.

**44** Vyatskikh, A., Delalande, S., Kudo, A., Zhang, X., Portela, C. M., & Greer, J. R. (2018). Additive manufacturing of 3D nano-architected metals. *Nature Communications*, 9(1), 1–8.

**45** Tumbleston, J. R., Shirvanyants, D., Ermoshkin, N., Janusziewicz, R., Johnson, A. R., Kelly, D., . . . DeSimone, J. M. (2015). Continuous liquid interface production of 3D objects. *Science*, 347(6228), 1349–1352.

**46** Han, D., Yang, C., Fang, N. X., & Lee, H. (2019). Rapid multi-material 3D printing with projection micro-stereolithography using dynamic fluidic control. *Additive Manufacturing*, 27, 606–615.

**47** Choi, J.-W., Kim, H.-C., & Wicker, R. (2011). Multi-material stereolithography. *Journal of Materials Processing Technology*, 211(3), 318–328.

**48** Hohnholz, A., Obata, K., Albrecht, D., Koch, J., Hohenhoff, G., Suttmann, O., . . . Overmeyer, L. (2019). Multimaterial bathless stereolithography using aerosol jet printing and UV laser based polymerization. *Journal of Laser Applications*, 31(2), 022301.

**49** Schwartz, J. J., & Boydston, A. J. (2019). Multimaterial actinic spatial control 3D and 4D printing. *Nature Communications*, 10(1), 791.

**50** Dolinski, N. D., Page, Z. A., Callaway, E. B., Eisenreich, F., Garcia, R. V., Chavez, R., . . . Hawker, C. J. (2018). Solution mask liquid lithography (SMaLL) for one-step, multimaterial 3D printing. *Advanced Materials*, 30(31), 1800364.

**51** Kuang, X., Wu, J., Chen, K., Zhao, Z., Ding, Z., Hu, F., . . . Qi, H. J. (2019). Grayscale digital light processing 3D printing for highly functionally graded materials. *Science Advances*, 5(5), eaav5790.

**52** Yin, H., Ding, Y., Zhai, Y., Tan, W., & Yin, X. (2018). Orthogonal programming of heterogeneous micro-mechano-environments and geometries in three-dimensional bio-stereolithography. *Nature Communications*, 9(1), 1–7.

**53** Camacho, P., Busari, H., Seims, K. B., Schwarzenberg, P., Dailey, H. L., & Chow, L. W. (2019). 3D printing with peptide–polymer conjugates for single-step fabrication of spatially functionalized scaffolds. *Biomaterials Science*, 7(10), 4237–4247.

**54** Kotz, F., Arnold, K., Bauer, W., Schild, D., Keller, N., Sachsenheimer, K., . . . Rapp, B. E. (2017). Three-dimensional printing of transparent fused silica glass. *Nature*, 544(7650), 337–339.

**55** Melchert, D. S., Collino, R. R., Ray, T. R., Dolinski, N. D., Friedrich, L., Begley, M. R., & Gianola, D. S. (2019). Flexible conductive composites with programmed electrical anisotropy using acoustophoresis. *Advanced Materials Technologies*, 4, 1900586.

**56** Lu, L., Zhang, Z., Xu, J., & Pan, Y. (2019). 3D-printed polymer composites with acoustically assembled multidimensional filler networks for accelerated heat dissipation. *Composites: Part B, Engineering*, 174, 106991.

**57** Yang, Y., Li, X., Chu, M., Sun, H., Jin, J., Yu, K., . . . Chen, Y. (2019). Electrically assisted 3D printing of nacre-inspired structures with self-sensing capability. *Science Advances*, 5(4), eaau9490.

**58** Martin, J. J., Fiore, B. E., & Erb, R. M. (2015). Designing bioinspired composite reinforcement architectures via 3D magnetic printing. *Nature Communications*, 6(1), 1–7.

**59** Song, S. Y., Park, M. S., Lee, D., Lee, J. W., & Yun, J. S. (2019). Optimization and characterization of high-viscosity ZrO2 ceramic nanocomposite resins for supportless stereolithography. *Materials and Design*, 180, 107960.

**60** Carbon, Inc. Technical data sheet for UMA90 resin. Retrieved from https://s3.amazonaws.com/carbon-staticassets/downloads/resin_data_sheets/UMA90_TDS.pdf.

**61** Zhang, K., Xie, C., Wang, G., He, R., Ding, G., Wang, M., . . . Fang, D. (2019). High solid loading, low viscosity photosensitive Al2O3 slurry for stereolithography based additive manufacturing. *Ceramics International*, 45(1), 203–208.

**62** de Beer, M. P., van der Laan, H. L., Cole, M. A., Whelan, R. J., Burns, M. A., & Scott, T. F. (2019). Rapid, continuous additive manufacturing by volumetric polymerization inhibition patterning. *Science Advances*, 5(1), eaau8723.

**63** Ebert, J. & Gmeiner, R. (2017). Stereolithography device having a heating unit. *United States US20170334129A1*, filed October 15, 2015 and issued November 23, 2017.

**64** Walker, D. A., Hedrick, J. L., & Mirkin, C. A. (2019). Rapid, large-volume, thermally controlled 3D printing using a mobile liquid interface. *Science*, 366(6463), 360–364.

**65** Luongo, A., Falster, V., Doest, M. B., Ribo, M. M., Eiriksson, E. R., Pedersen, D. B., & Frisvad, J. R. (2019). Microstructure control in 3D printing with digital light processing. *Computer Graphics Forum*, 39(1), 347–359.

**66** Yuan, C., Kowsari, K., Panjwani, S., Chen, Z., Wang, D., Zhang, B., . . . Ge, Q. (2019). Ultrafast three-dimensional printing of optically smooth microlens arrays by oscillation-assisted digital light processing. *ACS Applied Materials & Interfaces*, 11(43), 40662–40668.

**67** STEP Technology. Evolve additive solutions. Retrieved from https://www.evolveadditive.com/.

**68** Hanson, W. J., Sanders, J. R., Bacus, M. W., & Chillscyzn, S. A. (2015). Electrophotography-based additive manufacturing system with transfer-medium service loops. *United States US9141015B2*, filed June 7, 2013 and issued September 22, 2015.

**69** Batchelder, J. S., Chowdry, A., & Chillscyzn, S. A. (2017). Systems and methods for electrophotography-based additive manufacturing of parts. *United States US20170192377A1*, filed December 21, 2016 and issued July 6, 2017.

**70** Zhu, W., Qu, X., Zhu, J., Ma, X., Patel, S., Liu, J., . . . Chen, S. (2017). Direct 3D bioprinting of prevascularized tissue constructs with complex microarchitecture. *Biomaterials*, 124, 106–115.

**71** Grigoryan, B., Paulsen, S. J., Corbett, D. C., Sazer, D. W., Fortin, C. L., Zaita, A. J., . . . Miller, J. S. (2019). Multivascular networks and functional intravascular topologies within biocompatible hydrogels. *Science*, 364(6439), 458–464.

**72** Volumetric Bio. Retrieved from https://volumetricbio.com/

**73** Zhuang, P., Ng, W. L., An, J., Chua, C. K., & Tan, L. P. (2019). Layer-by-layer ultraviolet assisted extrusion-based (UAE) bioprinting of hydrogel constructs with high aspect ratio for soft tissue engineering applications. *PLoS ONE*, 14(6), e0216776.

**74** Ji, S., Almeida, E., & Guvendiren, M. (2019). 3D bioprinting of complex channels within cell-laden hydrogels. *Acta Biomaterialia*, 95, 214–224.

**75** Lee, A., Hudson, A. R., Shiwarski, D. J., Tashman, J. W., Hinton, T. J., Yerneni, S., . . . Feinberg, A. W. (2019). 3D bioprinting of collagen to rebuild components of the human heart. *Science*, 365(6452), 482–487.

**76** Senior, J. J., Cooke, M. E., Grover, L. M., & Smith, A. M. (2019). Fabrication of complex hydrogel structures using suspended layer additive manufacturing (SLAM). *Advanced Functional Materials*, 29(49), 1904845.

**77** Wang, Y., Huang, X., Shen, Y., Hang, R., Zhang, X., Wang, Y., . . . Tang, B. (2019). Direct writing alginate bioink inside pre-polymers of hydrogels to create patterned vascular networks. *Journal of Materials Science*, 54(10), 7883–7892.

**78** Skylar-Scott, M. A., Uzel, S. G. M., Nam, L. L., Ahrens, J. H., Truby, R. L., Damaraju, S., & Lewis, J. A. (2019). Biomanufacturing of organ-specific tissues with high cellular density and embedded vascular channels. *Science Advances*, 5(9), eaaw2459.

**79** Zhou, Y. (2016). The application of ultrasound in 3D bio-printing. *Molecules*, 21(5), 590.

**80** Chansoria, P., Narayanan, L. K., Schuchard, K., & Shirwaiker, R. (2019). Ultrasound-assisted biofabrication and bioprinting of preferentially aligned three-dimensional cellular constructs. *Biofabrication*, 11(3), 035015.

**81** Tam, K.-M. M., & Mueller, C. T. (2017). Additive manufacturing along principal stress lines. *3D Printing and Additive Manufacturing*, 4(2), 63–81.

**82** Coulter, F. B., Schaffner, M., Faber, J. A., Rafsanjani, A., Smith, R., Appa, H., . . . Studart, A. R. (2019). Bioinspired heart valve prosthesis made by silicone additive manufacturing. *Matter*, 1(1), 266–279.

**83** Kim, H., Jang, J., Park, J., Lee, K.-P., Lee, S., Lee, D.-M., . . . Cho, D.-W. (2019). Shear-induced alignment of collagen fibrils using 3D cell printing for corneal stroma tissue engineering. *Biofabrication*, 11(3), 035017.

**84** Piestun, R., & Shamir, J. (2002). Synthesis of three-dimensional light fields and applications. *Proceedings of the IEEE*, 90(2), 222–244.

**85** Zhang, J., Pégard, N., Zhong, J., Adesnik, H., & Waller, L. (2017). 3D computer-generated holography by non-convex optimization. *Optica*, 4(10), 1306–1313.

**86** Condliffe, J. (2017). This super-fast 3-D printer is powered by holograms. *MIT Technology Review*. Retrieved from https://www.technologyreview.com/s/603605/this-super-fast-3-d-printer-is-powered-by-holograms/

**87** Kamali, S. M., Arbabi, E., Kwon, H., & Faraon, A. (2019). Metasurface-generated complex 3-dimensional optical fields for interference lithography. *Proceedings of the National Academy of Sciences*, 116(43), 21379–21384.

**88** Shusteff, M., Browar, A. E. M., Kelly, B. E., Henriksson, J., Weisgraber, T. H., Panas, R. M., . . . Spadaccini, C. M. (2017). One-step volumetric additive manufacturing of complex polymer structures. *Science Advances*, 3(12), eaao5496.

89 Kelly, B. E., Bhattacharya, I., Heidari, H., Shusteff, M., Spadaccini, C. M., & Taylor, H. K. (2019). Volumetric additive manufacturing via tomographic reconstruction. *Science*, 363(6431), 1075–1079.

90 Kak, A., & Slaney, M. (2001). *Principles of computerized tomographic imaging*. Society for Industrial and Applied Mathematics.

91 Hounsfield, G. N. (1973). Method and apparatus for measuring x- or $\gamma$-radiation absorption or transmission at plural angles and analyzing the data. *United States US3778614A*, filed December 27, 1971 and issued December 11, 1973.

92 Bortfeld, T., Bürkelbach, J., Boesecke, R., & Schlegel, W. (1990). Methods of image reconstruction from projections applied to conformation radiotherapy. *Physics in Medicine and Biology*, 35(10), 1423–1434.

93 Sharpe, J., Ahlgren, U., Perry, P., Hill, B., Ross, A., Hecksher-Sørensen, J., . . . Davidson, D. (2002). Optical projection tomography as a tool for 3D microscopy and gene expression studies. *Science*, 296(5567), 541–545.

94 Choi, W., Fang-Yen, C., Badizadegan, K., Oh, S., Lue, N., Dasari, R. R., & Feld, M. S. (2007). Tomographic phase microscopy. *Nature Methods*, 4(9), 717–719.

95 Ramachandran, G. N., & Lakshminarayanan, A. V. (1971). Three-dimensional reconstruction from radiographs and electron micrographs: Application of convolutions instead of fourier transforms. *Proceedings of the National Academy of Sciences of the United States of America*, 68(9), 2236–2240.

96 Github Repository. Computed axial lithography. Retrieved from https://github.com/computed-axial-lithography/

97 Tretiak, O., & Metz, C. (1980). The exponential radon transform. *SIAM Journal on Applied Mathematics*, 39(2), 341–354.

98 Bernal, P. N., Delrot, P., Loterie, D., Li, Y., Malda, J., Moser, C., & Levato, R. (2019). Volumetric bioprinting of complex living-tissue constructs within seconds. *Advanced Materials*, 31, 1904209.

99 Luxexcel. Retrieved from https://www.luxexcel.com/

100 Poomathi, N., Singh, S., Prakash, C., Patil, R. V., Perumal, P. T., Barathi, V. A., . . . Maheshwari, N. U. (2019). Bioprinting in ophthalmology: Current advances and future pathways. *Rapid Prototyping Journal*, 25(3), 496–514.

101 Kewitsch, A. S., & Yariv, A. (1996). Self-focusing and self-trapping of optical beams upon photopolymerization. *Optics Letters*, 21(1), 24–26.

102 Bowyer, A. (2019) Electric 3D printing. Retrieved from https://reprapltd.com/electric-3d-printing/.

103 Ping, J., Gao, F., Chen, J. L., Webster, R. D., & Steele, T. W. J. (2015). Adhesive curing through low-voltage activation. *Nature Communications*, 6(1), 1–9.

104 Bharti, H., Patra, A. K., & Rakshit, S. K. (2018). 3D printing on the surface of an acoustic hologram. *United States US20180361680A1*, filed June 19, 2017 and issued Decenber 20, 2018.

105 http://www.spacefoundry.us/.

**106** Gandhiraman, R. P., Singh, E., Diaz-Cartagena, D. C., Nordlund, D., Koehne, J., & Meyyappan, M. (2016). Plasma jet printing for flexible substrates. *Applied Physics Letters*, 108(12), 123103.

**107** Nardiello, J. A., Christ, R. J., Crawford, J. A., & Madsen, J. S. (2019). Device and method for 3D printing with long-fiber reinforcement. *United States US10173410B2*, filed December 8, 2015 and issued January 8, 2019.

**108** An, Y., & Yu, W.-R. (2019). Three-dimensional printing of continuous carbon fiber-reinforced shape memory polymer composites. *AIP Conference Proceedings*, 2113(1), 020008.

**109** Buswell, R. A., Leal de Silva, W. R., Jones, S. Z., & Dirrenberger, J. (2018). 3D printing using concrete extrusion: A roadmap for research. *Cement and Concrete Research*, 112, 37–49.

**110** Favier, A., Scrivener, K., & Habert, G. (2019). Decarbonizing the cement and concrete sector: Integration of the full value chain to reach net zero emissions in Europe. *IOP Conference Series: Earth and Environmental Science*, 225, 012009.

**111** Pringle, A. M., Rudnicki, M., & Pearce, J. M. (2017). Wood furniture waste–based recycled 3-D printing filament. *Forest Products Journal*, 68(1), 86–95.

**112** Yuan, B., Guss, G. M., Wilson, A. C., Hau-Riege, S. P., DePond, P. J., McMains, S., . . . Giera, B. (2018). Machine-learning-based monitoring of laser powder bed fusion. *Advanced Materials Technologies*, 3(12), 1800136.

**113** Thomas-Seale, L. E. J., Kirkman-Brown, J. C., Attallah, M. M., Espino, D. M., & Shepherd, D. E. T. (2018). The barriers to the progression of additive manufacture: Perspectives from UK industry. *International Journal of Production Economics*, 198, 104–118.

# Part III

# 3D printing of energy devicest

# 8

# Current State of 3D Printing Technologies and Materials

*Poul Norby*

*Department of Energy Conversion and Storage, Technical University of Denmark, Lyngby, Denmark*

## 8.1 3D Printing of Energy Devices

### 8.1.1 Batteries

Storage of energy is crucial in the future society where sustainable energy sources will become dominating. The intermittent nature of energy sources as wind and solar makes it necessary to improve the energy storage solutions. Battery technology is one of several solutions for storing energy. For large scale energy storage (grid scale) batteries will play a role in short-term storage, up to maybe days, while longer term storage will be solved by other technologies (e.g., synthetic fuels, pumped hydro, etc.). In the transportation sector batteries will be crucial, maybe in combination with fuel cells. Presently, and in the foreseeable future, lithium ion batteries are the dominating technology for electrical cars, ferries, trains, and smaller trucks. Lithium ion batteries include a number of different chemistries, where especially the positive electrode material may vary. Another energy storage technology is the supercapacitors (see Chapter 9) which has the advantage of very fast charge/discharge rates and long cycle lifetime, but have lower specific energy and energy density than lithium ion batteries. The hybrid supercapacitors have elements from both batteries and supercapacitors and have a higher energy density than supercapacitors.

Most lithium ion batteries have a fixed format, either as a cylindrical battery (e.g., 18650 format) or prismatic batteries and are made in large scale. Production of these formats is optimized over a long period with significant focus on consistent and reproducible production. Quality control and homogeneous performance are essential parameters in production of batteries in order to avoid battery degradation and safety issues such as fire.

*3D Printing for Energy Applications*, First Edition. Edited by Albert Tarancón and Vincenzo Esposito.

In many applications it would be an advantage to have more freedom in designing batteries, for example when constructing devices with integrated batteries, custom made micro batteries and batteries for wearables. Here 3D printing of batteries would be of considerable interest as it is possible to design and realize new battery architectures. Traditional lithium ion battery technology is built from sandwich structures of coated electrodes which may be rolled or stacked to provide high capacity batteries. The electrodes are therefore basically 2-dimensional. The idea of building better batteries using 3-dimensional structuring of the electrodes has been investigated both experimentally and by modeling [1]. Early efforts related to 3D structuring were by deposition methods, for example using nanorod-patterns. It was only by the development of 3D printing techniques that it became possible to use additive manufacturing to produce 3D electrode structures. A recent example of the possibility of printing customized batteries intended for integration in drone technology can be found in Ref. [2] where a combination of 3D print and aerosol deposition is used for producing customizable, non-planar, integrated lithium-ion batteries [2].

Yoshima, Munakata, and Kanamura [3] published one of the very early reports on depositing a 3D interdigitated anode and cathode pattern in a functional lithium-ion battery. They used a flat $SiO_2$/Si substrate with a pre-deposited interdigitated pattern of gold current collectors and a pre-deposited polymer wall (30 μm thick) between the gold electrodes. The polymer wall was maintained during deposition of the anode (lithium titanium oxide, $Li_4Ti_5O_{12}$ (LTO)) and cathode (lithium cobalt oxide, $LiCoO_2$ (LCO)) interdigitated electrode structure, which helped in obtaining a high capacity of the final battery. The electrode slurries were deposited between the polymer walls using a micro-injection system with a 10 μm (i.d.) glass capillary nozzle. After drying the polymer wall was removed in acetone and a PMMA gel electrolyte was deposited between the electrodes. The electrolyte was finally thermally polymerized in situ. A high areal capacity of 0.270 $mAh/cm^2$ at 2C was obtained and good capacity retention (ca. 60% at 20C) was found, however with a significant polarization at higher C-rates.

One of the first real 3D-printed interdigitated lithium ion battery electrode structures was reported by Ke Sun et al. [4] Gold current collectors were deposited using lithographic methods on a glass substrate (Figure 8.1). An interdigitated pattern of LTO and LFP (lithium iron phosphate, $LiFePO_4$) nanoparticles as the negative and positive electrode, respectively, was printed on top of the gold current collectors using a slurry with 55–65 wt% solid loading (electrode materials and cellulose-based binder). Half-cells as well as full cells were produced and characterized. For the full cell with an 8-layer structure an areal capacity of 1.5 $mAh/cm^2$ at 1C and 1.2 $mAh/cm^2$ at 2C was found, significantly higher than for the earlier reported 3D battery structure by Yoshima et al. Both 8-layer and 16-layer electrode structures were investigated. No improvement in the capacity

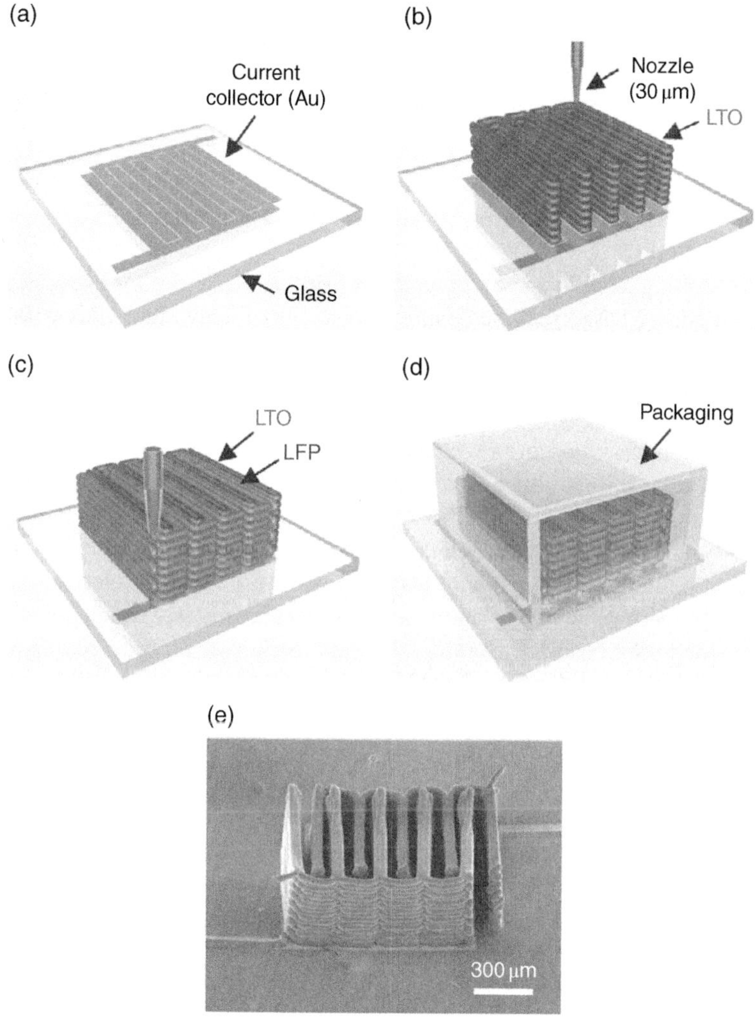


**Figure 8.1** Schematic illustration [4] of 3D printed interdigitated microbattery architectures fabricated on a gold current collector and a SEM image of the 3D printed interdigitated electrodes. *Source:* Reproduced with permission from Sun et al. [4]. © 2013 John Wiley & Sons, Inc.

was observed for the thicker electrodes, indicating that electronic conduction in the electrode structure is a limiting factor in the design.

Since then, there has been a significant amount of work on 3D printing various elements of a battery from interdigitated electrode and solid electrolyte structures to printing the entire batteries. Progress in 3D-printed devices and materials for energy technologies have been reviewed recently [5–12]. The reviews cover energy storage as well as other energy technologies where 3D-printing of functional materials is employed. In the following we will look closer at how 3D printing is employed in the different elements of a battery structure.

#### 8.1.1.1 3D Printing Structured Electrodes

Generally, 3D printed electrode structures can be divided into two classes reflecting the strategy for battery assembly: 3D structuring of 2-dimensional electrode surfaces and interdigitated patterns for *in situ* built 3-dimensional electrode/electrolyte structures.

Many 3D printing techniques can be used for printing electrodes for batteries, including ink/slurry based methods using dispersions of active materials with different viscosities (inkjet printing, drop-on-demand deposition, and slurry extrusion methods) as well as filament printing using, for example, PLA loaded with functional materials.

The formulation, viscosity, and rheology of developed inks and slurries for 3D printing are of utmost importance. Traditionally, slurries for electrode coating in lithium-ion batteries are made with NMP (*N*-methyl-2 pyrrolidon) as solvent and pvdf (Polyvinylidene fluorid) as a binder. NMP has very good properties for coating, but it has a high boiling point, leading to slow drying of the electrodes, and has toxic properties. Therefore significant resources are put into development of aqueous slurries for electrode coating in battery research and production. An example of development of a 3D-printable aqueous ink was published by Park et al. [13] and many of the 3D-printed electrodes in the examples below use aqueous inks.

***3D-Structured Electrodes*** Development of 3D-printed 3D-structured electrodes has shown a promising route for enhancing the performance of batteries which are then assembled using conventional battery assembly technology. The battery electrodes (positive and negative) are stacked using a separator membrane or solid polymer electrolyte in such a way that the electrodes are deposited on current collectors and are contacted to the external circuit from opposite sides, as is the case for traditional lithium-ion batteries. The electrodes are in essence patterned 2D electrodes where the improvement in the performance of the batteries is mainly governed by a specific 3D structure of the interface between electrode and electrolyte.

Using inkjet printing methods Delannoy et al. [14] prepared high-power positive electrodes of carbon coated $LiFePO_4$ from an aqueous ink with ca. 10 wt% solid content. The electrodes were printed on aluminum foil using an inkjet printer designed for R&D and feasibility studies. By testing several electrodes with different printing density and a thickness around 4 μm, it was found that the charge/discharge capacities for the printed electrodes were lower than for tape-casted electrodes. However, the polarization between discharge and charge for the inkjet printed electrodes was significantly lower for higher charge/discharge rates. This means that the rate capability was much better than for the coated electrodes resulting in only ca. 10% decrease in the capacity between 9C (total discharge in ca. 7 min) and 90C (total discharge in 40 s) rates, whereas there was virtually no capacity to be measured in the coated electrodes at 90C.

Ben-Barak et al. [15] prepared electrodes of carbon coated $LiFePO_4$ using drop-on-demand printing. Hundred layers of lines consisting of discrete droplets (thickness 150–170 μm) were deposited and they obtained capacities close to the theoretical value and a capacity retention of ca. 50% at 6C (total discharge in 10 min). Also filament printing [Fused filament fabrication (FFF) or Fused Deposition Modeling (FDM)] can be used to make electrodes for lithium ion batteries. Using this technique, a filament consisting of a printable polymer (e.g. PLA, polylactic acid) is blended with functional electrode materials, typically with a loading of >50 wt% content of functional materials. As an example Maurel et al. [16] prepared graphite loaded PLA filaments for printing negative electrodes for lithium-ion batteries. Using filament 3D-printing of functional materials is difficult as it may be hard to combine printability (floating and viscosity properties of the molten filament) with obtaining electrical and ionic conductivity percolation in the printed electrodes. By using a small amount (2 wt%) conducting additive with a PLA/graphite filament (30/70 weight ratio) and using PEGDME500 [Poly(ethylene glycol) dimethyl ether] as a plasticizer it was possible to print a 60 μm thick electrode, which was tested in an electrochemical half-cell with lithium metal and to obtain good electrochemical properties.

In order to improve the electrode properties, 3D structuring of 3D-printed electrodes has been investigated. It was shown, for example, by Hu et al. [17] that by printing a line structure of a $LiMn_{1-x}Fe_xPO_4$ positive electrode on an aluminum foil, Figure 8.2, using slurry extrusion from a micro nozzle, improved specific capacity and rate capability was obtained with discharge rates up to 100C (total discharge in 36 s).

Liu et al. used a slurry extrusion printing process (called Low Temperature Direct Writing, LTDW) where printing is performed a low temperature and freeze drying of the printed electrodes is used to obtain porous LFP electrodes [18, 19]. The electrodes were printed in a line structure or a cross-hatched pattern using an aqueous slurry and showed good specific capacity and reasonable rate capabilities. 3D printed electrodes prepared using DIW (Direct Ink Writing) and LTDW were compared to conventional roller coated electrode deposition.

In order to enhance the rate capability and areal capacity, 3D-structured electrodes with a hierarchical porosity may be used. Electrodes for metal-ion batteries require porosity in order for the electrolyte to penetrate the electrode and create ionic conductivity percolation, that is, ensure that the entire electrode is accessible to ions for intercalation and that mobility of ions between the electrodes

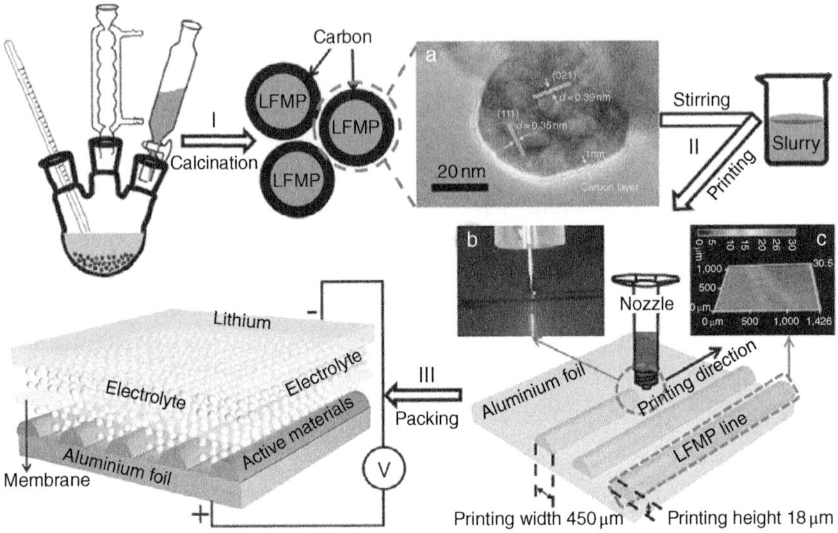


**Figure 8.2** Schematic of battery preparation using a $LiMn_{0.21}Fe_{0.79}PO_4$-based 3D-printed electrode. *Source:* Reproduced with permission from Hu et al. [17]. © 2016 John Wiley & Sons, Inc.

is not obstructed. Ionic diffusion is the important parameter determining how high charge/discharge rates can be obtained. Ion diffusion in solids (e.g., electrode particles) is slow and small particles will result in faster diffusion inside the solid particles. Ion mobility in the electrolyte is also of importance with smaller pores decreasing the diffusion rate. However, large pores means that there is less active material in the electrode, so a balance is important in the pore size distribution. Using a hierarchical pore structure having bimodal or varying pore size distribution may enhance the overall rate performance of the electrode. This is utilized by creating a pore size distribution during electrode preparation from a slurry, but a hierarchical pore size distribution could also be deliberately built into the electrode using 3D printing methods.

One example is from 3D printing of electrodes for sodium-ion batteries [20] where the porosity of the printed electrode is combined with large scale porosity created by 3D structuring the electrode. The electrodes printed by Ding, Shen, Du, Li, and Yang [20] were made from a slurry containing graphene oxide (GO) and NVP [$Na_3V_2(PO_4)_3$]. After freeze drying graphene oxide is thermally reduced to rGO, reduced graphene oxide, which has good electric conductivity and a disordered structure composed of graphene sheets. GO inks are often used for inkjet printing of conducting circuits or for 3D printing of high-surface area structures, for example supercapacitors (see Chapter 9). NVP is a sodium intercalation material which can be used as positive electrode in sodium ion batteries. The NVP particles were mixed with a GO dispersion which was concentrated to a printable ink or slurry by evaporation as shown in Figure 8.3. After printing and thermal reduction, continuous filaments with a porous structure are obtained. The pore structure consists of small pores (1–20 μm) created by the nanosheet structure and larger isolated pores (5–30 μm) created due to ice formation during the freeze drying process. By creating patterns with crosshatched (Figure 8.3) or otherwise structured electrodes (Figure 8.4) with different 3D patterns the electrochemical properties during sodium intercalation was investigated [20].

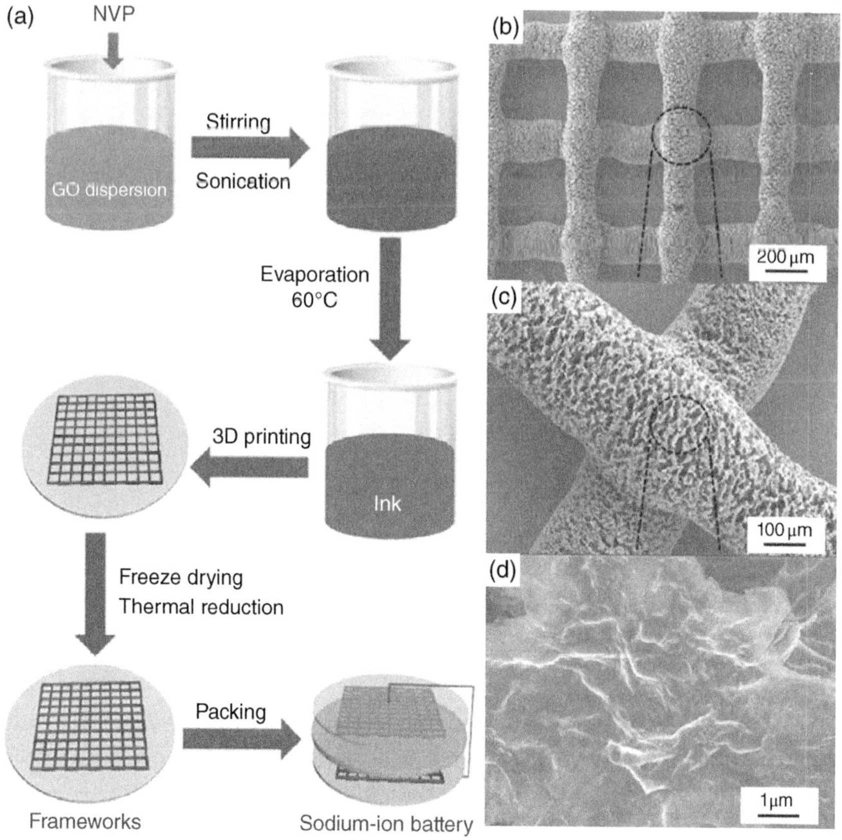


**Figure 8.3** Schematics and SEM images of 3D-printed hierarchical porous frameworks using NVP-GO inks. *Source:* Reproduced with permission from Ding et al. [20]. © 2017 American Chemical Society.

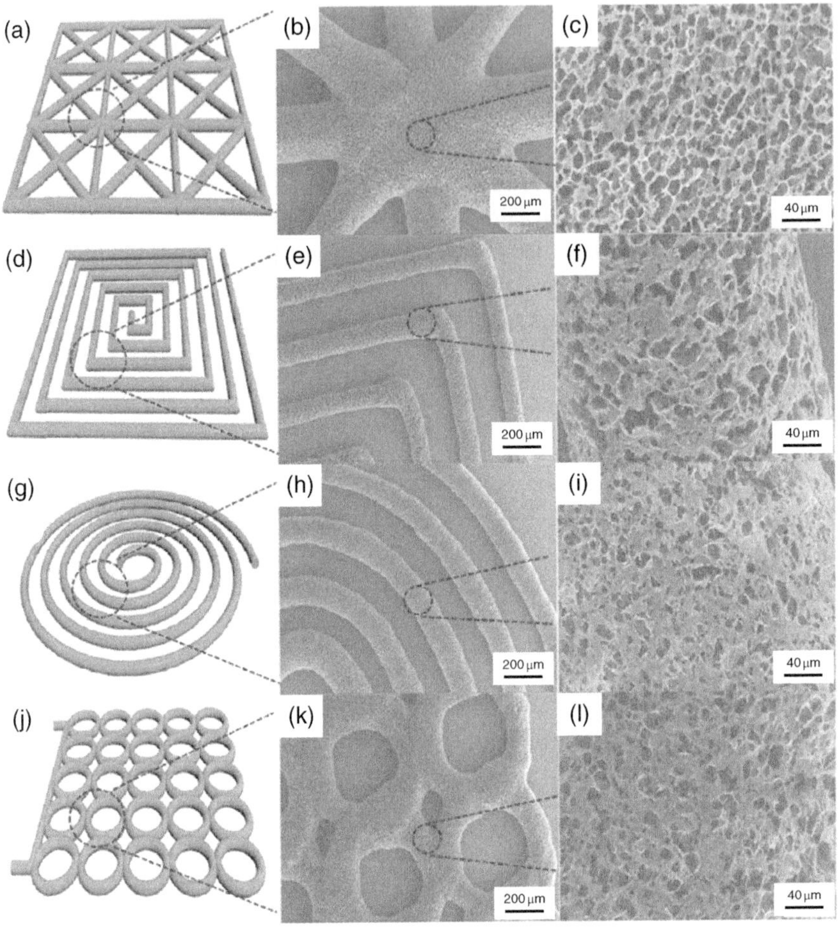


**Figure 8.4** Schematics and SEM images of four types of 3D-printed hierarchical porous NVP-GO electrode frameworks. *Source:* Reproduced with permission from Ding et al. [20]. © 2017 American Chemical Society.

A similar approach was employed by Wang et al. [21] where a series of patterns were generated using slurry deposition methods with LFP as the active lithium ion battery positive electrode material. Self-supported electrode were printed, Figure 8.5 and different patterns were investigated, Figure 8.6. The width of the printed line was ca. 150 μm and the height of each layer was around 200 μm. Multilayer electrodes with a total thickness of up to 1.6 mm were prepared, achieving a high area capacity while maintaining a high specific capacity, in contrast to

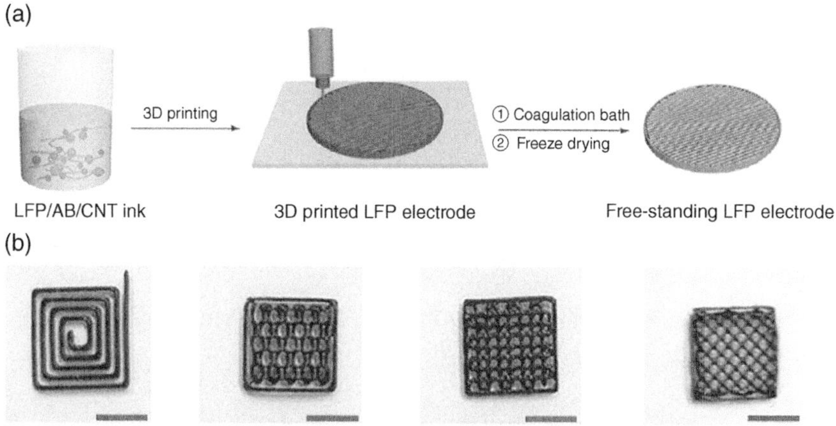


**Figure 8.5** Schematic illustration and optical images of 3D-printed self-supported LFP electrodes. *Source:* Reproduced with permission from Wang et al. [21]. © 2018 American Chemical Society.

conventional coated electrodes, where crack formation and poor adhesion often is observed with increased thickness. However, the cycle stability of the thick 3D printed electrodes is still limited and improvements are needed.

Extending the 3D printed structures further in thickness makes it possible to make larger open framework structures. This would further increase the energy and power density of the battery by allowing thicker electrodes to be used while maintaining fast ionic and electronic conduction. Some examples can be found in [Refs. 22–26] where 3-dimensional open frameworks are 3D printed. Saleh, Li, Park, and Panat [22] used an aerosol jet printing technique, to prepare complex porous wireframe structures of metallic silver resulting in very high specific capacity electrodes for Li-ion batteries, Figure 8.7.

Qiao et al. [23] used 3D extrusion printing of a GO ink to form a 3D open structure, which was then thermally reduced to rGO, impregnated with a $NiCl_2$ solution and shock-heated to form ultrafine Ni metal particles on the conducting high surface rGO structure, Figure 8.8.

A similar rGO structure was prepared by Lacey et al. as a porous air electrode for lithium-$O_2$ batteries [24]. They used holey graphene oxide to prepare electrodes with a tri-modal pore size distribution from the holes in the graphite sheet, the porosity of the printed material, and the macro-porosity obtained from the 3D printing pattern. 3D-printed mesh patterns have also been reported for high capacity quantum dot based anodes for lithium-ion batteries [25] and sulfur copolymer/graphene structures for lithium–sulfur batteries [26].

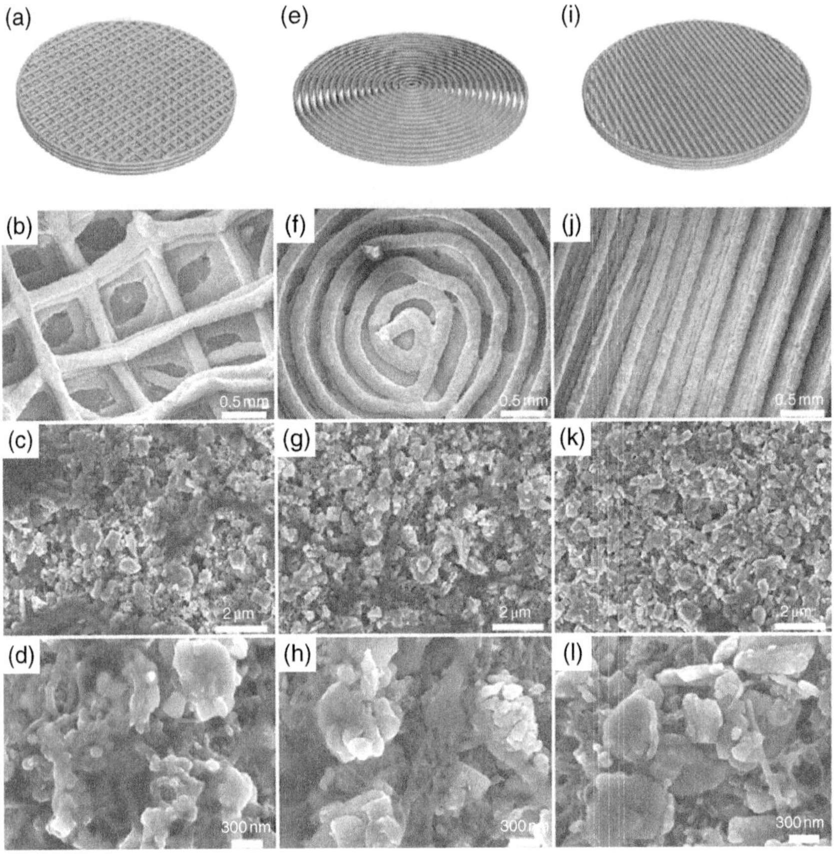


**Figure 8.6** Schematic and SEM images of three types of 3D-printed LFP electrode structures. *Source:* Reproduced with permission from Wang et al. [21]. © 2018 American Chemical Society.

***Interdigitated Electrode Structures*** In the examples from Section "3D-Structured Electrodes" only one of the battery electrodes (negative or positive) are printed, and the battery is assembled mainly as a half-cells with lithium as the counter electrode. Another approach is to continue the development of an interdigitated electrode pattern seen in Refs. [3, 4], Figure 8.1, where the positive and negative electrodes are printed together. Contacting of the negative and positive electrode is then in the same plane with, for example, 3D printed or deposited current collectors. More intricate and 3D interaction between electrodes and electrolyte can be obtained by these interdigitated patterns as described, for example, by Miranda, Costa, Almeida, and Lanceros-Méndez [27], where the performance of

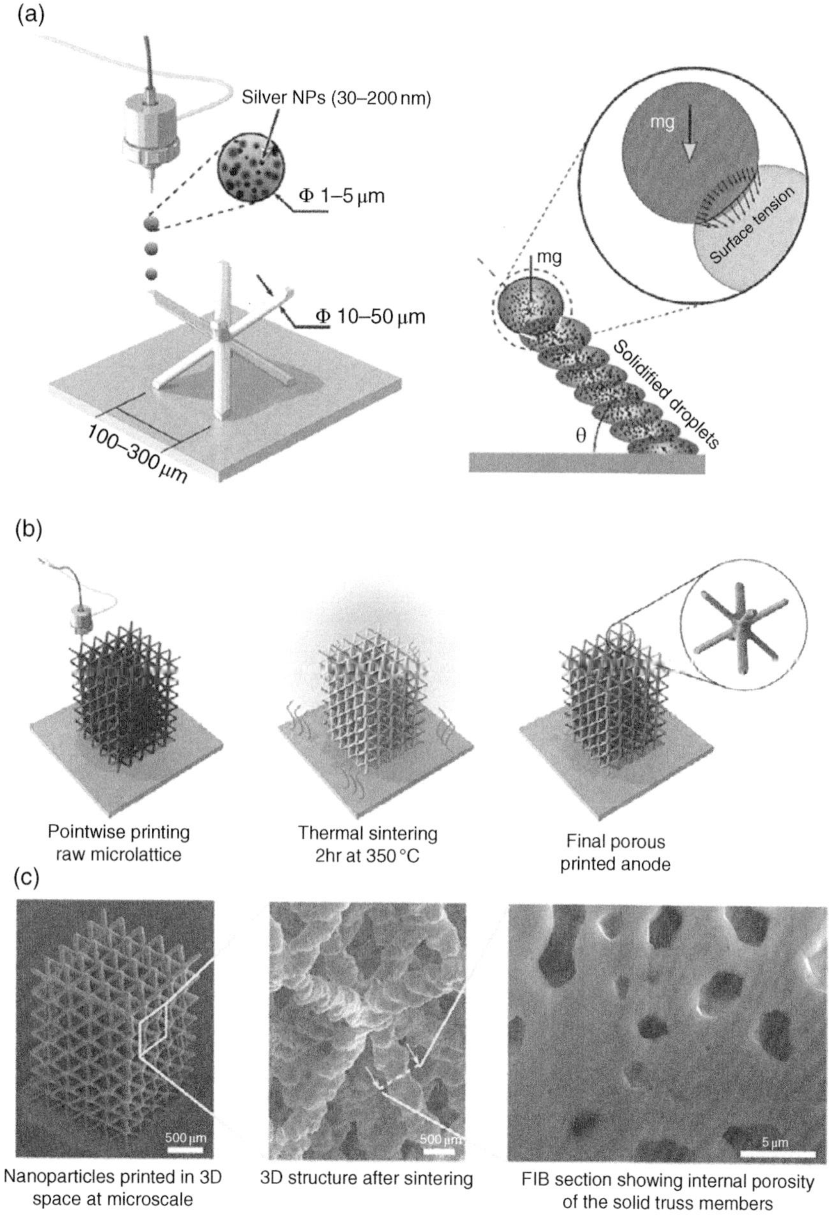


**Figure 8.7** Schematic of the fabrication of 3D micro-architected battery electrodes by assembling nanoparticles in 3D space using an aerosol jet printing process together with SEM images of the printed microstructure. *Source:* Reproduced with permission from Saleh et al. [22]. © 2018 Elsevier.

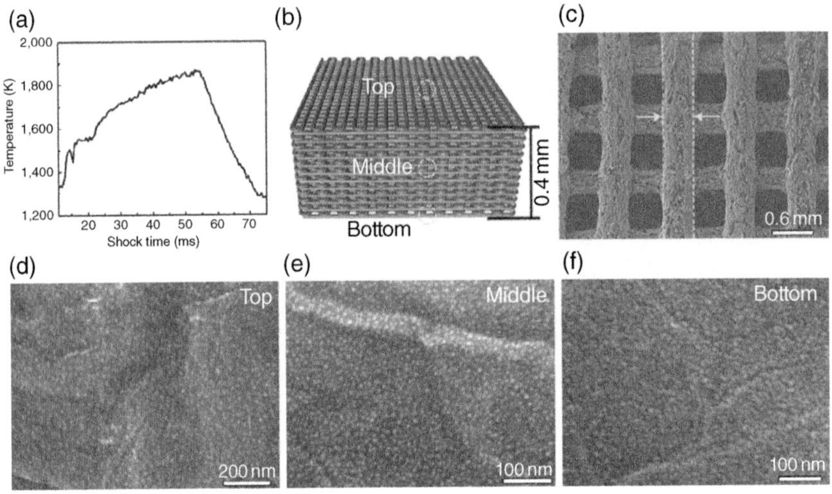


**Figure 8.8** Schematic illustration and SEM images of 3D-printed Ni/r-GO framework. *Source:* Reproduced with permission from Qiao et al. [23]. © 2018 John Wiley & Sons, Inc.

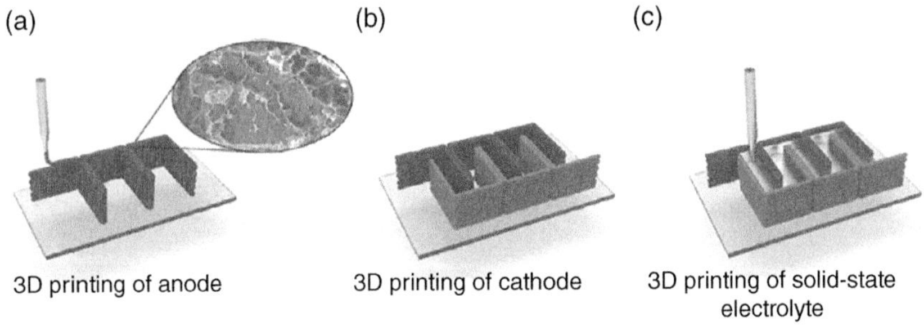


**Figure 8.9** Schematic of the 3D-printed interdigitated LTO and LFP electrodes. *Source:* Reproduced with permission from Fu et al. [29]. © 2016 John Wiley & Sons, Inc.

2D interdigitated batteries are investigated using computer modeling. By using interdigitated patterns, it is possible to optimize the contact between the electrodes, and also to print entire battery structures by including 3D-printed solid electrolytes. Interdigitated-like patterns were used by Li, Leu, Panat, and Park [28] as single electrodes in conventional coin cells, giving improved properties compared to coated electrodes. However, as demonstrated in Refs. [3, 4], the true potential of interdigitated electrodes is released when the entire electrode structure is printed together. Fu et al. [29] demonstrated 3D printing of interdigitated patterns of LFP/GO and LTO/GO electrodes with a solid state electrolyte. The positive (cathode) and negative (anode) electrodes were printed in a consecutive manner,

**Figure 8.10** SEM images of the annealed LFP/rGO electrodes. *Source:* Reproduced with permission from Fu et al. [29]. © 2016 John Wiley & Sons, Inc.

Figure 8.9, and high temperature treatment was used to reduce GO before a gel-type solid electrolyte was deposited between the electrodes. A porous structure built from rGO nanosheets was obtained after thermal reduction, Figure 8.10. For electrochemical tests of full cells, simplified two-prong microbatteries were prepared, showing promising capacities and rate behavior.

#### 8.1.1.2 3D Printing Solid Electrolytes

Solid state batteries, especially lithium-ion batteries, are becoming increasingly interesting due to the potential of higher energy densities, high power density, thermal stability, and improved safety due to the lack of flammable organic solvents which are present in conventional lithium-ion batteries. When 3D-printing an entire battery the goal is to prepare an all-solid-state battery, and one of the key requirements is therefore to be able to print a solid electrolyte. Ionic conductivity in solids is significantly lower than in liquids, and even as solid electrolytes with improved conductivities are found the conductivity, especially at ambient or low temperature, is still orders of magnitude smaller than the commonly used liquid electrolytes (e.g., lithium ion batteries). This can be remedied if it is possible to prepare very thin electrolytes, but the requirement is that the electrolyte is mechanically stable and without defects such as pin holes. Another challenge with using solid electrolytes is the ionic conductivity of the coated or printed electrodes. Conventional electrodes are porous and soaked with liquid electrolyte forming a percolating ionic transport route. Ion transport is therefore predominantly taking place in the liquid phase, while solid state conductivity is limited to transport in and out of the small electrode particles. In a solid state battery with a dense electrode, the obtainable thickness of the electrode is limited by the necessity of the ions being conducted from particle to particle due to the low ionic conductivity in the solid electrode material. This could be alleviated by mixing the electrode material with a solid electrolyte (having a higher ion diffusion coefficient that the electrode material) to enhance ionic transport. In addition, the interfaces between the electrodes and the solid electrolyte should be stable and allow easy transport of ions.

Solid electrolytes for solid state lithium-ion batteries can be divided into organic (polymer based) and inorganic materials, but could also be an inorganic/organic composite. The electrolyte has to be chemically and electrochemically stable during battery operation and maintain its electrically insulating properties during extended use. Recent reviews of solid state polymer and composite electrolytes can be found [30, 31].

When printing electrodes for traditional liquid electrolyte lithium-ion batteries, porosity is created, for example, by evaporation (by heating or freeze drying) of the dispersant solvent in the ink or slurry. When printing solid state batteries, the electrode structure should be non-porous with intrinsic ion conductivity. Several strategies for printing solid state batteries can be applied, high temperature solidification, mixing the polymer precursors into the slurry and perform *in situ* polymerization using thermal or UV driven polymerization processes or by printing a porous electrode, soak it with polymer precursors and perform polymerization within the electrode.

Cheng et al. [32] developed a new method for high temperature printing of a solid electrolyte in an interdigitated battery structure using a hybrid composite electrolyte. They mixed ionic liquids, polymer (PVDF-based) and $TiO_2$ nanoparticles as filler and evaporated the solvent at 90 °C forming a soft material which could be extruded at 120 °C through a thin nozzle. 3D printed interdigitated batteries were prepared using LFP and LTO as the positive and negative electrode materials, respectively. Figure 8.11 from Ref. [32] shows a simple interdigitated battery structure printed with solid electrolyte as well as a more intricate Hilbert curved

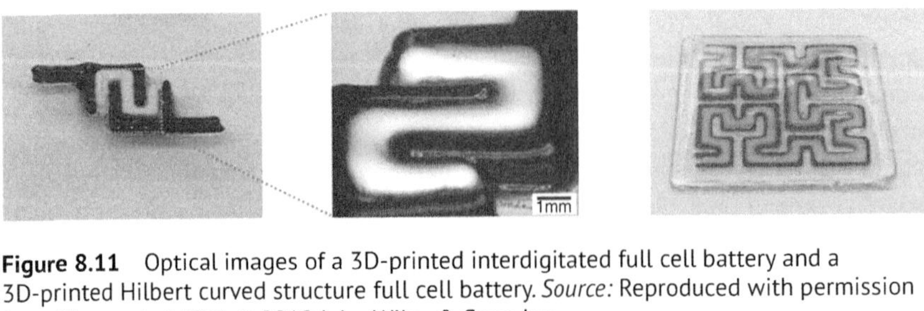


**Figure 8.11** Optical images of a 3D-printed interdigitated full cell battery and a 3D-printed Hilbert curved structure full cell battery. *Source:* Reproduced with permission from Cheng et al. [32]. © 2018 John Wiley & Sons, Inc.

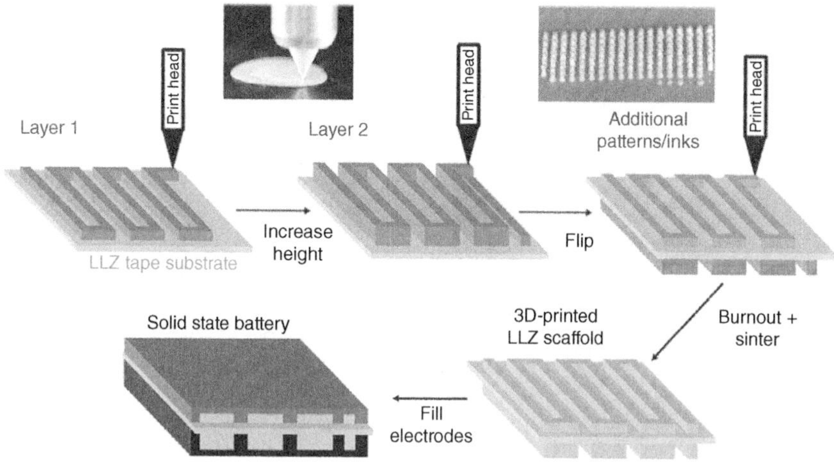


**Figure 8.12** Schematic of a process to 3D-print solid electrolyte structures. *Source:* Reproduced with permission from McOwen et al. [34]. © 2018 John Wiley & Sons, Inc.

structure which is a fully functioning battery sealed with poly-dimethylsiloxane gel. Other approaches include development of ceramic/polymer 3D-printable electrolytes based on, for example, pvdf/alumina blends [33].

A different approach is taken by McOwen et al. [34] where an inorganic garnet type LLZ ($Li_7La_3Zr_2O_{12}$) solid electrolyte structure was manufactured. An LLZ thin substrate was patterned on both sides and consolidated by high temperature firing, Figure 8.12. An electrode can then be infiltrated on each side of the electrolyte. The solid state electrolyte structure was tested in a symmetrical configuration using lithium metal on both sides.

#### 8.1.1.3 3D Printed Full Batteries

One of the goals for 3D printing of solid state batteries is to be able to print the entire battery, including current collectors and packaging. This would allow for instance real integration between 3D-printed devices and batteries by combining device and power source in the same manufacturing process. Yu et al. [2] used aerosol printing to produce non-planar printing of lithium-ion batteries with arbitrary shape. They demonstrated non-planar conformal aerosol printing of a lithium-ion battery and one of the visions was to be able to print a fully functioning lithium-ion battery on a drone structure, Figure 8.13.

3D print of wearables with integrated lithium-ion batteries is of great interest, as it gives a freedom to design and shape the battery after the device. Fused filament printing techniques have been applied to print a complete lithium-ion

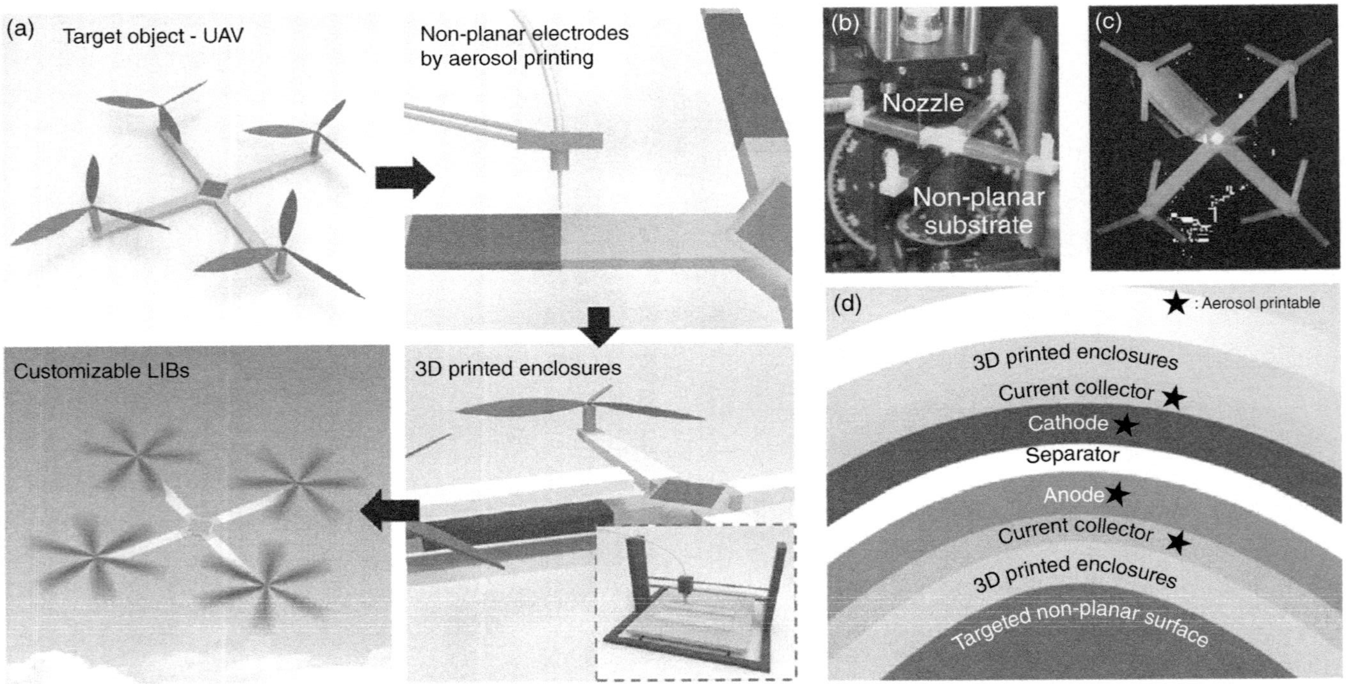


**Figure 8.13** Conceptual illustration and optical images of the 3D printing of the customizable nonplanar lithium-ion batteries on the four arms of an unmanned aerial vehicle. *Source:* Reproduced with permission from Yu et al. [2]. © 2019 John Wiley & Sons, Inc.

battery integrated with a device [35]. As an example they printed LCD driven switchable sunglasses with lithium-ion batteries integrated into the side temples and a bangle battery with an integrated LED light, Figure 8.14.

Using extrusion techniques, Wang et al. [36] fabricated a flexible, all-fiber lithium-ion battery intended for integration into textiles for wearables. They prepared LFP and LTO fibers, which were twisted using a quasi-solid gel electrolyte to form a battery with fairly high specific capacity and good cyclability. Other approaches use post-assembly of 3D-printed parts to form a fully functioning battery. As an example, Down et al. [37] printed a functioning cylindrical sodium-ion battery using fused filament printing techniques. A soluble polymer was used in printing the electrodes in order to create the needed porosity. The battery uses a liquid electrolyte, but the 3D-printed structure of the electrode assembly means that it

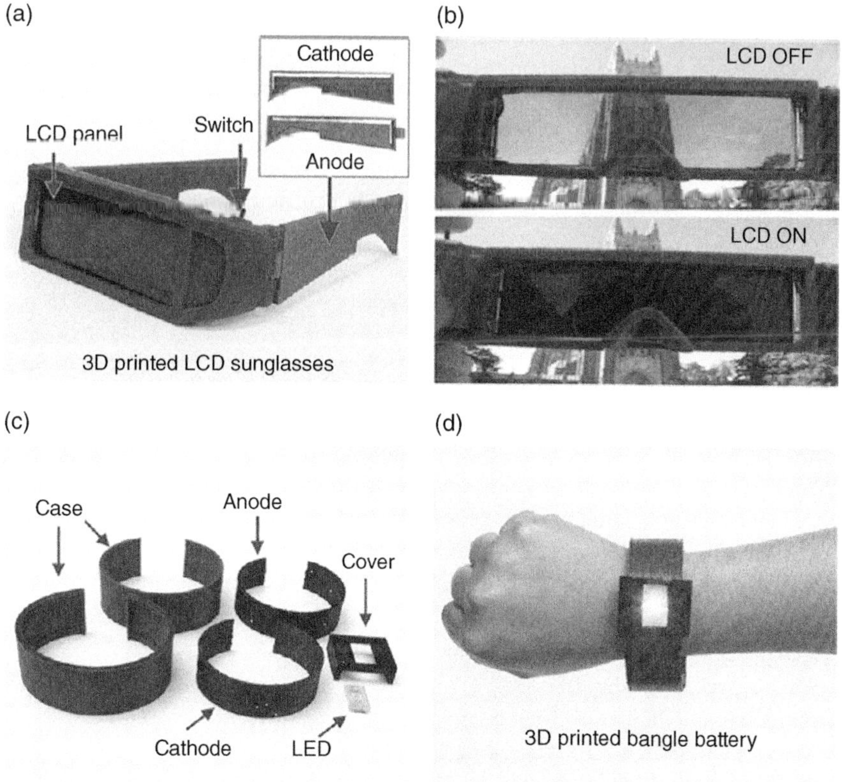


**Figure 8.14** Three-dimensional printed glasses with an electronic darkening LCD lens and a 3D printed bangle battery powering an LED. *Source:* Reproduced with permission from Reyes et al. [35]. © 2108 American Chemical Society.

does not need a separator membrane. Using slurry extrusion methods Wei, Ahn, Grotto, and Lewis [38] printed a fully assembled LFP/LTO based lithium-ion battery with a polymer electrolyte/liquid electrolyte mixture and an UV curable epoxy based sealant ink for packaging.

#### 8.1.1.4 Conclusion and Outlook

Using 3D printing methods for battery fabrication is a field where significant progress can be expected in the coming years. The efforts so far include fabrication of various elements in a battery, for example, 3D printed electrodes which are used together with a liquid electrolyte or 3D printed solid electrolyte patterns, which are then assembled using conventional electrode casting methods.

There is little doubt that interdigitated electrode/electrolyte structures will be of importance in future battery development. Advantages include the possibility of obtaining higher charge/discharge rates and a more efficient packing of the battery structure. Present day interdigitated patterns are basically 2-dimensional patterns, which have been extended in the third dimension. A further development into true 3-dimensional interdigitated patterns could be one of the next steps in development of 3D-printed batteries. This is a development, which would only be possible to realize using advanced 3D-printing techniques and which would take full advantage of the unique opportunities in 3D-printing.

## References


1 Arthur, T. S., Bates, D. J., Cirigliano, N., Johnson, D. C., Malati, P., Mosby, J. M., . . . Dunn, B. (2011). Three-dimensional electrodes and battery architectures. *MRS Bulletin*, 36(7), 523–531. doi:10.1557/mrs.2011.156

2 Yu, X., Liu, Y., Pham, H., Sarkar, S., Ludwig, B., Chen, I., . . . Pan, H. (2019). Customizable nonplanar printing of lithium-ion batteries. *Advanced Materials Technologies*, 1900645. doi:10.1002/admt.201900645

3 Yoshima, K., Munakata, H., & Kanamura, K. (2012). Fabrication of micro lithium-ion battery with 3D anode and 3D cathode by using polymer wall. *Journal of Power Sources*, 208, 404–408. doi:10.1016/j.jpowsour.2012.02.045

4 Sun, K., Wei, T. S., Ahn, B. Y., Seo, J. Y., Dillon, S. J., & Lewis, J. A. (2013). 3D printing of interdigitated Li-ion microbattery architectures. *Advanced Materials*, 25(33), 4539–4543. doi:10.1002/adma.201301036

5 Chang, P., Mei, H., Zhou, S., Dassios, K. G., & Cheng, L. (2019). 3D printed electrochemical energy storage devices. *Journal of Materials Chemistry A*, 7(9), 4230–4258. doi:10.1039/c8ta11860d

6 Deiner, L. J., Gomes Bezerra, C. A., Howell, T. G., & Powell, A. S. (2019). Digital printing of solid state lithium ion batteries. *Advanced Engineering Materials*, 1900737, 1–18. doi:10.1002/adem.201900737

7 Zhu, C., Liu, T., Qian, F., Chen, W., Chandrasekaran, S., Yao, B., . . . Li, Y. (2017). 3D printed functional nanomaterials for electrochemical energy storage. *Nano Today*, 15, 107–120. doi:10.1016/j.nantod.2017.06.007

8 Wei, M., Zhang, F., Wang, W., Alexandridis, P., Zhou, C., & Wu, G. (2017). 3D direct writing fabrication of electrodes for electrochemical storage devices. *Journal of Power Sources*, 354, 134–147. doi:10.1016/j.jpowsour.2017.04.042

9 Ruiz-Morales, J. C., Tarancón, A., Canales-Vázquez, J., Méndez-Ramos, J., Hernández-Afonso, L., Acosta-Mora, P., . . . Fernández-González, R. (2017). Three dimensional printing of components and functional devices for energy and environmental applications. *Energy and Environmental Science*, 10(4), 846–859. doi:10.1039/c6ee03526d

10 Ambrosi, A., & Pumera, M. (2016). 3D-printing technologies for electrochemical applications. *Chemical Society Reviews*, 45(10), 2740–2755. doi:10.1039/c5cs00714c

11 Tian, X., Jin, J., Yuan, S., Chua, C. K., Tor, S. B., & Zhou, K. (2017). Emerging 3D-printed electrochemical energy storage devices: A critical review. *Advanced Energy Materials*, 7(17), 1–17. doi:10.1002/aenm.201700127

12 Zhang, F., Wei, M., Viswanathan, V. V., Swart, B., Shao, Y., Wu, G., & Zhou, C. (2017). 3D printing technologies for electrochemical energy storage. *Nano Energy*, 40(May), 418–431. doi:10.1016/j.nanoen.2017.08.037

13 Park, S., Nenov, N. S., Ramachandran, A., Chung, K., Hoon Lee, S., Yoo, J., . . . Bae, C. J. (2018). Development of highly energy densified ink for 3D printable batteries. *Energy Technology*, 6(10), 2058–2064. doi:10.1002/ente.201800279

14 Delannoy, P. E., Riou, B., Brousse, T., Le Bideau, J., Guyomard, D., & Lestriez, B. (2015). Ink-jet printed porous composite $LiFePO_4$ electrode from aqueous suspension for microbatteries. *Journal of Power Sources*, 287, 261–268. doi:10.1016/j.jpowsour.2015.04.067

15 Ben-Barak, I., Kamir, Y., Menkin, S., Goor, M., Shekhtman, I., Ripenbein, T., . . . Peled, E. (2019). Drop-on-demand 3D printing of lithium iron phosphate cathodes. *Journal of The Electrochemical Society*, 166(3), A5059–A5064. doi:10.1149/2.0091903jes

16 Maurel, A., Courty, M., Fleutot, B., Tortajada, H., Prashantha, K., Armand, M., . . . Dupont, L. (2018). Highly loaded graphite-polylactic acid composite-based filaments for lithium-ion battery three-dimensional printing. *Chemistry of Materials*, 30(21), 7484–7493. doi:10.1021/acs.chemmater.8b02062

17 Hu, J., Jiang, Y., Cui, S., Duan, Y., Liu, T., Guo, H., . . . Pan, F. (2016). 3D-Printed cathodes of $LiMn_{1-x}Fe_xPO_4$ nanocrystals achieve both ultrahigh rate and high

capacity for advanced lithium-ion battery. *Advanced Energy Materials*, 6(18), 1–8. doi:10.1002/aenm.201600856

**18** Liu, C., Cheng, X., Li, B., Chen, Z., Mi, S., & Lao, C. (2017). Fabrication and characterization of 3D-printed highly-porous 3D $LiFePO_4$ electrodes by low temperature direct writing process. *Materials*, 10(8). doi:10.3390/ma10080934

**19** Liu, C., Xu, F., Cheng, X., Tong, J., Liu, Y., Chen, Z., . . . Ma, J. (2019). Comparative study on the electrochemical performance of LiFePO 4 cathodes fabricated by low temperature 3D printing, direct ink writing and conventional roller coating process. *Ceramics International*, 45(11), 14188–14197. doi:10.1016/j.ceramint.2019.04.124

**20** Ding, J., Shen, K., Du, Z., Li, B., & Yang, S. (2017). 3D-Printed hierarchical porous frameworks for sodium storage. *ACS Applied Materials and Interfaces*, 9(48), 41871–41877. doi:10.1021/acsami.7b12892

**21** Wang, J., Sun, Q., Gao, X., Wang, C., Li, W., Holness, F. B., . . . Sun, X. (2018). Toward high areal energy and power density electrode for li-ion batteries via optimized 3D printing approach. *ACS Applied Materials and Interfaces*, 10(46), 39794–39801. doi:10.1021/acsami.8b14797

**22** Saleh, M. S., Li, J., Park, J., & Panat, R. (2018). 3D printed hierarchically-porous microlattice electrode materials for exceptionally high specific capacity and areal capacity lithium ion batteries. *Additive Manufacturing*, 23(April), 70–78. doi:10.1016/j.addma.2018.07.006

**23** Qiao, Y., Liu, Y., Chen, C., Xie, H., Yao, Y., He, S., . . . Hu, L. (2018). 3D-Printed graphene oxide framework with thermal shock synthesized nanoparticles for Li-$CO_2$ batteries. *Advanced Functional Materials*, 28(51). doi:10.1002/adfm.201805899

**24** Lacey, S. D., Kirsch, D. J., Li, Y., Morgenstern, J. T., Zarket, B. C., Yao, Y., . . . Hu, L. (2018). Extrusion-based 3D printing of hierarchically porous advanced battery electrodes. *Advanced Materials*, 30(12), 1–9. doi:10.1002/adma.201705651

**25** Zhang, C., Shen, K., Li, B., Li, S., & Yang, S. (2018). Continuously 3D printed quantum dot-based electrodes for lithium storage with ultrahigh capacities. *Journal of Materials Chemistry A*, 6(41), 19960–19966. doi:10.1039/c8ta08559e

**26** Shen, K., Mei, H., Li, B., Ding, J., & Yang, S. (2018). 3D Printing sulfur copolymer-graphene architectures for Li-S batteries. *Advanced Energy Materials*, 8(4), 1–6. doi:10.1002/aenm.201701527

**27** Miranda, D., Costa, C. M., Almeida, A. M., & Lanceros-Méndez, S. (2016). Computer simulation evaluation of the geometrical parameters affecting the performance of two dimensional interdigitated batteries. *Journal of Electroanalytical Chemistry*, 780, 1–11. doi:10.1016/j.jelechem.2016.08.031

**28** Li, J., Leu, M. C., Panat, R., & Park, J. (2017). A hybrid three-dimensionally structured electrode for lithium-ion batteries via 3D printing. *Materials and Design*, 119, 417–424. doi:10.1016/j.matdes.2017.01.088

29 Fu, K., Wang, Y., Yan, C., Yao, Y., Chen, Y., Dai, J., . . . Hu, L. (2016). Graphene oxide-based electrode inks for 3D-printed lithium-ion batteries. *Advanced Materials*, 28, 2587–2594. doi:10.1002/adma.201505391

30 Yeganeh Ghotbi, M. (2019). Solid state electrolytes for electrochemical energy devices. *Journal of Materials Science: Materials in Electronics*, 30(15), 13835–13854. doi:10.1007/s10854-019-01749-4

31 Yao, P., Yu, H., Ding, Z., Liu, Y., Lu, J., Lavorgna, M., . . . Liu, X. (2019). Review on polymer-based composite electrolytes for lithium batteries. *Frontiers in Chemistry*, 7(August), 1–17. doi:10.3389/fchem.2019.00522

32 Cheng, M., Jiang, Y., Yao, W., Yuan, Y., Deivanayagam, R., Foroozan, T., . . . Shahbazian-Yassar, R. (2018). Elevated-temperature 3D printing of hybrid solid-state electrolyte for li-ion batteries. *Advanced Materials*, 30(39), 1–10. doi:10.1002/adma.201800615

33 Blake, A. J., Kohlmeyer, R. R., Hardin, J. O., Carmona, E. A., Maruyama, B., Berrigan, J. D., . . . Durstock, M. F. (2017). 3D Printable ceramic–polymer electrolytes for flexible high-performance li-ion batteries with enhanced thermal stability. *Advanced Energy Materials*, 7, 1602920. doi:10.1002/aenm.201602920

34 McOwen, D. W., Xu, S., Gong, Y., Wen, Y., Godbey, G. L., Gritton, J. E., . . . Wachsman, E. D. (2018). 3D-Printing electrolytes for solid-state batteries. *Advanced Materials*, 30(18), 1–7. doi:10.1002/adma.201707132

35 Reyes, C., Somogyi, R., Niu, S., Cruz, M. A., Yang, F., Catenacci, M. J., . . . Wiley, B. J. (2018). Three-dimensional printing of a complete lithium ion battery with fused filament fabrication. *ACS Applied Energy Materials*, 1(10), 5268–5279. doi:10.1021/acsaem.8b00885

36 Wang, Y., Chen, C., Xie, H., Gao, T., Yao, Y., Pastel, G., . . . Hu, L. (2017). 3D-Printed all-fiber li-ion battery toward wearable energy storage. *Advanced Functional Materials*, 27(43), 1–8. doi:10.1002/adfm.201703140

37 Down, M. P., Martínez-Periñán, E., Foster, C. W., Lorenzo, E., Smith, G. C., & Banks, C. E. (2019). Next-generation additive manufacturing of complete standalone sodium-ion energy storage architectures. *Advanced Energy Materials*, 9(11). doi:10.1002/aenm.201803019

38 Wei, T. S., Ahn, B. Y., Grotto, J., & Lewis, J. A. (2018). 3D Printing of customized li-ion batteries with thick electrodes. *Advanced Materials*, 30(16), 1–7. doi:10.1002/adma.201703027

# 9

# Capacitors

*Lukas Fieber and Patrick S. Grant*

*Department of Materials, University of Oxford, Oxford, UK*

## 9.1 Introduction

**Rising demand for electricity** Ever-increasing demands in global energy needs (>1.6×1,017 Wh/year, +2.1% in 2017) have been met through the unsustainable exploitation of greenhouse gas emitting fossil fuels including oil, coal, and gas (72% of the annual increase, 81% of total energy in 2017) [1, 2]. Meanwhile, electricity has become an indispensable form of energy, following the electrification of leading industrialized nations in the 1880s and global industrialization. However, strengthening global environmental awareness and concern [3] requires us to radically rethink how we generate, store, and use electricity [4, 5]. Technological advancements, including photovoltaics and nuclear energy, aim to make electricity generation carbon-neutral [6, 7], whereas consumer electronics and ubiquitous digitization are providing us with more ways to consume electricity [8].

**Capacitors** To bridge both temporal and geographic *gaps* in electricity generation and consumption, various energy storage solutions (ESSs) have been developed [9]. For applications with short charge/discharge times (ms to min) and long cycle-lives (≫1,000 cycles), capacitors have dominated. Developments have included new and improved dielectrics [10, 11], electrolytes [12], and nanostructured electroactive electrode materials [9, 13, 14]; however, only recently (since 2010) have transformative approaches to the manufacture of capacitors, as a way to enhance performance, received significant research and development (R & D) attention [9, 15–18].

**This review** This chapter reviews the progress that has been made in using emerging additive manufacture (AM), or *3D printing*, technologies to fabricate

*3D Printing for Energy Applications*, First Edition. Edited by Albert Tarancón and Vincenzo Esposito.

capacitors. First, it outlines different capacitor types and their current manufacture, identifying the principal functionality of capacitor components, and performance reporting standards and limitations in current manufacturing approaches (Section 9.2). The case for AM of capacitors is made (Section 9.3) and the operating principles as well as materials considerations for common AM techniques (e.g., inkjet printing (IJP), direct ink writing (DIW)) are outlined (Section 9.4). Examples of how individual components (e.g., electrodes, current collectors) and novel devices (e.g., coplanar, 3D interdigitated) have been fabricated using *additive* techniques are provided. Finally, a summary and outlook on future research needs in the field is presented (Section 9.5).

## 9.2 Capacitors and Their Current Manufacture

### 9.2.1 Capacitor Classifications, Operating Principles, Applications, and Current Manufacture

**Classifications** Capacitors are passive, two-terminal electrical devices that store, filter, and regulate electrical energy. Many capacitor types have been developed since the original Leyden Jar in 1745 (see [19, 20] for a brief history of capacitors and supercapacitors). They are commonly classified by their: (a) internal electric field distribution (linear, asymmetric, or localized – see Table 9.1), (b) size (dimensions, capacitance, energy, or power performance), (c) architecture (hybrid, symmetric, or asymmetric), (d) primary charge storage mechanism (electrostatic, pseudocapacitive, or intercalation pseudocapacitive), (e) design (coin-, wound-, or pouch-cell; cylindrical; or prismatic), or (f) a combination thereof (a–e).

#### 9.2.1.1 Electrostatic Capacitors

**Architecture and operating principle** A simple, electrostatic capacitor consists of (at least) two electronic conductors (*electrodes*) separated by an insulator (*dielectric*). Upon external polarization of the electrodes, energy is reversibly stored in the form of an electrostatic field across the dielectric. The electrostatic charge storage mechanism is fundamentally different from the reduction and oxidation processes in ion batteries, leading to a distinctly different, linear, charge/discharge behavior (i.e., current vs. voltage), as schematically depicted in Figure 9.1. The capacitor's ability to store charge, that is, the *capacitance*, is proportional to the electrode surface area ($m^2$) as well as absolute permittivity of the dielectric (F/m), and inversely proportional to the polarization plane separation distance (m). In an ideal capacitor, the capacitance is a constant, independent of, for example, frequency, voltage, or temperature. The capacitance is limited by the dielectric strength or *breakdown voltage*.

**Table 9.1** Summary of capacitor types showing schematics of device architecture and their internal electric potential $\phi_e$ distribution, the associated primary charge storage mechanisms, device examples, device components and common materials, as well as the manufacturing process.

| | | | |
|---|---|---|---|
| **Schematic and internal electric potential distribution** | 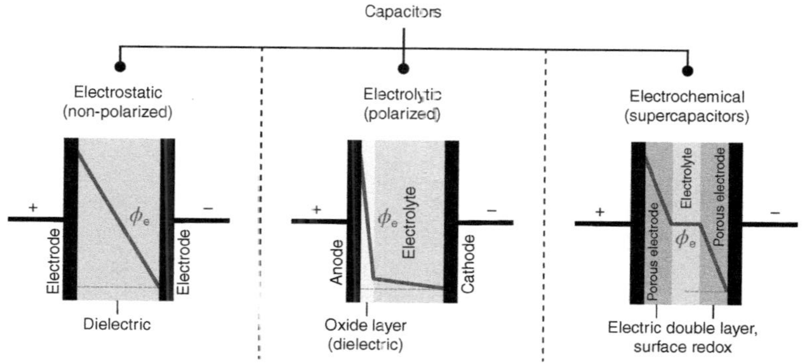 | | |
| **Charge storage mechanism** | Electrostatic across dielectric | Electrostatic across oxide layer | Electrostatic within EDL<br>(Pseudo-capacitive/surface redox)<br>(Intercalation pseudo-capacitive) |
| **Examples** | Film (metalized polymer)<br>Foil (ceramic) | Aluminum oxide<br>Tantalum oxide<br>Niobium oxide (solid electrolyte) | Electro-chemical double-layer (EDLC)<br>Pseudo-capacitor<br>Hybrid |



*(Continued)*

**Table 9.1** (Continued)

| | | | |
|---|---|---|---|
| **Components and common materials** | *Electrodes*: Al foil or metalized film<br>*Dielectric*: ceramic or polymer | *Electrodes*: etched Al foil<br>*Dielectric*: metal oxide $Al_2O_3$<br>*Electrolyte*: solid or non-solid<br>*Separator*: cellulose | *Current collector*: etched Al, Ni or stainless steel foil<br>*Porous electrode*: active material (activated carbon), conductive additive (carbon black), binder (PVDF/PTFE)<br>*Electrolyte*: aqueous, organic or ionic<br>*Separator*: cellulose or glass-fibre |
| **Conventional manufacturing process** | Film stretching/foil casting<br>Metallization/roll-screen printing<br>Film/foil slitting<br>Winding/stacking (slightly offset)<br>Compression/thermal post-treatment<br>Attachment of terminals<br>Assembly/insulation | Anode foil etching<br>Oxide forming (anodic oxidation)<br>Foil slitting/attachment of terminals<br>Winding/coiling<br>Electrolyte impregnation (vacuum)<br>Assembly/sealing/insulation<br>Artificial aging | Foil etching/primer<br>Slurry casting of electrode<br>Calendaring/drying<br>Foil slitting/punching<br>Winding/stacking<br>Electrolyte impregnation<br>Assembly/sealing/insulation<br>Artificial aging |

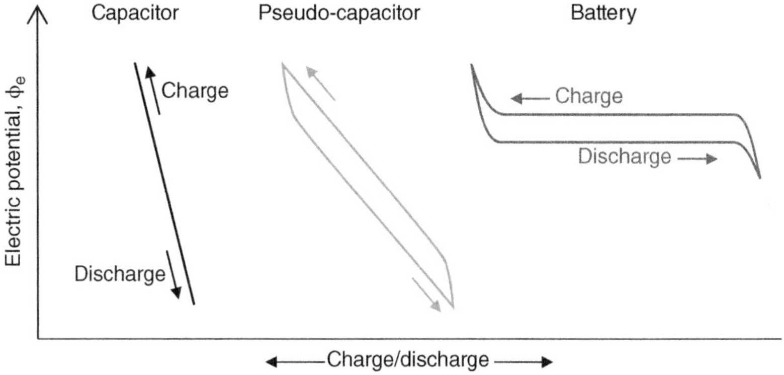


**Figure 9.1** Charge/discharge profiles for ideal electrostatic capacitors, pseudo-capacitive *supercapacitors*, and rechargeable ion batteries. *Source:* Adapted from Conway [21].

**Conventional manufacturing process** Some of the most common electrostatic capacitors used today, in virtually all devices with microelectronics, are multilayer ceramic capacitors. They are fabricated in a multistep process including [22, 23] (a) ceramic tape-casting (e.g., 1 μm $BaTiO_3$ particles with a binder), (b) roll-screen printing or metallization of electrodes (e.g., >0.02 μm Zn or Al) on ceramic tape, (c) tape slitting and lamination, (d) stacking (slightly offset), (e) compression and thermal post-treatment (bake-out and firing inducing volume reduction of up to 50%), (f) termination dipping and plating, and finally (g) coating and packaging.

**Classifications, characteristics, examples, and applications** Electrostatic capacitors are commonly classified by their dielectric material (polymer film, glass, mica, or polymer) or class (1: high stability, low loss, and low capacitance; 2: high capacitance, temperature dependent). Besides the aforementioned multilayer ceramic capacitors, polymer film and Al foil capacitors are very commonly used. These simple, electrostatic capacitors are a mature technology and available for a wide range of cost-effective high-frequency applications (e.g., microelectronics, resonators, or communications systems [24]). However, depending on the dielectric material used, their performance may be temperature sensitive (e.g., $BaTiO_3$).

#### 9.2.1.2 Electrolytic Capacitors

**Architecture and operating principle** Electrolytic capacitors consist of (at least) two electrical conductors (*anode* and *cathode*) separated by an ionically conductive liquid or gel (*electrolyte*). The anode is plated with a metal oxide, acting as a very thin (>1.0 nm/V) electric insulator (*dielectric*). Due to the small polarization plane separation distance, across the metal oxide layer, the achievable capacitance of electrolytic capacitors exceeds electrostatic capacitors (although energy is still being stored in an electrostatic manner).

**Conventional manufacturing process** The manufacturing process for electrolytic capacitors includes [25–28]: (a) anode foil etching or sintering to increase surface area (e.g., Al, Ta, or Nb/$NbO_2$ foil), (b) metal oxide forming (anodic oxidation, e.g., $Al_2O_3$, $Ta_2O_5$ or $Nb_2O_5$), (c) slitting of etched Al foil, (d) lead attachment (e.g., crimping or ultrasonic welding), (e) coiling or winding of anode, paper separator, and cathode foils, (f) impregnation of electrolyte (e.g., vacuum method), (g) assembly and sealing, (h) accelerated aging at elevated temperatures, and (i) inspection and testing.

**Classifications, characteristics, examples, and applications** Electrolytic capacitors are commonly classified by their anode material (Al, Ta, or Nb) and electrolyte (organic or aqueous; liquid, gel, or solid state). A wide range of devices in terms of size, shape, and capacitance (0.1 μF to 3 F) are available commercially. Al-based electrolytic capacitors are among the most common, in particular due to their low cost and low impedance at low frequencies (<5 mΩ at 100 kHz). As such, electrolytic capacitors find many uses in power systems. However, most electrolytic capacitors suffer from a limited lifetime due to evaporation of the electrolyte (~20 years), and may have unexpected and dramatic short-circuit (rapid discharge) failures.

#### 9.2.1.3 Electrochemical Capacitors

**Architecture and operating principle** Electrochemical capacitors, *ultracapacitors* or *supercapacitors*, consist of (at least) two, typically symmetric (identical anode and cathode), ultra-high surface area electrical conductors (*porous electrodes*) separated by an ion permeable but electrically insulating membrane (*separator*). The porous electrodes are connected to an external circuit by electric conductors (*current collectors*) and the entire cell is immersed in an ion-conducting liquid or gel (*electrolyte*), allowing ions to shuttle between the porous electrodes during charge/discharge. Supercapacitors make use of various charge-storage mechanisms, including (a) electrostatic: physical adsorption of electrolyte ions in an electric double layer (EDL) on the electrode surface; (b) pseudocapacitive: reversible, fast faradaic surface reactions; and (c) intercalation-pseudocapacitive: fast intercalation of cations in the bulk, active electrode material but not limited by diffusion of cations within the crystalline framework of active material [29]. Comprehensive reviews of the underlying physical electrochemistry [13, 21, 30] and recent advances [29, 31, 32] have been published elsewhere. It should be noted that many supercapacitors make use of a combination of charge-storage mechanisms (so-called *hybrid supercapacitors*), frequently resulting in misleading naming conventions [33]. A widely accepted publication clearly defines the boundaries [34].

**Conventional manufacturing process** Supercapacitors that store energy in an electrostatic EDL only are referred to as electrochemical double layer capacitors

(EDLCs). Their highly mature, roll-to-roll mass-manufacturing process is similar to how Li-ion batteries are fabricated, and includes [13, 35] (a) current collector foil (e.g., Al foil 15–40 μm for organic electrolytes and Ni or stainless steel (SS) for aqueous electrolytes) etching alongside application of a conductive primer (~1 μm carbon black) to increase surface area and reduce electrical series resistance (ESR), respectively; (b) doctor blading or slurry casting of active material (e.g., activated carbon from wood [36] or coconut [37] precursor, activated by water vapor with up to 2,000 $m^2/g$), conductive additive (e.g., ultrafine carbon black particulate), binder (e.g., 10wt% polyvinylidene fluoride (PVDF) or polytetrafluoroethylene (PTFE)), and a fugitive solvent (e.g., *N*-Methyl-2-pyrrolidone (NMP) and/or isopropyl alcohol (IPA)) to a preset, constant thickness (~50–150 μm) by a passing blade; (c) drying (e.g., along a drying tunnel on a conveyor) and calendering (i.e., compression through heat pressing) of the porous electrodes; (d) electrode and current collector foil slitting or punching; (e) winding or stacking of current collector, electrodes, and porous (50–90% porosity) separator (e.g., 15–50 μm paper of solvent-spun cellulose or recycled cellulose fibers for organic [38], or 30 μm glass-fiber-based for aqueous electrolyte [39]); (f) impregnation of the cell with liquid or gel electrolyte (organic, aqueous, or ionic); (g) multistep assembly and sealing involving welding; and finally (h) artificial aging at elevated temperatures before testing and packaging.

**Classifications, characteristics, examples, and applications** Supercapacitors are commonly classified by their electrolyte (organic, aqueous, or ionic); electrode materials (one dimensional (1D), two dimensional (2D), or three dimensional (3D) carbons [40]; endohedral, planar, or exohedral carbons; metal oxides; conducting polymers or composites [14, 41–44]); charge storage mechanism (pure EDLC, pseudo-capacitor, or hybrid-supercapacitor); application (e.g., consumer electronics, automotive, or aerospace); and configuration (symmetric, asymmetric, or hybrid). An effective polarization plane separation distance, or inner Helmholtz plane (IHP), of 0.2–0.8 nm gives rise to significantly higher capacitance (up to 5,000 F, approximately ×100 more specific energy than electrolytic capacitors) compared with other capacitors. EDLCs can operate over a wide range of temperatures (−40 to 85 °C) that is difficult to achieve for other electrochemical devices such as ion batteries. The supercapacitor market has matured significantly over the last years, in particular for applications such as regenerative braking in electric or hybrid vehicles, smart grid applications, rapid transit rail or buses [20, 45, 46] which make use of the high power and reasonable energy density capabilities of EDLCs. An overview of the performance of commercially available supercapacitors may be found elsewhere [47].

### 9.2.2 Capacitor Components: Function and Requirements

The principal functionality of capacitor components (dielectric, current collector, housing, electrolyte/separator, and porous electrode), their requirements, and general AM processing considerations are now considered. Table 9.2 outlines which functional components are present in the different capacitor types outlined in Section 9.2.1. Overarching, general component requirements include health and safety (H&S) concerns (e.g., toxicity of embodied materials), environmental concerns (ecological sourcing, end of life disposal), regulatory adherence (compatibility with current legislation for product manufacture), process-ability, chemical stability, and shelf life.

**Dielectric** The dielectric (insulator) electrically separates conductive electrode plates in electrostatic capacitors. The primary requirements include a high dielectric constant ($\varepsilon_r$), high breakdown voltage (V), or dielectric strength (MV/m), and the ability to allow for a small capacitor polarization plane separation distance (nm).

**Current collector** The current collector (electrical conductor) supplies electrons to and from an external circuit to polarize the capacitor. The primary requirements include (a) excellent intrinsic electron conductivity (S/m) and, in electrochemical capacitors, (b) good adhesion to the porous electrode to reduce internal resistances, as well as (c) chemical inertness to the electrolyte.

**Housing** The housing (encasing) provides mechanical protection and prevents electrolyte leakage. The primary requirements for this passive component include (a) sufficient mechanical strength to endure environmental stresses and (b) degradation resistance over the entirety of intended shelf and operation life. The exact definition of these depends on the type of electrolyte used and the environment of the capacitor. For example, solid-state electrochemical capacitors (i.e., supercapacitors with a solid electrolyte) do not require the same robust, rigid packaging requirements of liquid electrolyte systems, potentially providing more design flexibility and geometric freedom [48].

**Electrolyte** The electrolyte (salt + solvent) provides ions to satisfy electrostatic equilibrium at the electrode–electrolyte interface in electrolytic and electrochemical capacitors. The primary requirements include (a) high ionic concentration (mol/m$^3$) with high intrinsic ionic mobility, (b) a wide operating window, or electrochemical stability window (ESW), in terms of potential (V) and temperature (°C), and (c) a small solvated ion radius (nm) to increase surface accessibility into porous electrodes [44]. There are a wide range of electrolytes including aqueous (e.g., polyvinyl alcohol (PVA) hydrogels [49, 50]), organic [51], and ionic liquids [52]; however, safety concerns related to nonaqueous liquid and some gel-based electrolytes create an imperative for the use of solid electrolytes in some energy storage applications. Reviews of electrolyte types [12] and specifically hydrogel polymer electrolytes (frequently used in AM processes) can be found elsewhere [53].

**Separator** The separator (porous membrane) electrically insulates electrodes from one another while allowing electrolyte ions to shuttle between the electrodes

**Table 9.2** Functional components in different types of capacitor devices.

| | Component | | | | |
|---|---|---|---|---|---|
| **Capacitor type** | **Dielectric** | **Current collector** | **Housing** | **Electrolyte/ separator** | **Porous electrode** |
| Electrostatic | X | X | X | | |
| Electrolytic | | X | X | X | |
| Electrochemical | | X | X | X | X |

during charge/discharge. In capacitor systems fabricated by AM, solid or gel electrolytes that simultaneously act as the separator are common. The separator should provide [13]: (a) mechanical stability to allow for cell assembly, (b) high porosity (>50%) for facile electrolyte ion mobility, (c) high thermal stability, and (d) be as thin as is practical to reduce the length of ion pathways. Coating of the electrolyte/separator layer is often considered one of the most difficult stages in capacitor manufacture [54]. A study on the influence of separator properties on electrochemical performance of supercapacitors can be found in [55].

**Porous electrode** The porous electrode (active material + conductive additive + binder) aims to provide the maximum accessible active surface area for electrolyte ions to be adsorbed onto (in electrochemical capacitors) and is ultimately how/where charge is stored. The binder provides mechanical cohesion and stability between active particles and the current collector. Electrode requirements include (a) high specific electrochemical (accessible) surface area ($m^2/g$), influenced by the porosity, pore tortuosity, and wettability of active material particles; and (b) good intra- and inter-particle electron conductivity. The binder may have a significant effect on performance in electrochemical capacitors [56, 57].

### 9.2.3 Performance

**Key performance parameters** Capacitor performance parameters include capacitance (F), operating voltage (V), impedance (Ω), time constant (s), energy stored (J), power (W), dissipation factor (%), cycling stability (capacitance retention [%] after >10,000 cycles), charge transfer resistance or ESR (Ω), self-discharge behavior (V/min), and cost [$ $F^{-1}$]. Where applicable, absolute performance is normalized to active material weight ($g^{-1}$), electrode area ($cm^{-2}$), or device volume ($mm^{-3}$). Electrochemical parameters are frequently evaluated by cyclic voltammetry (CV), constant current charge/discharge (CCCD), and electrochemical impedance spectroscopy (EIS). An excellent overview of the instruments and measurements required to estimate key performance parameters, alongside common inconsistencies in evaluating these, is given in [58]. Further guidelines outlining conventional experimental methods and how to interpret data are given [58–64].

**Lack of standardization** Unfortunately, performance parameters, and evaluation thereof, are not well standardized. Therefore, data in many publications can only be used to inform what AM techniques are being explored in the context of energy storage devices and there are difficulties in making robust quantitative comparisons, particularly between research and commercial performance. International Electrotechnical Commission (IEC) norms for the standardized testing of capacitors have been published [65]; however, access to these and their adoption, especially in academic environments, is often limited.

### 9.2.4 The Challenge of Manufacturing Capacitors

**The market** The global capacitor market is expected to grow significantly, at ~22%/year, to USD 4 billion by 2022 [46]. It is a competitive, high-volume mass market with many suppliers including Asahi Glass (Japan), Batscap (France), Maxwell Technologies (USA), Murata Manufacturing Co. Ltd. (Japan), NEC Corporation (Japan), Nesscap (South Korea, Canada), Nippon Chemicon (Japan), Panasonic (Japan), and Samsung Electro-Mechanics (South Korea). Manufacturing challenges arise from the perpetual desire for increased performance (increased capacitance, lifetime, and energy/power density) at reduced cost. Meanwhile, the environmental requirements (operating temperatures, exposure to stresses, and device size restrictions) and regulatory environment (ecological sourcing of materials, manufacturing health, safety, and end of life) are becoming more stringent/demanding.

**Principal requirements** The many different types of capacitors and corresponding manufacturing approaches (outlined in Section 9.2.1) each bring their own challenges, from effective electrolyte impregnation in supercapacitors to crack-free fabrication of thin ceramic films in multilayer electrostatic capacitors. However, the principal challenge in capacitor manufacture remains the same: to engineer macroscopic components (*devices*), with designed and highly resolved, multi-material, electro-active substructures that cost-effectively meet the required performance [9]. The manufacturing technologies must be scalable and financially attractive to significantly penetrate the price-sensitive energy storage device market.

**Limitations of current manufacturing methods** Some of the limitations of current, commercial capacitor fabrication techniques include (a) limited geometric design freedom – the inherent planar form-factor restrictions of *slurry casting* techniques for electrochemical capacitor electrode fabrication restrict the device architectures possible (flat, wound, or coin-cell configurations) [66]; (b) homogeneous electrode morphology – the intrinsically stochastic microporosity and tortuous pore morphology of electrodes fabricated using conventional *slurry* methods hinders optimal use of electro-active materials (especially at ultrafast charging rates) due to limited ion mobility [9]; and (c) expensive cell assembly – the multistep assemblies (outlined in Table 9.1) required to fabricate functioning capacitor devices induce high up-front capital costs and long product lead-times [67].

## 9.3 The Promise of Additive Manufacturing

**Overview** Emerging AM technologies and concepts are having a fundamental and wide-reaching impact on modern manufacturing [68]. While not yet penetrating significantly into the energy storage sector or applied specifically to the manufacture of capacitors, AM ideas and approaches may provide new opportunities to address the limitations (outlined in Section 9.2.4) inherent within conventional device manufacture. While the full potential of AM technologies for the fabrication of capacitors is still unclear, possible advantages and disadvantages are now described.

**Advantages of AM for capacitor fabrication** With respect to the fabrication of capacitors devices and components, the spatial and temporal material manipulation abilities of AM techniques may allow for:

a) **Increased geometric design freedom** – the sequential, *layer-by-layer* deposition of functional materials allows for the fabrication of geometrically complex, multi-material (ceramic, plastic, and metal) devices such as capacitors [69]. Such devices may demonstrate novel geometric form factors (e.g., periodic or interdigitated electrode capacitor configurations) for improved performance (through, e.g., shortened ion diffusion pathways) and allow for the utilization of irregular volumes for energy storage [70].
b) **Nano- to macro-scale control of material properties** – AM technologies may provide an economical approach to control material properties over a wide range of length scales (from, e.g., aligned carbon nanotubes (CNTs) in extrudates to large hierarchically structured devices). Nano- to micro-scale architectures have already been demonstrated for other applications [71–73].
c) **Functionally gradient devices** – the spatial control of material properties (e.g., porosity and morphology) may allow AM processes to provide additional functionality of capacitors through functional gradients [74] or heterogeneous designs. The potential benefits of graded electrodes in energy storage devices [75] and Li-ion batteries specifically [76] as well as supercapacitors [77] have begun to be explored.

d) **Direct manufacturing integration** – the direct integration of capacitor manufacture in product manufacture may facilitate novel, more compact electronic device designs. Integrated manufacture may allow for on-chip printing [78] and direct manufacture of energy storage devices on crowded micro-devices [48].
e) **Cost reduction** – AM processes may provide opportunities for new, decentralized manufacturing business models [79]. They may provide environmental benefits (waste reduction) and reduce or eliminate the need for expensive part-specific tooling.
f) **Process flexibility** – in terms of material feedstock properties (e.g., paste, powder, laminate, filament, and suspension), AM processes may provide unprecedented versatility. The shortened overall manufacturing lead times achievable with AM techniques may promote rapid prototyping for capacitor designs and allow the testing of novel materials combinations at an unprecedented pace.

**Disadvantages of AM** Some of the main drawbacks of AM technologies for the fabrication of capacitors include (a) relatively high cost (current high-throughput, roll-to-roll techniques are highly efficient, optimized systems), (b) intrinsic anisotropy of material properties (strength, conductivity, etc.) resulting from the line-by-line or layer-by-layer deposition and sequential integration/application of materials, (c) limited scalability (discrete machines with preset working envelopes), (d) immature technology (AM process dynamics have not yet been fully understood and in particular feedback control approaches are generally absent), and (e) low material throughput (e.g., [$m^2$/hr] of electrode).

**Existing reviews of AM technologies** AM technologies have been reviewed frequently, outlining in detail the operating principles of common technologies and giving examples of what applications may benefit from their use [68, 80–82]. Alongside these, common materials and materials issues for individual AM technologies have been reviewed [83, 84]. However, only a few recent top-level publications review AM for energy storage applications, for example [9, 15–18].

**General selection criteria for AM processes** AM processes are free from some conventional manufacturing constraints but are still developing and evolving quickly, making the selection of a suitable technology for a specific application difficult. Knowledge based decision support systems (kb-DSSs) are still in their infancy for AM processes [85, 86]. Nevertheless, the following aspects of manufacture may be considered when choosing an appropriate AM technology for the fabrication of capacitor components [87, 88]: (a) physical (resolution/precision, accuracy, speed, linear or logarithmic print cycle, build size, surface finish, and repeatability), (b) design (support requirements, workflow, and available materials palette), (c) cost (initial investment, running costs, and waste), (d) post-processing requirements (thermal, pressure, ultraviolet (UV) curing, sintering, and debinding), (e) multi-material, multi-technology (MMMT) or *hybrid-AM* ability, and (f) throughput.

## 9.4 Additive Manufacturing Technologies: Considerations for Capacitor Fabrication

### 9.4.1 AM Process Categories

AM processes are commonly divided into the following process categories, jointly defined by the American Society for Testing and Materials (ASTM) and International Organization for Standardization (ISO) [89]: material extrusion, vat polymerization, powder bed fusion, material jetting, binder jetting, direct energy deposition, and sheet lamination. The operating principle, process considerations and feedstock requirements, benefits as well as challenges relating to how individual technologies have been used to fabricate capacitor components *additively* are discussed hereafter, with exclusion of direct energy deposition and sheet lamination technologies. As yet uncommon AM technologies such as flexographic printing, 3D screen printing, electrophoretic deposition, electrode conversion, and electrolytic deposition are also excluded. Table 9.3 provides a summary of examples where AM techniques have been reported for the fabrication of individual capacitor components.

#### 9.4.1.1 Material Extrusion – Fused Filament Fabrication

**Operating principle** In fused filament fabrication (FFF), a thermoplastic filament, such as amorphous acrylonitrile butadiene styrene (ABS) or semicrystalline polylactic acid (PLA), is heated above its glass transition temperature in a *hot-end*, that is, *print-head* (180–250°C) and pushed through a *nozzle* (Ø = 0.2–1.0 mm). The extrudate solidifies (through polymer crystallization and chain entanglement [18]) upon deposition onto a temperature-controlled *build platform* (20–150°C) to form 3D objects in a layer-by-layer manner. Commercial systems are mature and readily available (e.g., from Makerbot, Ultimaker, and Stratasys).

**Process considerations and feedstock requirements** FFF process parameters include print speed (mm/s), extrusion temperature (°C), and print path. Feedstock material properties include melt viscosity (Pa s) and elastic modulus (Pa). Process and material parameters influence the quality (tolerance,

**Table 9.3** Examples of AM technologies, split by process categories, used for the fabrication of capacitor components (dielectric, current collector, housing, electrolyte, and porous electrode).

| | | Capacitor component | | | | |
|---|---|---|---|---|---|---|
| **AM technology** | | **Dielectric** | **Current collector** | **Housing** | **Electrolyte/separator** | **Porous electrode** |
| **Material extrusion** | Fused filament fabrication (FFF) | [90] Polylactic acid (PLA)<br>[91] PLA$^{\dagger}$<br>[92] ceramic$^{\dagger}$ | [90] Thermoplastics<br>[93, 94] graphene-acrylonitrile butadiene styrene (ABS)$^{\dagger}$/PLA$^{\dagger}$ | [95] Silicone<br>[96] PLA<br>[69], polypropylene (PP) | [97] Nafion precursor<br>[98] ion exchange membrane$^{*}$ | N/A Too low mass loading achievable |
| | DIW<br>e-3D | [99] Rheology of ceramic pastes$^{\dagger}$ | [69, 100] Silver (Ag) ink<br>[101] embedded 3D printing$^{\dagger}$ | [100] Silicon | [69, 100] Gel<br>[102] ionic liquid$^{*}$<br>[54] solid-state printed electrode membrane assembly (PEMA)$^{*}$ | [103] Review<br>[104–107] graphene<br>[69, 95] activated carbon (C)<br>[108, 109] CNT |
| **Vat polymerization** | SLP<br>PμSL<br>HL<br>DLP | [110] PμSL: polymer-ceramic | [111] SLP: metal-UV resin$^{\dagger}$ | N/A no literature | [112] PμSL: macro-structured polymer$^{*}$<br>[113] PμSL: micro-structured ceramic$^{*}$<br>[114] DLP: porous$^{\dagger}$ | [115] HL: lithium ions (Li-ion)$^{*}$ |

| | | | | | | |
|---|---|---|---|---|---|---|
| **Powder bed fusion** | SLM/DMLS<br>SLS<br>EBM | [116] Ceramics† | [117] SLM: steel<br>[118] SLM: titanium (Ti)<br>[119] DMLS: steel | [120] Polymer†<br>[121] polyether ether ketone (PEEK)† | [122] EBM: liquid/gas diffusion layer* | [123] Tool electrode† |
| **Material jetting** | IJP spray | [124, 125] Barium titanate ($BaTiO_3$)<br>[126, 127] polymer<br>[128, 129] boron nitride (BN)<br>[130] iron oxide ($Fe_3O_4$)† | [131, 132] Reviews†<br>[124, 125, 133] Ag<br>[128, 129] graphene<br>[126] polymer | [134] Printed hydraulics† | [135] PVA, solid<br>[136] solid-state<br>[137] ionogel*<br>[138] spray* | [139] Graphene<br>[135, 136, 140, 141] CNT<br>[142] activated C<br>[143] lithium ferrophosphate ($LiFePO_4$)* |
| **Binder jetting** | IJP | [144] $BaTiO_3$<br>[145] ceramics† | [146] Copper† | N/A Generally too porous without post-processing | N/A No literature | [147] Graphene<br>[148] carbon nanofiber (CNF) plaster† |

Notes: * indicates that the example given is in another energy storage-related application and † indicates the example is for applications further removed from capacitors but of potential relevance.

pores, surface finish, distortion, etc.) of fabricated parts in a complex, and usually difficult to predict, manner [69, 149, 150]. Reviews of the FFF process include design considerations and process models [149]; materials, dimensional accuracy, and surface roughness [151]; as well as approaches to optimizing process parameters [150, 152–154]. FFF feedstock materials usually come in the form of a *filament* spool (Ø = 1.75 or 3.0mm). The primary requirement for the filament is to allow for a mechanical feed mechanism to push the filament through the nozzle without breaking and to reduce the likelihood of nozzle clogging, over print periods typically of several hours.

**Benefits** Principal benefits of FFF technologies for the fabrication of capacitors include:

a) **An expanding palette of processable materials** – commercial thermoplastics can be doped with functional materials, for example, $BaTiO_3$ for use as the dielectric [92]; metallic nano-platelets [90] or conductive graphene [94] for use as a current collector. Recently, an ion-conducting thermoplastic based on a Nafion precursor has been printed for applications in supercapacitors [97].
b) **Facile hybridization** – FFF processes do not rely on a vat or powder-bed system, and as such are easily compatible with MMMT processes, that is, FFF processes can be easily combined with other, complementing AM processes. A *hybrid-AM* approach has been used to fabricate functioning, form-factor free electrochemical capacitors [69].
c) **Simple, fast, and economical operation** – the low capital and running costs of FFF processes provide an excellent breeding ground for manufacturing innovations related to capacitor fabrication. FFF technologies are mature, commercially available, and there are growing online support communities/ resources (forums and cogeneration platforms) [155]. Control of FFF processes is fairly simple and supported by vibrant open-source software projects (RepRap, Sailfish, Repetier, etc.). Furthermore, FFF is a fairly quick process (print speed ~0.01 kg/hr [156], near instantaneous solidification of extrudate) providing potential benefits in terms of scalability.
d) **Minimal heat input and post-processing required** – the low glass transition temperature of most thermoplastics does not require large heat inputs during processing (<250°C). Generally, there is no need for chemical post-treatment of printed parts.
e) **Mechanical strength** – parts fabricated by FFF have demonstrated sufficient mechanical strength using electrically insulating materials (e.g., polycarbonate (PC), polypropylene (PP), PLA, ABS, and PC-ABS blends [80]) to be applicable for fabricating capacitor housing elements.

**Challenges** Technological challenges associated with FFF of capacitor components include:

a) **Low mass loading** – composite functional filaments rarely exceed 30wt% of active material. The thermoplastic *carrier* (e.g., 70wt% ABS or PLA) may significantly hinder device performance. Furthermore, the process of developing functional filaments is not trivial and may require many steps [93]. Long-time, robust printing without breaks or clogging can be problematic.
b) **Limited understanding of the underlying processing science** – current models of FFF processes suffer from excessive oversimplifications and lack insight into the dynamics between melt and extrusion processes, liquifier or *hot-end* dynamics (e.g., pressure drop within the heated nozzle), as well as bead spreading (wetting and rheology) and bonding processes between successive layers [149].
c) **Geometric restrictions** – *overhangs* of more than 30° may require support structures [88], limiting the ability to fabricate geometrically unconstrained devices.
d) **Dimensional accuracy** – thermal effects including shrinking and warpage make it difficult to predict the dimensional accuracy. The coarse feature resolution (>200 μm) achievable with current FFF technologies significantly limits the polarization plane separation distance and associated capacitance for FFF fabricated capacitors.
e) **Compatibility** – some of the more mechanically desirable polymers are *difficult to print* because bonding even between like layers is challenging, and bonding to chemically different layers, even other polymers, can be extremely variable.

**Conclusion** The popular material extrusion additive manufacture (MEAM) technology FFF may provide an excellent candidate for fabricating thermoplastic, insulating housing components for capacitors. Although in principle it is possible to fabricate other functional capacitor components using FFF, including an ion-permeable separator [97] or current collector [90], the low active mass loading of FFF-compatible filaments and coarse feature-resolution suggests these components may be better fabricated by other AM approaches.

#### 9.4.1.2 Material Extrusion – Direct Ink Writing

**Operating principle** In DIW or *robocasting*, a particle-loaded ink (e.g., concentrated colloidal/nanoparticle suspension) is micro-dispensed (pneumatic, hydraulic, or electronic actuation) through a syringe onto the build platform. DIW as a technique to fabricate functional devices has been discussed [157, 158] and has been reviewed extensively, for example [159, 160]. Common lab-scale arrangements include modified FFF printers, for example, Discovery [95]. Modified DIW technologies in which the ink is deposited in a liquid/gel supporting reservoir have been reported recently and are referred to as embedded 3D printing (e-3D) [101].

**Process considerations and feedstock requirements** DIW process parameters include print speed (mm/s), extrusion speed (mL/s), nozzle diameter (Ø mm), and nozzle geometry (tapered and step). Feedstock material properties include mass loading (vol%), curing parameters (thermal, UV, etc.), and rheological considerations (typically viscoelastic gels) such as viscosity (mPa s) or yield stress (Pa) [99]. Functional inks should allow for the formulation of long-term stable suspensions without particle agglomeration or sedimentation. In particular, ink-wetting properties with the substrate should be considered (i.e., wetting substrates including paper invoke spreading of inks due to uncontrollable capillary forces whereas non-wetting substrates such as polyethylene terephthalate (PET) may lead to a *coffee-ring* effect) [161, 162].

**Benefits** Principal benefits of DIW technologies for the fabrication of capacitors include:

a) **High mass loading** – the formulation of functional inks with a high proportion of active/functional materials is possible, for example, through the use of phase inversion, freeze gelation [108], or a fugitive solvent which evaporates upon deposition [69]. Conductive inks (e.g., Ag nanoparticles) that sinter without external heating, at room temperature, have been reported and may be good candidates for current collectors in *printed* capacitors [163]. Ceramic-polymer electrolytes/separators for applications in Li-ion batteries have been demonstrated and may be extended for the similar requirements of electrochemical capacitors [54].
b) **Liquid and gel materials** – DIW technologies allow the direct deposition of liquid and gel electrolytes without the need for subsequent phase transformation. Liquid/gel electrolytes intrinsically have higher ion mobility than solid counterparts and may lead to superior electrochemical performance in capacitors. The deposition of ionic liquid electrolytes has been demonstrated in [102], as well as the fabrication of highly flexible capacitor components using DIW [100].
c) **Flexible ink properties** – relative to other ink-based AM processes such as IJP, DIW technologies are forgiving in terms of the rheology of inks and may operate reliably with a relatively wide range of particle size, viscosity, surface tension, and so forth [104]. Highly concentrated inks based on thermally reduced graphene oxide (GO) [104, 105] or CNTs [109] (both difficult to fabricate using conventional methods) have been demonstrated for applications in porous electrodes for electrochemical capacitors.
d) **Facile hybridization** – similar to FFF processes, the absence of a vat or powder-bed make DIW technologies good candidates for a MMMT or *hybrid-AM* approach.

**Challenges** Challenges associated with DIW of capacitor components include:

a) **Post-processing** – many DIW processes require post-processing [107] (e.g., high-temperature sintering at 600°C or UV curing [164, 165]) to achieve the desired material properties. The post-processing requirements increase cost and may prohibit MMMT or hybrid-AM approaches.
b) **Complex deposition/drying process** – the somewhat unpredictable nature of depositing liquids (spreading, wetting, evaporation, etc.) makes accurately predicting the dimensions (e.g., layer thickness) of DIW fabricated parts difficult in all but the most well-characterized systems. However, if DIW processes could be controlled sufficiently well, physical phenomena such as the coffee-ring effect might be exploited to fabricate, for example, very thin separators [127].
c) **Stable suspension** – developing stable suspensions (no material separation/sedimentation) may be difficult due to the large difference in material densities (e.g., dense metallic particulates in an organic solvent). In particular, inks prepared with carbon black (common conductive additive for capacitors) require additional, performance-hindering polymers to achieve good stability, dispersion, and printability [103]. Low concentrations may be needed for suspension stability/non-clogging, leading to high fugitive carrier loss and low active material throughput.

**Conclusion** The high mass loading achievable with DIW technologies, facile hybridization with other AM processes, and ongoing development of functional inks make DIW technologies excellent candidates for the fabrication of the current collector, electrolyte/separator, and porous electrode in capacitors [103]. However, the relatively poor spatial resolution (> 100 μm) makes DIW slightly less suitable for fabricating thin dielectric layers. The relatively poor mechanical properties of components fabricated by DIW make such technologies unsuitable for fabricating housing components.

#### 9.4.1.3 Vat Polymerization

**Operating principle** In vat polymerization AM technologies, a liquid photopolymer *resin* (monomer + oligomer + photoinitiator) in a vat is selectively cured (laser, projector, or display) by typically light-activated, polymerization to fabricate 3D parts layer-by-layer [88]. Both top-down (part is built facing up) and bottom-up (part is built facing upside down) vat polymerization technologies are commercially available (e.g., Formlabs or 3D Systems). Vat polymerization technologies that have been reported for the fabrication of capacitor components include stereolithography (SLA), projection micro stereo-lithography (PμSL), and selective laser printing (SLP). The photo-curable resin can also be loaded with a second-phase particulate (metal and ceramic) up to a loading of ~60vol%.

**Process considerations and feedstock requirements** Process parameters in vat polymerization technologies include the source energy (W), exposure time (s), part orientation (°), and scanning speed (mm/s). Feedstock parameters include the vat viscosity (mP s), mass loading (vol%), and curing parameters (i.e., sensitivity of the photopolymer to light). Dispersions of functional materials (e.g., ceramics or metals) in photopolymers for the direct fabrication of capacitor components have rarely been reported; however, the fabrication of components for energy storage devices with similar requirements (e.g., porous separator in Li-ion batteries) is available [113].

**Benefits** The principal benefits of vat polymerization technologies for the fabrication of capacitors include:

a) **High resolution** – vat polymerization technologies allow for the fabrication of fine detail (<10 μm) such as conducting ceramic and insulating polymer microchannels for applications in solid electrolytes for energy storage devices [113]. Patterned polymer electrolytes making use of the high spatial resolution of vat polymerization technologies have also been reported [112].
b) **Low cost** – relatively mature optical imaging techniques (e.g., laser or liquid crystal) and good process-ability of functional dispersions may make vat polymerization technologies an economical option for fabricating capacitor components [110].

**Challenges** Challenges associated with vat polymerization technologies for the fabrication of capacitor components include:

a) **Limited range of materials** – the materials available for vat polymerization technologies are limited to the use of curable, thermo-set plastics. The resins may be relatively hazardous due to the use of organic solvents. The high proportion of photo-curable, insulating thermoset required for process-ability may inhibit the fraction of the minority active phase, for example, the electrical conductivity of metal components [111, 114] and fabricated parts often exhibit poor mechanical behavior (e.g., low elongation to failure). The uncured materials must also transmit light to some extent.
b) **Cross-contamination** – the *wet*, vat-based nature of vat polymerization technologies make these inherently unsuitable for MMMT or *hybrid-AM* approaches. Furthermore, an entire vat of material needs to be available when using vat polymerization technologies, which may lead to a low conversion rate of feedstock to component.
c) **Slow and complex curing** – the kinetics of photopolymerization limit the fabrication speed (~10–50 mm/hr) of vat polymerization technologies. Additionally, photopolymerization reactions in dispersions with active/functional material

particulates may be complex so the optimum polymerization approach is bespoke and developed by trial-and-error [82].

d) **Post-processing** – to achieve the desired material properties, post-processing in the form of heat treatment or extended photocuring (e.g. >6 hr) is often required. For parts fabricated with suspended ceramic or metal particles, additional debinding and sintering steps may be required [111]. A complicated, multistep fabrication process based on holographic patterning has been reported for electrodes with energy storage applications [115].

**Conclusion** Vat polymerization technologies are not common for the fabrication of capacitor components. However, the high resolution, controllable porosity, and range of functional dispersions for vat polymerization AM technologies may make PμSL, DLP, and related processes suitable for fabricating electrolyte/separator [113] and dielectric components. However, the vat-based nature and post-processing requirements of such technologies are inherently incompatible with MMMT or *hybrid-AM* approaches and as such, single-operation fabrication of complete capacitor devices may remain unfeasible for vat polymerization-based approaches. Furthermore, vat polymerization technologies are among the slowest (<50 mm/hr) of all additive techniques.

#### 9.4.1.4 Powder Bed Fusion

**Operating principle** In powder bed fusion AM technologies, thermal energy selectively fuses (solid/liquid-state sintering or partial/full-melting) regions of a closely packed powder bed (10–100 μm ceramic, metal, or polymer powders) [88]. Post-processing including detailing, coating, sintering, surface finishing, and infiltration is often required [82]. Technologies reported for the use in fabricating capacitor components include selective laser melting (SLM), selective laser sintering (SLS), direct metal laser sintering (DMLS), and electron beam melting (EBM). Commercial powder bed fusion machines are available (e.g., EOS or 3D systems).

**Process considerations and feedstock requirements** Process parameters for powder bed fusion technologies include fusing energy ($W/mm^{-3}$), spot size and scan spacing (Ø mm), scan speed (mm/s), and pre-heating conditions (°C). The heat distribution and thermal control of printed components needs to be considered to reduce thermal effects such as distortion due to inhomogeneous thermal expansion/contraction. Feedstock parameters include powder size distribution (Ø mm), packing density (vol%), and powder morphology.

**Benefits** The principal benefits of powder bed fusion technologies for the fabrication of capacitors include:

a) **Design freedom and speed** – throughout fabrication, a part is supported by the surrounding powder bed and there is generally no need for substantial support structures. Complex, 3D helical SS current collectors [117] and interdigitated Ti electrodes [118] for applications in electrochemical capacitors have been reported. Powder bed technologies are relatively fast and span a wide range of printing speeds (~0.005–0.5 kg/hr [156]).

b) **Binder free** – performance-hindering additives such as polymeric binders or support structures frequently found in other AM (e.g., IJP or FFF) processes may not be required for powder bed fusion technologies. The high *purity* of fabricated devices may promote superior performance [118], for example, ceramic dielectrics fabricated by SLS [116].
c) **Porous structures** – a powder bed with relatively low packing density may provide an opportunity to fabricate controlled porous electrode structures. Although not yet demonstrated for capacitors, porous electrodes for tooling applications have been reported [123].
d) **Mechanical properties** – components fabricated by powder bed fusion technologies may exhibit good mechanical properties approaching those of fully dense, conventionally processed equivalents. This may be useful for the fabrication of structural electrodes for capacitors [116].

**Challenges** Challenges associated with the fabrication of capacitor components using powder bed fusion technologies include:

a) **Defects** – lack of consistent fusing of powders is a common and enduring problem, and defects or voids in the interior of bulk materials are difficult to identify and control. They arise from variability of the *green powder* density in each coated layer of powder. In some cases, overheating is also a problem leading to excessive flow of liquid followed by shrinkage-induced pores and residual stresses. The high specific surface area of powders can promote oxide and other contaminations when using reactive powders.
b) **Thermal effects** – to fuse/sinter together particles, in particular ceramics or metal, high temperatures are required inducing strong thermal gradients and associated stresses. Furthermore, the high temperatures required to sinter metal or ceramic powders may prevent powder bed fusion technologies from being useful for MMMT or *hybrid-AM* approaches.
c) **Hazardous powders** – the fine powder (10–100 μm) used in powder bed fusion technologies may pose H&S concerns (airborne nano-particulates generated during aggressive heating/evaporation under the laser or electron-beam).
d) **Limited spatial resolution** – the spatial resolution (>100 μm) of powder bed fusion technologies is limited by the particle size of the powder used and cannot approach that of other AM technologies such as SLA [120]. The use of very small particles (Ø < 50 μm) may make spreading the powder between layers difficult due to static forces [120]. The achievable polarization separation planes or relatively low spatial resolution of other capacitor features may inhibit overall performance. The resolution of, for example, polymer components using

powder bed fusion technologies is often limited and may not be sufficient for capacitor component requirements (e.g., a leak-proof housing/encasing) [121].

e) **Post-processing** – to improve the spatial resolution and surface finish of parts fabricated, subtractive post-processing steps are often necessary.

**Conclusion** Powder bed fusion technologies are not common for the fabrication of capacitor components. However, the absence of binders or fugitive solvents/carriers and, for example, the associated good electrical conductivity of metal components fabricated by powder bed fusion technologies may make powder bed approaches attractive for fabricating capacitor current collectors. Nevertheless, the poor spatial resolution and post-processing requirements significantly limit the potential of powder bed fusion technologies for the fabrication of other capacitor components.

#### 9.4.1.5 Material Jetting

**Operating principle** In material jetting AM technologies, droplets of a functional ink (a suspension of active/functional particulates + fugitive solvent or carrier) are selectively deposited and cured (e.g., thermal, UV) on a build plate to fabricate 3D components layer-by-layer [88]. Among the most commonly used AM technologies used for fabricating capacitor components are IJP processes. Modified commercial IJP printers (e.g., from Stratasys, Fujifilm Dimatix) are used for lab-scale exploration of novel device fabrication.

**Process considerations and feedstock requirements** Material jetting process parameters include jetting velocity (m/s) and droplet size (μm). To prevent nozzle-clogging, material jetting feedstock properties are stringent and include mass loading (vol%), dynamic viscosity (e.g., <20 μPa s for IJP [166]), surface tension (e.g., <80 μN/m for IJP [167]), and wetting properties of the ink on the chosen substrate [161, 162]. The required flow behavior of the ink and associated rheological tuning may limit the options for optimizing the ink chemistry [48]. Similar to DIW processes, the wetting properties of the substrate may play a critical role in the quality of material jetted parts [161, 162].

**Benefits** Principal benefits of material jetting technologies for the fabrication of capacitors include:

a) **High resolution** – the micro-dispensing of functional inks, for example, in IJP processes, can allow for the fabrication of high-resolution components (~30 μm), with very thin layer heights (<15 μm) that are attractive for applications in capacitor fabrication [135].

b) **Versatility** – a wide range of functional inks has already been developed for use in material jetting technologies. The fabrication of dielectrics ($BaTiO_3$-based [124, 125], polymeric [126, 127], BN-based [128, 129]), current collectors (Ag-based [124, 125, 133], graphene-based [128, 129]), electrolyte/separator (PVA hydrogel [135], solid-state [136]), and porous electrodes (based on graphene [139], CNTs [135, 136, 140, 141] and activated C [142]) have all been demonstrated.

c) **Mature printer technology** – material jetting machines are readily available commercially and can be modified for R & D. The capital and operational costs associated with material jetting technologies are relatively low.

**Challenges** Challenges associated with the fabrication of capacitor components using material jetting techniques include:

a) **Clogging of nozzles** – process instability associated with agglomerates in functionally loaded inks may lead to nozzle clogging in material jetting processes. To prevent clogging, electrochemical performance-hindering binders or surfactants may be used and/or functional inks are diluted significantly. It should be noted that the fugitive solvents used are often hazardous and expensive.
b) **Post-processing** – some functional inks require thermal post-processing, adding an additional step to the manufacturing process.
c) **Throughput** – relatively low material throughput due to diluted inks (~1,000 [168] or 0.5 cm/hr [169]).

**Conclusion** Among AM processes reported for the fabrication of capacitor components, material jetting technologies and in particular IJP are common. The versatile functional inks, high resolution, and low cost of mature material jetting printers make material jetting technologies good candidates for fabricating all capacitor components (dielectric, current collector, housing, electrolyte/separator, and porous electrode). However, the low throughput of jetting suggests it will be restricted to certain, high value and low mass fraction parts of the assembly such as the current collectors.

#### 9.4.1.6 Binder Jetting

**Operating principle** In binder jetting AM technologies, a liquid bonding agent selectively binds regions of a tightly packed powder bed (typically sand, gypsum, or metal) to fabricate 3D components layer-by-layer [88]. Binder jetting processes may require post-processing steps such as thermal or UV curing, debinding, and sintering as well as multiple steps of infiltration to achieve dense parts. Binder jetting technologies usually involve a dispensing system similar to IJP, with a bed of powder similar to powder bed fusion technologies. Commercial systems are available (e.g., ExOne).

**Process considerations and feedstock requirements** Binder jetting process considerations are similar to material jetting and powder bed considerations and include jetting velocity (m/s), droplet size (μm), and sintering profile (for parts requiring post-processing). There are a variety of excellent binders already developed, most notably furan (does not require heat to fully cure), phenolic (for high-temperature applications), silicate (environmentally friendly), and aqueous (easy to burn out). The rheological properties and wetting properties of binding agents in relation to the powder bed are key considerations.

**Benefits** Principal benefits of binder jetting technologies for the fabrication of capacitors include:

a) **Geometric design freedom** – binder jetting technologies do not rely on a support structure for more complex geometries. The fabrication of crack-free thick-electrodes based on GO for applications in electrochemical capacitors has been reported [147].
b) **Room temperature** – most binder jetting processes allow for the bulk powder to be at room temperature, preventing, for example, *warping* as may be induced in FFF or SLM processes and *curling* in SLA processes [88].
c) **Economical** – relative to AM processes such as SLS, SLM, or DMLS, binder jetting technologies may be relatively economical due to the achievable throughput (among the faster additive techniques ~0.01 kg/hr [156] or 1 cm/hr [169]). Furthermore, binder jetting technologies allow for the fabrication of relatively large parts (>50 cm in all dimensions) and so scalability is to some extent already demonstrated.

**Challenges** Challenges associated with the fabrication of capacitor components using binder jetting technologies include:

a) **Post-processing** – the poor mechanical properties of so-called *green-state* parts (as printed) may require multiple post-processing steps including UV or thermal curing. Some binders require baking (e.g., 1,600°C for 16 hr) to fully cure and the parts may suffer from anisotropic shrinking [145]. Any post-processing steps reduce the ease of hybridizing the technology.
b) **Powder bed** – H&S concerns regarding the fine particulates in the powder bed may restrict development. Furthermore, the size of the particles in the powder bed limit the spatial resolution (~80 μm) and associated polarization plane separation distance [146].

**Conclusions** Binder jetting technologies are uncommon for fabricating capacitor components, primarily due to the restrictions on hybridization and poor spatial resolution inherent to powder systems. They are however relatively mature and productive, and further development may enhance their applicability to capacitor manufacture.

### 9.4.2 Multi-technology or Hybrid Printing

**Overview** Capacitors are multi-material devices and as such, developing a MMMT or *hybrid-AM* approach may be desirable and arguably necessary to fabricate complete devices. MMMT concepts, machines, and processes are emerging and have been reviewed for general applications elsewhere [170, 171]. Hybrid approaches allow each component to be fabricated in a preferred, optimized way, and even for entire devices to be fabricated in a single operation [69]. Critically, they rely on a good understanding and careful study of how different AM processes interact, especially the interactions of the different materials, for example, at interfaces. General requirements for MMMT processes include (a) materials should form good interparticle bonds, once deposited, without the need for additional compression such as calendering (rolling), (b) fabrication routes should not require any post-processing (e.g., thermal curing) which may degrade some already printed parts of the assembly, and (c) processes should not be vulnerable to contamination of feedstock through the introduction of a multi-material component (e.g., vat-based systems). It follows that, in general, powder-bed or vat systems are intrinsically/systematically unsuited for MMMT approaches as they are mono-material, and that nozzle/dispensing technologies such as MEAM or material jetting processes are suitable. Challenges in MMMT approaches are often associated with the sequential, layer-by-layer, deposition of compositionally discrete layers, and ensuring the right behavior/nature of the associated material interfaces.

### 9.4.3 Complete Capacitor Devices Fabricated by Additive Manufacturing

**Capacitor devices** First attempts to fabricate complete capacitor devices using only *additive* technologies have been made. Figure 9.2 provides an overview of capacitors (electrolytic and electrochemical) that have been fabricated exclusively using *additive* techniques such as IJP, DIW, FFF, and direct laser write (DLW). Given the current immature research-scape of MMMT AM processes, the performance of reported capacitor devices is surprisingly impressive and promising, and there are many obvious routes for optimization to be explored. Currently the fabrication of complete capacitor devices is principally restricted to proof-of-concept work, and devices suffer from poor long-term cycling performance as well as the need for additional encapsulation (to prevent electrolyte from leaking) [136].

**Geometric form factor and graded devices** A variety of geometric form factors, with different objectives (e.g., maximum volumetric or areal performance), have been proposed. Figure 9.3 provides a schematic overview of common configurations. These include in-plane (e.g., interdigitated electrodes [173–175]) and parallel-plate (e.g., multilayer [90] or interdigitated [118]). The exploration of novel geometric form factors, in both experimental work and simulations, as well as the benefits of functionally graded electrodes remains in their infancy, but

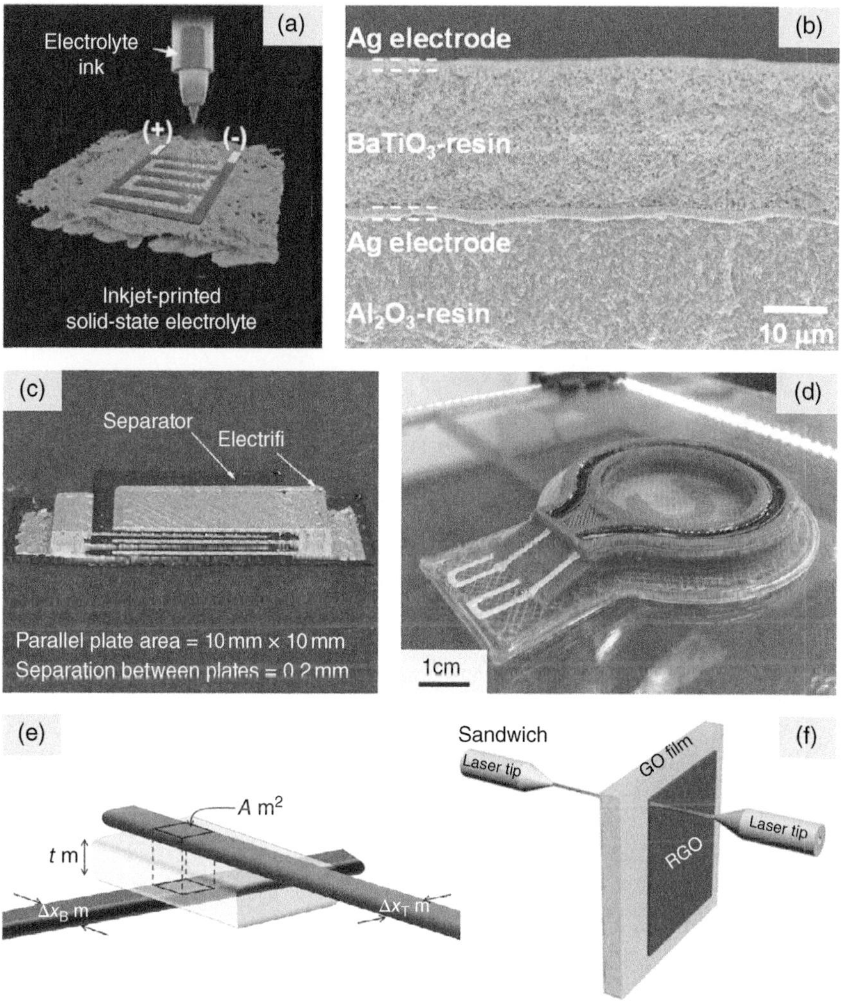


**Figure 9.2** Complete capacitor devices fabricated only by *additive* methods: (a) interdigitated solid-state electrolyte capacitor – IJP. Source: From Choi et al. [136]. © 2016 Royal Society of Chemistry; (b) thin film capacitor – IJP. Source: Form Lim et al. [125]. © 2012 Elsevier; (c) stacked non-polarized capacitor – FFF. Source: From Flowers et al. [90]. © 2017 Elsevier; (d) ring-shaped electrochemical capacitor – FFF and DIW Source: From Fieber et al. [69]. © 2019 Elsevier; (e) 2D material thin film capacitor – IJP. Source: From Worsley et al. [129]. © 2019 American Chemical Society; and (f) hydrated GO micro-supercapacitor – DLW. Source: From Gao et al. [172]. © 2011 Springer Nature.

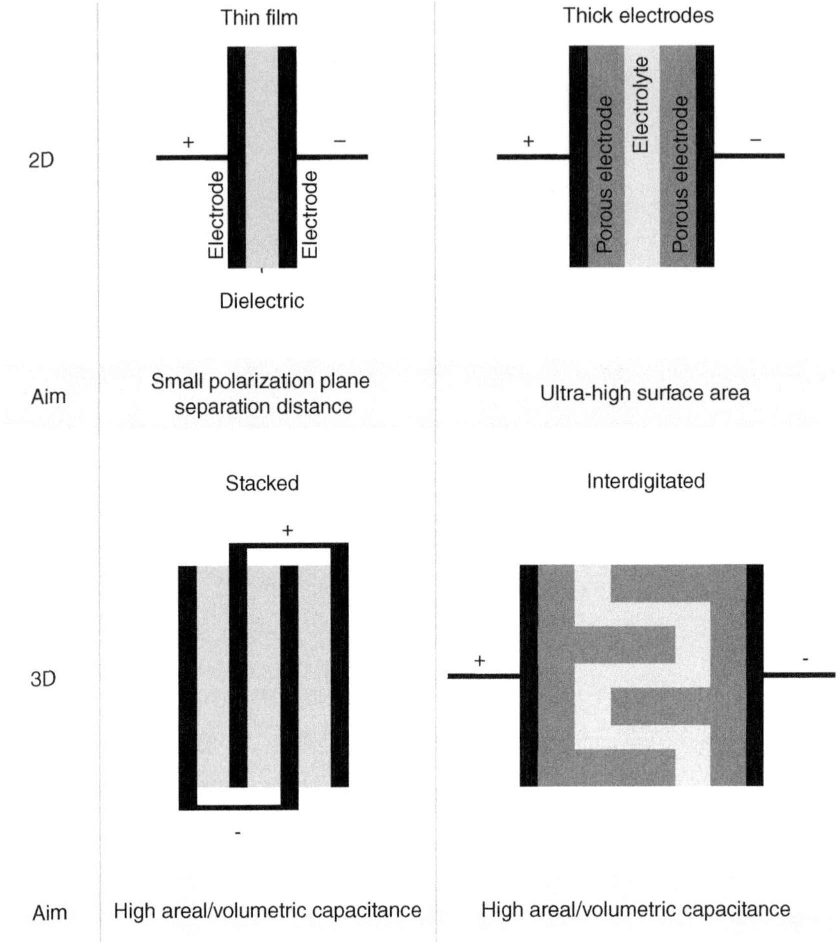


**Figure 9.3** Schematic overview of capacitor configurations in 2D and next-generation 3D configurations.

activity is increasing. Promising results indicating the benefits of functionally graded electrochemical energy storage devices have been reported [77, 141].

## 9.5 Summary and Outlook

**Summary** A wide range of capacitors (electrostatic, electrolytic, and electrochemical) have been developed to temporarily store electrical energy in an electrostatic and/or pseudo-capacitive manner (Section 9.2). Technological advances have tended to focus on the development and incorporation of high-performance materials (e.g., hierarchical activated carbons and nano-materials) rather than new manufacturing approaches to capacitors. While the full range of potential benefits of emerging AM technologies for the fabrication of capacitors (relative to traditional high-volume *slurry casting* approaches for EDLCs or polymer films for electrostatic capacitors) is still unclear, *additive* techniques may offer benefits including (Section 9.3) (a) increased geometric design freedom, (b) nano- to macro-scale control of material properties, (c) the ability to fabricate functionally gradient devices, (d) direct manufacturing integration, (e) cost reduction, and (f) process flexibility. Research interest in AM of energy storage devices and in particular capacitors is increasing rapidly. Reports of capacitor components fabricated by AM technologies, in particular IJP, are common (Table 9.3). Pioneering work principally demonstrating the feasibility and benefits of *additively* fabricating entire capacitor devices by AM have been published (Section 9.4). DIW and IJP technologies have been identified as particularly suitable for the fabrication of capacitor components and devices; however, in practice, the achieved performance when normalized by price and considering scalability issues cannot yet rival current mature mass-production methods. Nevertheless, the speed and ease of prototyping using AM technologies may allow for the development of novel materials arrangements. AM is unlikely to displace conventional manufacturing approaches for mass market capacitors but is likely to gain market share in high-value, low-volume applications where geometric flexibility is valued and only short production runs/low productivity is required, for example, med-tech and defense sectors.

**Outlook** AM of capacitor components (dielectric, current collector, housing, electrolyte/separator, and porous electrode) and capacitor devices is in its infancy. Further research is required in particular in the fields of (a) hybrid-AM – an improved understanding of the interactions of different AM processes to allow for multi-material assemblies in a single operation and potentially the integration of in-operation quality control measures to facilitate *reactive* manufacturing or on-demand repair; (b) new materials – the incorporation of advanced functional materials (e.g., nano-materials) and materials combinations (interaction of surfactants, binders, active material, conductive additives, and electrolytes) to demonstrate good electrical and other performance; (c) applications – flexible, conformal devices for new markets and applications (e.g., wearable electronics) may be developed; and (d) functional gradients – the exploration of functionally graded devices may allow for improved absolute performance [77, 82].

## Acronyms

**1D** one dimensional
**2D** two dimensional

| | |
|---|---|
| **3D** | three dimensional |
| **ABS** | acrylonitrile butadiene styrene |
| **Ag** | silver |
| **Al** | aluminum |
| **AM** | additive manufacture |
| **ASTM** | American Society for Testing and Materials |
| **$BaTiO_3$** | barium titanate |
| **BN** | boron nitride |
| **C** | carbon |
| **CCCD** | constant current charge/discharge |
| **CNF** | carbon nanofiber |
| **CNT** | carbon nanotube |
| **CV** | cyclic voltammetry |
| **DIW** | direct ink writing |
| **DLP** | digital light processing |
| **DLW** | direct laser write |
| **DMLS** | direct metal laser sintering |
| **e-3D** | embedded 3D printing |
| **EBM** | electron beam melting |
| **EDL** | electric double layer |
| **EDLC** | electrochemical double layer capacitor |
| **EIS** | electrochemical impedance spectroscopy |
| **ESR** | electrical series resistance |
| **ESS** | energy storage solution |
| **ESW** | electrochemical stability window |
| **$Fe_3O_4$** | iron oxide |
| **FFF** | fused filament fabrication |
| **GO** | graphene oxide |
| **H&S** | health and safety |
| **HL** | holographic lithography |
| **IEC** | International Electrotechnical Commission |
| **IHP** | inner Helmholtz plane |
| **IJP** | inkjet printing |
| **IPA** | isopropyl alcohol |
| **ISO** | International Organization for Standardization |
| **kb-DSS** | knowledge-based decision support system |
| **$LiFePO_4$** | lithium ferrophosphate |
| **Li-ion** | lithium ions |
| **MEAM** | material extrusion additive manufacture |
| **MMMT** | multi-material, multi-technology |
| **Ni** | nickel |

| | |
|---|---|
| **NMP** | *N*-Methyl-2-pyrrolidone |
| **PC** | polycarbonate |
| **PEEK** | polyether ether ketone |
| **PEMA** | printed electrode membrane assembly |
| **PET** | polyethylene terephthalate |
| **PLA** | polylactic acid |
| **PP** | polypropylene |
| **PTFE** | polytetrafluoroethylene |
| **PVA** | polyvinyl alcohol |
| **PVDF** | polyvinylidene fluoride |
| **PμSL** | projection micro stereo-lithography |
| **R & D** | research and development |
| **SLA** | stereolithography |
| **SLM** | selective laser melting |
| **SLP** | selective laser printing |
| **SLS** | selective laser sintering |
| **SS** | stainless steel |
| **Ti** | titanium |
| **UV** | ultraviolet |
| **Zn** | zinc |

## References


1 International Energy Agency. (2017). *Global energy and $CO_2$ status report 2017*. Tech. Rep. International Energy Agency. Retrieved from https://www.iea.org/publications/freepublications/publication/GECO2017.pdf.

2 Enerdata. (2018). *Global energy trends, 2018 edition. A step backward for the energy transition?* Tech. Rep. Enerdata – Intelligence + Consulting. Retrieved from https://www.enerdata.net/publications/reports-presentations/2018-world-energy-trends-projections.html.

3 Ham, M., Strossmayer, J. J., Mrčela, D., & Horvat, M. (2016). Insights for measuring environmental awareness. *Ekonomski Vjesnik Econviews*, 159–176.

4 Klein, N. (2014). *This changes everything*. Simon & Schuster.

5 Aneke, M., & Wang, M. (2016). Energy storage technologies and real life applications A state of the art review. *Applied Energy*, 179, 350–377. doi:10.1016/j.apenergy.2016.06.097

6 Dincer, I., & Acar, C. (2015). A review on clean energy solutions for better sustainability. *International Journal of Energy Research*, 39(5), 585–606. doi:10.1002/er.3329

7 Mohtasham, J. (2015). Review article-renewable energies. *Energy Procedia*, 74, 1289–1297. doi:10.1016/j.egypro.2015.07.774

8 JEITA. (2017). *Production forecasts for the global electronics and information technology industries*. Tech. Rep. Japan Electronics and Information Technology Industries Association. Retrieved from https://www.jeita.or.jp/english/topics/2017/1219{_}en.pdf.

9 Zhu, C., Liu, T., Qian, F., Chen, W., Chandrasekaran, S., Yao, B., . . . Li, Y. (2017). 3D printed functional nanomaterials for electrochemical energy storage. *Nano Today*, 1–14. doi:10.1016/j.nantod.2017.06.007

10 Tan, Q., Irwin, P., & Cao, Y. (2006). Advanced dielectrics for capacitors. *IEEJ Transactions on Fundamentals and Materials*, 126(11), 1153–1159. doi:10.1541/ieejfms.126.1153

11 Huan, T. D., Boggs, S., Teyssedre, G., Laurent, C., Cakmak, M., Kumar, S., & Ramprasad, R. (2016). Advanced polymeric dielectrics for high energy density applications. *Progress in Materials Science*, 83, 236–269. doi:10.1016/j.pmatsci.2016.05.001

12 Zhong, C., Deng, Y., Hu, W., Qiao, J., Zhang, L., & Zhang, J. (2015). A review of electrolyte materials and compositions for electrochemical supercapacitors. *Chemical Society Reviews*, 44(21), 7484–7539. doi:10.1039/C5CS00303B

13 Beguin, F. (2013). *Supercapacitors*. Weinheim: Wiley. doi:10.1002/9783527646661

14 Zhang, Y., Feng, H., Wu, X., Wang, L., Zhang, A., Xia, T., . . . Zhang, L. (2009). Progress of electrochemical capacitor electrode materials: A review. *International Journal of Hydrogen Energy*, 34(11), 4889–4899. doi:10.1016/j.ijhydene.2009.04.005

15 Ambrosi, A., & Pumera, M. (2016). 3D-printing technologies for electrochemical applications. *Chemical Society Reviews*, 45(10), 2740–2755. doi:10.1039/C5CS00714C

16 Tian, X., Jin, J., Yuan, S., Chua, C. K., Tor, S. B., & Zhou, K. (2017). Emerging 3D-printed electrochemical energy storage devices: A critical review. *Advanced Energy Materials*, 7(17), 1700127. doi:10.1002/aenm.201700127

17 Zhang, F., Wei, M., Viswanathan, V. V., Swart, B., Shao, Y., Wu, G., & Zhou, C. (2017). 3D printing technologies for electrochemical energy storage. *Nano Energy*, 40(May), 418–431. doi:10.1016/j.nanoen.2017. 08.037

18 Chang, P., Mei, H., Zhou, S., Dassios, K. G., & Cheng, L. (2019). 3D printed electrochemical energy storage devices. *Journal of Materials Chemistry A*, 7(9), 4230–4258. doi:10.1039/C8TA11860D

19 Ho, J., Jow, T., & Boggs, S. (2010). Historical introduction to capacitor technology. *IEEE Electrical Insulation Magazine*, 26(1), 20–25. arXiv:arXiv:1011.1669v3. doi:10.1109/MEI.2010.5383924

20 Miller, J. (2007). A brief history of supercapacitors. *Battery + Energy Technologies*, 21(52), 61–78.

**21** Conway, B. E. (1999). *Electrochemical supercapacitors, scientific fundamentals and technological applications.* New York: Kluwer Academic, Plenum Publishers.

**22** Murata. (2019). *How multilayer ceramic capacitors are made.* Retrieved from https://www.murata.com/en-eu/products/emiconfun/capacitor/2011/06/28/en-20110628-p1.

**23** Kahn, M. (1990). *Multilayer ceramic capacitors – Materials and Manufacture* (Tech. Rep. S-MLCMM00M900-R). Myrtle Beach, SC: Mlc, AVX Corporation.

**24** Choi, T., Kaneshiro, R., Brodersen, R., Gray, P., Jett, W., & Wilcox, M. (1983). High-frequency CMOS switched-capacitor filters for communications application. *IEEE Journal of Solid-State Circuits*, 18(6), 652–664. doi:10.1109/JSSC.1983.1052015

**25** Panasonic. (2014). *Production of aluminum electrolytic capacitors.* Tech. Rep. Panasonic. Retrieved from https://web.archive.org/web/20141214154719/http://industrial.panasonic.com/www-data/pdf/ABA0000/ABA0000TE3.pdf

**26** TDK. (2016). *Aluminum electrolytic capacitors: Quality and environment* (Tech. Rep.). TDK. Retrieved from https://www.tdk-electronics.tdk.com/download/530714/6542c9c7cf9cd55cf1ab2a2d36b44f51/pdf-qualityandenvironment.pdf

**27** Elna. (2019). *Capacitors – Fabrication.* Retrieved from http://www.elna.co.jp/en/capacitor/alumi/manufacture.html.

**28** Kemet. (2019). *How we make capacitors.* Retrieved from https://ec.kemet.com/how-we-make-capacitors.

**29** Wang, Y., Song, Y., & Xia, Y. (2016). Electrochemical capacitors: mechanism, materials, systems, characterization and applications. *Chemical Society Reviews*, 45(21), 5925–5950. doi:10.1039/C5CS00580A

**30** Forse, A. C., Merlet, C., Griffin, J. M., & Grey, C. P. (2016). New perspectives on the charging mechanisms of supercapacitors. *Journal of the American Chemical Society*, 138(18), 5731–5744. doi:10.1021/jacs. 6b02115

**31** Sharma, P., & Bhatti, T. (2010). A review on electrochemical double-layer capacitors. *Energy Conversion and Management*, 51(12), 2901–2912. doi:10.1016/j.enconman.2010.06.031

**32** González, A., Goikolea, E., Barrena, J. A., & Mysyk, R. (2016). Review on supercapacitors: Technologies and materials. *Renewable and Sustainable Energy Reviews*, 58, 1189–1206. doi:10.1016/j.rser.2015.12.249

**33** Simon, P., Gogotsi, Y., & Dunn, B. (2014). Where do batteries end and supercapacitors begin? *Science*, 343(6176), 1210–1211. doi:10.1126/science.1249625

**34** Brousse, T., Bélanger, D., & Long, J. W. (2015). To be or not to be pseudocapacitive? *Journal of the Electrochemical Society*, 162(5), A5185–A5189. doi:10.1149/2.0201505jes

**35** Erlichman, I. (1976). Method for manufacturing flat batteries. *U.S. Patent No. 3993508A.*

**36** Alford, J. A. & Creek, G. (1997). High power density double layer energy storage devices. *U.S. Patent No. 5926361A*.

**37** Ishida, S., Takenaka, H., NishimuraY, S., Egawa, Y., & Otsuka, K. (1997). Activated carbon and process for production thereof, nonaqueous type polarizable electrodes and electric double-layer capacitors. *European Patent No. 1027716A1*.

**38** Mizobuchi, T., Shinsenji, T., & Yanase, M. (2001). Separator for electrochemical device and method for its production and electrochemical device. *German Patent No. DE10196273T1*.

**39** Tsushima, M., Morimoto, T., Hiratsuka, K., Kawasato, T., & Suhara, M. (1998). Electric double layer capacitor and process for its production. *U.S. Patent No. 6402792B1*.

**40** Yu, Z., Tetard, L., Zhai, L., & Thomas, J. (2015). Supercapacitor electrode materials: Nanostructures from 0 to 3 dimensions. *Energy & Environmental Science*, 8(3), 702–730. doi:10.1039/C4EE03229B

**41** Frackowiak, E. (2007). Carbon materials for supercapacitor application. *Physical Chemistry Chemical Physics*, 9(15), 1774. doi:10.1039/b618139m

**42** Zhang, L. L., Zhou, R., & Zhao, X. S. (2010). Graphene-based materials as supercapacitor electrodes. *Journal of Materials Chemistry*, 20(29), 5983. arXiv:0706.1062v1. doi:10.1039/c000417k

**43** Yuan, C. Z., Gao, B., Shen, L. F., Yang, S. D., Hao, L., Lu, X. J., . . . Zhang, X. G. (2011). Hierarchically structured carbon-based composites: Design, synthesis and their application in electrochemical capacitors. *Nanoscale*, 3(2), 529–545. doi:10.1039/C0NR00423E

**44** Wang, G., Zhang, L., & Zhang, J. (2012). A review of electrode materials for electrochemical supercapacitors. *Chemical Society Reviews*, 41(2), 797–828. arXiv:1005.0853. doi:10.1039/C1CS15060J

**45** Emmerich, J. (2017). *Technical summary: Supercapacitors*. Retrieved from https://www.design-concepts.com/insights/supercapacitors.

**46** Market Research Future. (2019). *Super capacitor market research report – Forecast 2022*. Tech. Rep. Market Research Future, Global. Retrieved from https://www.marketresearchfuture.com/reports/super-capacitor-market-2548.

**47** Yassine, M., & Fabris, D. (2017). Performance of commercially available supercapacitors. *Energies*, 10(9), 1340. doi:10.3390/en10091340

**48** Choi, K.-H., Ahn, D. B., & Lee, S.-Y. (2018). Current status and challenges in printed batteries: Toward form factor-free, monolithic integrated power sources. *ACS Energy Letters*, 3(1), 220–236. doi:10.1021/acsenergylett.7b01086

**49** Hassan, C. M., & Peppas, N. a. (2000). Structure and applications of poly(vinyl alcohol) hydrogels produced by conventional crosslinking or by freezing/thawing methods. In *Biopolymers · PVA hydrogels, anionic polymerisation nanocomposites* (Vol. 153, pp. 37–65). Berlin Heidelberg: Springer. doi:10.1007/3-540-46414-X_2

**50** Sun, G., An, J., Chua, C. K., Pang, H., Zhang, J., & Chen, P. (2015). Layer-by-layer printing of laminated graphene-based interdigitated microelectrodes for flexible planar micro-supercapacitors. *Electrochemistry Communications*, 51, 33–36. doi:10.1016/j.elecom.2014.11.023

**51** Kovalska, E., & Kocabas, C. (2016). Organic electrolytes for graphene-based supercapacitor: Liquid, gel or solid. *Materials Today Communications*, 7, 155–160. doi:10.1016/j.mtcomm.2016.04.013

**52** Mousavi, M. P. S., Wilson, B. E., Kashefolgheta, S., Anderson, E. L., He, S., Bühlmann, P., & Stein, A. (2016). Ionic liquids as electrolytes for electrochemical double-layer capacitors: Structures that optimize specific energy. *ACS Applied Materials & Interfaces*, 8(5), 3396–3406. doi:10.1021/acsami.5b11353

**53** Choudhury, N. A., Sampath, S., & Shukla, A. K. (2009). Hydrogel-polymer electrolytes for electrochemical capacitors: An overview. *Energy & Environmental Science*, 2(1), 55–67. doi:10.1039/B811217G

**54** Blake, A. J., Kohlmeyer, R. R., Hardin, J. O., Carmona, E. A., Maruyama, B., Berrigan, J. D., . . . Durstock, M. F. (2017). 3D printable ceramic-polymer electrolytes for flexible high-performance Li-ion batteries with enhanced thermal stability. *Advanced Energy Materials*, 7(14), 1602920. doi:10.1002/aenm.201602920

**55** Tõnurist, K., Thomberg, T., Jänes, A., Romann, T., Sammelselg, V., & Lust, E. (2013). Influence of separator properties on electrochemical performance of electrical double layer capacitors. *Journal of Electroanalytical Chemistry*, 689, 8–20. doi:10.1016/j.jelechem.2012.11.024

**56** Zhu, Z. (2016). Effects of various binders on supercapacitor performances. *International Journal of Electrochemical Science*, 11(10), 8270–8279. doi:10.20964/2016.10.04

**57** Maarof, H. I., Wan Daud, W. M. A., & Aroua, M. K. (2017). Effect of varying the amount of binder on the electrochemical characteristics of palm shell activated carbon. *IOP Conference Series: Materials Science and Engineering*, 210(1), 012011. doi:10.1088/1757-899X/210/1/012011

**58** Zhang, S., & Pan, N. (2015). Supercapacitors performance evaluation. *Advanced Energy Materials*, 5(6), 1401401. doi:10.1002/aenm.201401401

**59** Burke, A. (2009). *Testing of supercapacitors: Capacitance, resistance, and energy energy and power capacity*. Proceedings of ISEE'Cap09, Nantes, France, 1–42.

**60** Stoller, M. D., & Ruoff, R. S. (2010). Best practice methods for determining an electrode material's performance for ultracapacitors. *Energy & Environmental Science*, 3(9), 1294. arXiv:1005.0805. doi:10.1039/c0ee00074d

**61** Stryzhakova, D. (2012). *SC testing methodology manual*. Tech. Rep. IAPP Energy Caps. Retrieved from http://www.energycaps.eu/wp-content/uploads/2013/09/Testing-Methodologies-Manual.pdf.

**62** Allagui, A., Freeborn, T. J., Elwakil, A. S., & Maundy, B. J. (2016). Reevaluation of performance of electric double-layer capacitors from constant-current charge/discharge and cyclic voltammetry. *Scientific Reports*, 6(1), 38568. doi:10.1038/srep38568

**63** Balducci, A., Belanger, D., Brousse, T., Long, J. W., & Sugimoto, W. (2017). Perspective – A guideline for reporting performance metrics with electrochemical capacitors: From electrode materials to full devices. *Journal of the Electrochemical Society*, 164(7), A1487–A1488. doi:10.1149/2.0851707jes

**64** Mei, B.-A., Munteshari, O., Lau, J., Dunn, B., & Pilon, L. (2018). Physical interpretations of Nyquist plots for EDLC electrodes and devices. *The Journal of Physical Chemistry C*, 122(1), 194–206. doi:10.1021/acs.jpcc.7b10582

**65** B. S. I. (BSI). (2015). *BSI standards publication fixed resistors for use in electronic equipment part 1: Generic specification* (Tech. Rep. BSI-BS EN 60115-1). BSI.

**66** Zhang, K., Gao, F., Xu, J., Wang, L., Zhang, Q., & Kong, J. (2017). Fabrication and dielectric properties of $Ba_{0.6}Sr_{0.4}TiO_3$ / acrylonitrilebutadienestyrene resin composites. *Journal of Materials Science: Materials in Electronics*, 28(12), 8960–8968. doi:10.1007/s10854-017-6626-y

**67** Chadalavada, H., Samuel Raj, D., Ashok Kumar, R., & Sankar, K. (2015). Production lead time reduction in a battery manufacturing unit using lean manufacturing. *International Journal of Engineering Research & Technology*, 4(04), 842–847.

**68** Bikas, H., Stavropoulos, P., & Chryssolouris, G. (2016). Additive manufacturing methods and modelling approaches: a critical review. *The International Journal of Advanced Manufacturing Technology*, 83(1–4), 389–405. doi:10.1007/s00170-015-7576-2

**69** Fieber, L., Evans, J. D., Huang, C., & Grant, P. S. (2019). Single-operation, multi-phase additive manufacture of electro-chemical double layer capacitor devices. *Additive Manufacturing*, 28, 344–353.

**70** Zhang, J., & Zhao, X. S. (2012). On the configuration of supercapacitors for maximizing electrochemical performance. *ChemSusChem*, 5(5), 818–841. doi:10.1002/cssc.201100571

**71** Ariga, K., Yamauchi, Y., Rydzek, G., Ji, Q., Yonamine, Y., Wu, K. C.-W., & Hill, J. P. (2014). Layer-by-layer nanoarchitectonics: Invention, innovation, and evolution. *Chemistry Letters*, 43(1), 36–68. doi:10.1246/cl.130987

**72** Xu, Q., Lv, Y., Dong, C., Sreeprased, T. S., Tian, A., Zhang, H., . . . Li, N. (2015). Three-dimensional micro/nanoscale architectures: Fabrication and applications. *Nanoscale*, 7(25), 10883–10895. doi:10.1039/C5NR02048D

**73** Raney, J. R., & Lewis, J. A. (2015). Printing mesoscale architectures. *MRS Bulletin*, 40(11), 943–950. doi:10.1557/mrs.2015.235

**74** Chiu, W., & Yu, K. (2008). Direct digital manufacturing of three-dimensional functionally graded material objects. *Computer-Aided Design*, 40(12), 1080–1093. doi:10.1016/j.cad.2008.10.002

**75** Li, Y., Fu, Z.-Y., & Su, B.-l. (2012). Hierarchically structured porous materials for energy conversion and storage. *Advanced Functional Materials*, 22(22), 4634–4667. doi:10.1002/adfm.201200591

**76** Qi, Y., Jang, T., Ramadesigan, V., Schwartz, D. T., & Subramanian, V. R. (2017). Is there a benefit in employing graded electrodes for lithium-ion batteries? *Journal of the Electrochemical Society*, 164(13), A3196–A3207. doi:10.1149/2.1051713jes

**77** Cheng, C., Drummond, R., Duncan, S. R., & Grant, P. S. (2019). Micro-scale graded electrodes for improved dynamic and cycling performance of Li-ion batteries. *Journal of Power Sources*, 413, 59–67. doi:10.1016/j.jpowsour.2018.12.021

**78** Kyeremateng, N. A., Brousse, T., & Pech, D. (2016). Microsupercapacitors as miniaturized energy-storage components for on-chip electronics. *Nature Nanotechnology*, 12(1), 7–15. doi:10.1038/nnano.2016.196

**79** Attaran, M. (2017). Additive manufacturing: the most promising technology to alter the supply chain and logistics. *Journal of Service Science and Management*, 10(03), 189–206. doi:10.4236/jssm.2017.103017

**80** Wong, K. V., & Hernandez, A. (2012). A review of additive manufacturing. *ISRN Mechanical Engineering*, 2012, 1–10. arXiv:208760. doi:10.5402/2012/208760

**81** Vaezi, M., Seitz, H., & Yang, S. (2013). A review on 3D micro-additive manufacturing technologies. *The International Journal of Advanced Manufacturing Technology*, 67(5-8), 1721–1754. doi:10.1007/ s00170-012-4605-2

**82** T. D. Ngo, A. Kashani, G. Imbalzano, K. T. Nguyen, D. Hui, Additive manufacturing (3D printing): A review of materials, methods, applications and challenges, *Composites Part B: Engineering* 143 (December 2017) (2018) 172–196. doi:10.1016/j.compositesb.2018.02.012.

**83** Singh, S., Ramakrishna, S., & Singh, R. (2017). Material issues in additive manufacturing: A review. *Journal of Manufacturing Processes*, 25, 185–200. doi:10.1016/j.jmapro.2016.11.006

**84** Bourell, D., Kruth, J. P., Leu, M., Levy, G., Rosen, D., Beese, A. M., & Clare, A. (2017). Materials for additive manufacturing. *CIRP Annals*, 66(2), 659–681. doi:10.1016/j.cirp.2017.05.009

**85** Klahn, C., Leutenecker, B., & Meboldt, M. (2015). Design strategies for the process of additive manufacturing. *Procedia CIRP*, 36, 230–235. doi:10.1016/j.procir.2015.01.082

**86** Wang, Y., Blache, R., & Xu, X. (2017). Selection of additive manufacturing processes. *Rapid Prototyping Journal*, 23(2), 434–447. doi:10.1108/RPJ-09-2015-0123

**87** Z.-X. Low, Y. T. Chua, B. M. Ray, D. Mattia, I. S. Metcalfe, D. A. Patterson, Perspective on 3D printing of separation membranes and comparison to related unconventional fabrication techniques, *Journal of Membrane Science* 523 (October 2016) (2017) 596–613. doi:10.1016/j. memsci.2016.10.006.

**88** Redwood, B., Schöffer, F., Garret, B., Fadell, T., & Hubs, D. (2018). *The 3D printing handbook: Technologies, design and applications*. 3D Hubs. Retrieved from https://books.google.co.uk/books?id=R3OvswEACAAJ

**89** ASTM International. (2013). F2792-12a – Standard terminology for additive manufacturing technologies. Rapid Manufacturing Association (2013), 10–12. arXiv:1011.1669v3. doi:10.1520/F2792-12A.

**90** P. F. Flowers, C. Reyes, S. Ye, M. J. Kim, B. J. Wiley, 3D printing electronic components and circuits with conductive thermoplastic filament, *Additive Manufacturing* 18 (2017) (2017) 156–163. doi:10.1016/j.addma.2017.10.002.

**91** Dichtl, C., Sippel, P., & Krohns, S. (2017). Dielectric properties of 3D printed polylactic acid. *Advances in Materials Science and Engineering*, 2017, 1–10. doi:10.1155/2017/6913835

**92** Wu, Y., Isakov, D., & Grant, P. (2017). Fabrication of composite filaments with high dielectric permittivity for fused deposition 3D printing. *Materials*, 10(10), 1218. doi:10.3390/ma10101218

**93** Wei, X., Li, D., Jiang, W., Gu, Z., Wang, X., Zhang, Z., & Sun, Z. (2015). 3D printable graphene composite. *Scientific Reports*, 5(1), 11181. doi:10.1038/srep11181

**94** Zhang, D., Chi, B., Li, B., Gao, Z., Du, Y., Guo, J., & Wei, J. (2016). Fabrication of highly conductive graphene flexible circuits by 3D printing. *Synthetic Metals*, 217, 79–86. doi:10.1016/j.synthmet.2016.03.014

**95** Areir, M., Xu, Y., Zhang, R., Harrison, D., Fyson, J., & Pei, E. (2017). A study of 3D printed active carbon electrode for the manufacture of electric double-layer capacitors. *Journal of Manufacturing Processes*, 25, 351–356. doi:10.1016/j.jmapro.2016.12.020

**96** Tanwilaisiri, A., Xu, Y., Zhang, R., Harrison, D., Fyson, J., & Areir, M. (2018). Design and fabrication of modular supercapacitors using 3D printing. *Journal of Energy Storage*, 16, 1–7. doi:10.1016/j.est.2017.12.020

**97** Carrico, J. D., Traeden, N. W., Aureli, M., & Leang, K. K. (2015). Fused filament 3D printing of ionic polymer-metal composites (IPMCs). *Smart Materials and Structures*, 24(12), 125021. doi:10.1088/0964-1726/24/12/125021

**98** Philamore, H., Rossiter, J., Walters, P., Winfield, J., & Ieropoulos, I. (2015). Cast and 3D printed ion exchange membranes for monolithic microbial fuel cell fabrication. *Journal of Power Sources*, 289, 91–99. doi:10.1016/j.jpowsour.2015.04.113

**99** King, B. H., Morissette, S. L., Denham, H., Cesarano, J., & Dimos, D. (1998, August). Influence of rheology on deposition behavior of ceramic pastes in direct fabrication systems. Solid Freeform Fabrication Proceedings, 391–398.

**100** M. Areir, Y. Xu, D. Harrison, J. Fyson, 3D printing of highly flexible supercapacitor designed for wearable energy storage, *Materials Science and Engineering B* 226 (August) (2017) 29–38. doi:10.1016/j.mseb.2017.09.004.

**101** Muth, J. T., Vogt, D. M., Truby, R. L., Mengüç, Y., Kolesky, D. B., Wood, R. J., & Lewis, J. A. (2014). Embedded 3D printing of strain sensors within highly stretchable elastomers. *Advanced Materials*, 26(36), 6307–6312. arXiv:1111.4970. doi:10.1002/adma.201400334

**102** Ho, C. C., Evans, J. W., & Wright, P. K. (2010). Direct write dispenser printing of a zinc microbattery with an ionic liquid gel electrolyte. *Journal of Micromechanics and Microengineering*, 20(10), 104009. doi:10.1088/0960-1317/20/10/104009

**103** Wei, M., Zhang, F., Wang, W., Alexandridis, P., Zhou, C., & Wu, G. (2017). 3D direct writing fabrication of electrodes for electrochemical storage devices. *Journal of Power Sources*, 354, 134–147. doi:10.1016/j.jpowsour.2017.04.042

**104** Nathan-Walleser, T., Lazar, I.-M., Fabritius, M., Tölle, F. J., Xia, Q., Bruchmann, B., . . . Mülhaupt, R. (2014). 3D micro-extrusion of graphene-based active electrodes: Towards high-rate AC line filtering performance electrochemical capacitors. *Advanced Functional Materials*, 24(29), 4706–4716. doi:10.1002/adfm.201304151

**105** Yang, Z., Chabi, S., Xia, Y., & Zhu, Y. (2015). Preparation of 3D graphene-based architectures and their applications in supercapacitors. *Progress in Natural Science: Materials International*, 25(6), 554–562. doi:10.1016/j.pnsc.2015.11.010

**106** Zhu, C., Liu, T., Qian, F., Han, T. Y.-J., Duoss, E. B., Kuntz, J. D., . . . Li, Y. (2016). Supercapacitors based on three-dimensional hierarchical graphene aerogels with periodic macropores. *Nano Letters*, 16(6), 3448–3456. doi:10.1021/acs.nanolett.5b04965

**107** Yao, B., Chandrasekaran, S., Zhang, J., Xiao, W., Qian, F., Zhu, C., . . . Li, Y. (2019). Efficient 3D printed pseudocapacitive electrodes with ultrahigh $MnO_2$ loading. *Joule*, 3(2), 459–470. doi:10.1016/j.joule.2018.09.020

**108** Lin, Y., Liu, F., Casano, G., Bhavsar, R., Kinloch, I. A., & Derby, B. (2016). Pristine graphene aerogels by room-temperature freeze gelation. *Advanced Materials*, 28(36), 7993–8000. doi:10.1002/adma.201602393

**109** Yu, W., Zhou, H., Li, B. Q., & Ding, S. (2017). 3D printing of carbon nanotubes-based microsupercapacitors. *ACS Applied Materials & Interfaces*, 9(5), 4597–4604. doi:10.1021/acsami.6b13904

**110** Yang, Y., Chen, Z., Song, X., Zhu, B., Hsiai, T., Wu, P.-I., . . . Shung, K. K. (2016). Three dimensional printing of high dielectric capacitor using projection based stereolithography method. *Nano Energy*, 22, 414–421. doi:10.1016/j.nanoen.2016.02.045

**111** Bartolo, P., & Gaspar, J. (2008). Metal filled resin for stereolithography metal part. *CIRP Annals*, 57(1), 235–238. doi:10.1016/j.cirp.2008.03.124

**112** Chen, Q., Xu, R., He, Z., Zhao, K., & Pan, L. (2017). Printing 3D gel polymer electrolyte in lithium-ion microbattery using stereolithography. *Journal of the Electrochemical Society*, 164(9), A1852–A1857. doi:10.1149/2.0651709jes

**113** Zekoll, S., Marriner-Edwards, C., Hekselman, A. K. O., Kasemchainan, J., Kuss, C., Armstrong, D. E. J., . . . Bruce, P. G. (2018). Hybrid electrolytes with 3D bicontinuous ordered ceramic and polymer microchannels for all-solid-state batteries. *Energy & Environmental Science*, 11(1), 185–201. doi:10.1039/C7EE02723K

**114** Cooperstein, I., Layani, M., & Magdassi, S. (2015). 3D printing of porous structures by UV-curable O/W emulsion for fabrication of conductive objects. *Journal of Materials Chemistry C*, 3(9), 2040–2044. doi:10.1039/C4TC02215G

**115** Ning, H., Pikul, J. H., Zhang, R., Li, X., Xu, S., Wang, J., . . . Braun, P. V. (2015). Holographic patterning of high-performance on-chip 3D lithium-ion microbatteries. *Proceedings of the National Academy of Sciences*, 112(21), 6573–6578. doi:10.1073/pnas.1423889112

**116** Sing, S. L., Yeong, W. Y., Wiria, F. E., Tay, B. Y., Zhao, Z., Zhao, L., . . . Yang, S. (2017). Direct selective laser sintering and melting of ceramics: A review. *Rapid Prototyping Journal*, 23(3), 611–623. doi:10.1108/RPJ-11-2015-0178

**117** Ambrosi, A., Moo, J. G. S., & Pumera, M. (2016). Helical 3D-printed metal electrodes as custom-shaped 3D platform for electrochemical devices. *Advanced Functional Materials*, 26(5), 698–703. doi:10.1002/ adfm.201503902

**118** Zhao, C., Wang, C., Gorkin, R., Beirne, S., Shu, K., & Wallace, G. G. (2014). Three dimensional (3D) printed electrodes for interdigitated supercapacitors. *Electrochemistry Communications*, 41, 20–23. doi:10.1016/j.elecom.2014.01.013

**119** Liu, X., Jervis, R., Maher, R. C., Villar-Garcia, I. J., Naylor-Marlow, M., Shearing, P. R., . . . Wu, B. (2016). 3D-printed structural pseudocapacitors. *Advanced Materials Technologies*, 1(9), 1600167. doi:10.1002/admt.201600167

**120** Goodridge, R., Tuck, C., & Hague, R. (2012). Laser sintering of polyamides and other polymers. *Progress in Materials Science*, 57(2), 229–267. doi:10.1016/j.pmatsci.2011.04.001

**121** Schmidt, M., Pohle, D., & Rechtenwald, T. (2007). Selective laser sintering of PEEK. *CIRP Annals*, 56(1), 205–208. doi:10.1016/j.cirp.2007.05.097

**122** Mo, J., Steen, S. M., Zhang, F.-Y., Dehoff, R. R., Peter, W. H., Toops, T. J., & J. B. G. Jr. (2014). *Development of liquid/gas diffusion layers with low-cost additive manufacturing* (Tech. Rep.).

**123** Koyano, T., Hosokawa, A., Igusa, R., & Ueda, T. (2017). Electrochemical machining using porous electrodes fabricated by powder bed fusion additive manufacturing process. *CIRP Annals*, 66(1), 213–216. doi:10.1016/j.cirp.2017.04.127

**124** Kang, B. J., Lee, C. K., & Oh, J. H. (2012). All-inkjet-printed electrical components and circuit fabrication on a plastic substrate. *Microelectronic Engineering*, 97(4023), 251–254. doi:10.1016/j.mee.2012.03.032

**125** Lim, J., Kim, J., Yoon, Y. J., Kim, H., Yoon, H. G., Lee, S.-N., & Kim, J. (2012). All-inkjet-printed metal-insulator-metal (MIM) capacitor. *Current Applied Physics*, 12(Suppl. 1), e14–e17. doi:10.1016/j.cap.2011.04.035

**126** Liu, Y., Cui, T., & Varahramyan, K. (2003). All-polymer capacitor fabricated with inkjet printing technique. *Solid-State Electronics*, 47(9), 1543–1548. doi:10.1016/S0038-1101(03)00082-0

**127** Graddage, N., Chu, T.-Y., Ding, H., Py, C., Dadvand, A., & Tao, Y. (2016). Inkjet printed thin and uniform dielectrics for capacitors and organic thin film transistors enabled by the coffee ring effect. *Organic Electronics*, 29, 114–119. doi:10.1016/j.orgel.2015.11.039

**128** Kelly, A. G., Finn, D., Harvey, A., Hallam, T., & Coleman, J. N. (2016). All-printed capacitors from graphene-BN-graphene nanosheet heterostructures. *Applied Physics Letters*, 109(2), 023107. doi:10.1063/1.4958858

**129** Worsley, R., Pimpolari, L., McManus, D., Ge, N., Ionescu, R., Wittkopf, J. A., . . . Casiraghi, C. (2019). All-2D material inkjet-printed capacitors: Toward fully printed integrated circuits. *ACS Nano*, 13(1), 54–60. doi:10.1021/acsnano.8b06464

**130** Saleh, E., Woolliams, P., Clarke, B., Gregory, A., Greedy, S., Smartt, C., . . . Tuck, C. (2017). 3D inkjet-printed UV-curable inks for multi-functional electromagnetic applications. *Additive Manufacturing*, 13, 143–148. doi:10.1016/j.addma.2016.10.002

**131** Perelaer, J., Smith, P. J., Mager, D., Soltman, D., Volkman, S. K., Subramanian, V., . . . Schubert, U. S. (2010). Printed electronics: the challenges involved in printing devices, interconnects, and contacts based on inorganic materials. *Journal of Materials Chemistry*, 20(39), 8446. doi:10.1039/c0jm00264j

**132** Venkata, K. R. R., Venkata, A. K., Karthik, P. S., & Singh, S. P. (2015). Conductive silver inks and their applications in printed and flexible electronics. *RSC Advances*, 5(95), 77760–77790. doi:10.1039/C5RA12013F

**133** Correia, V., Mitra, K., Castro, H., Rocha, J., Sowade, E., Baumann, R., & Lanceros-Mendez, S. (2018). Design and fabrication of multilayer inkjet-printed passive components for printed electronics circuit development. *Journal of Manufacturing Processes*, 31, 364–371. doi:10.1016/j.jmapro.2017.11.016

**134** MacCurdy, R., Katzschmann, R., Kim, Y., & Rus, D. (2016). *Printable hydraulics: A method for fabricating robots by 3D co-printing solids and liquids*. 2016 IEEE International Conference on Robotics and Automation (ICRA) (pp. 3878–3885). IEEE. arXiv: 1512.03744. doi:10.1109/ICRA.2016.7487576.

**135** Kaempgen, M., Chan, C. K., Ma, J., Cui, Y., & Gruner, G. (2009). Printable thin film supercapacitors using single-walled carbon nanotubes. *Nano Letters*, 9(5), 1872–1876. doi:10.1021/nl8038579

**136** Choi, K.-H., Yoo, J., Lee, C. K., & Lee, S.-Y. (2016). All-inkjet-printed, solid-state flexible supercapacitors on paper. *Energy & Environmental Science*, 9(9), 2812–2821. doi:10.1039/C6EE00966B

**137** Delannoy, P.-E., Riou, B., Lestriez, B., Guyomard, D., Brousse, T., & Le Bideau, J. (2015). Toward fast and cost-effective ink-jet printing of solid electrolyte for lithium microbatteries. *Journal of Power Sources*, 274, 1085–1090. doi:10.1016/j.jpowsour.2014.10.164

**138** Singh, N., Galande, C., Miranda, A., Mathkar, A., Gao, W., Reddy, A. L. M., . . . Ajayan, P. M. (2012). Paintable battery. *Scientific Reports*, 2(1), 481. doi:10.1038/srep00481

**139** Li, J., Ye, F., Vaziri, S., Muhammed, M., Lemme, M. C., & Östling, M. (2013). Efficient inkjet printing of graphene. *Advanced Materials*, 25(29), 3985–3992. doi:10.1002/adma.201300361

**140** Huang, C., & Grant, P. S. (2013). One-step spray processing of high power all-solid-state supercapacitors. *Scientific Reports*, 3(1), 2393. doi:10.1038/srep02393

**141** Huang, C., Zhang, J., Young, N. P., Snaith, H. J., & Grant, P. S. (2016). Solid-state supercapacitors with rationally designed heterogeneous electrodes fabricated by large area spray processing for wearable energy storage applications. *Scientific Reports*, 6(1), 25684. doi:10.1038/srep25684

**142** Pech, D., Brunet, M., Taberna, P.-L., Simon, P., Fabre, N., Mesnilgrente, F., . . . Durou, H. (2010). Elaboration of a microstructured inkjet-printed carbon electrochemical capacitor. *Journal of Power Sources*, 195(4), 1266–1269. doi:10.1016/j.jpowsour.2009.08.085

**143** Delannoy, P.-E., Riou, B., Brousse, T., Le Bideau, J., Guyomard, D., & Lestriez, B. (2015). Ink-jet printed porous composite $LiFePO_4$ electrode from aqueous suspension for microbatteries. *Journal of Power Sources*, 287, 261–268. doi:10.1016/j.jpowsour.2015.04.067

**144** Gaytan, S., Cadena, M., Karim, H., Delfin, D., Lin, Y., Espalin, D., . . . Wicker, R. (2015). Fabrication of barium titanate by binder jetting additive manufacturing technology. *Ceramics International*, 41(5), 6610–6619. doi:10.1016/j.ceramint.2015.01.108

**145** Gonzalez, J., Mireles, J., Lin, Y., & Wicker, R. (2016). Characterization of ceramic components fabricated using binder jetting additive manufacturing technology. *Ceramics International*, 42(9), 10559–10564. doi:10.1016/j.ceramint.2016.03.079

**146** Bai, Y., Williams, C. B. (2014). *An exploration of binder jetting of copper* (Tech. Rep.). doi:10.1108/RPJ-12-2014-0180.

**147** Azhari, A., Marzbanrad, E., Yilman, D., Toyserkani, E., & Pope, M. A. (2017). Binder-jet powder-bed additive manufacturing (3D printing) of thick graphene-based electrodes. *Carbon*, 119, 257–266. doi:10.1016/j.carbon.2017.04.028

**148** Czyżewski, J., Burzyński, P., Gaweł, K., & Meisner, J. (2009). Rapid prototyping of electrically conductive components using 3D printing technology. *Journal of Materials Processing Technology*, 209(12–13), 5281–5285. doi:10.1016/j.jmatprotec.2009.03.015

**149** Turner, B. N., Strong, R., & Gold, S. A. (2014). A review of melt extrusion additive manufacturing processes: I. Process design and modeling. *Rapid Prototyping Journal*, 20(3), 192–204. doi:10.1108/RPJ-01-2013-0012

**150** Mohamed, O. A., Masood, S. H., & Bhowmik, J. L. (2015). Optimization of fused deposition modeling process parameters: A review of current research and future prospects. *Advances in Manufacturing*, 3(1), 42–53. arXiv:1011.1669v3. doi:10.1007/s40436-014-0097-7

**151** Turner, B. N., & Gold, S. A. (2015). A review of melt extrusion additive manufacturing processes: II. Materials, dimensional accuracy, and surface roughness. *Rapid Prototyping Journal*, 21(3), 250–261. arXiv:1011.1669v3. doi:10.1108/RPJ-02-2013-0017

**152** Anitha, R., Arunachalam, S., & Radhakrishnan, P. (2001). Critical parameters influencing the quality of prototypes in fused deposition modelling. *Journal of Materials Processing Technology*, 118(1-3), 385–388. doi:10.1016/S0924-0136(01)00980-3

**153** Ćwikła, G., Grabowik, C., Kalinowski, K., Paprocka, I., & Ociepka, P. (2017). The influence of printing parameters on selected mechanical properties of FDM/FFF 3D-printed parts. *IOP Conference Series: Materials Science and Engineering*, 227(1), 012033. doi:10.1088/1757-899X/227/1/012033

**154** Alafaghani, A., Qattawi, A., Alrawi, B., & Guzman, A. (2017). Experimental optimization of fused deposition modelling processing parameters: A design-for-manufacturing approach. *Procedia Manufacturing*, 10, 791–803. doi:10.1016/j.promfg.2017.07.079

**155** Bowyer, A. & Olliver, V. (2019). RepRap. Retrieved from https://reprap.org/

**156** Gutowski, T., Jiang, S., Cooper, D., Corman, G., Hausmann, M., Manson, J. A., . . . Sekulic, D. P. (2017). Note on the rate and energy efficiency limits for additive manufacturing. *Journal of Industrial Ecology*, 21, S69–S79. doi:10.1111/jiec.12664

**157** Lewis, J. A., & Gratson, G. M. (2004). Direct writing in three dimensions. *Materials Today*, 7(7–8), 32–39. doi:10.1016/S1369-7021(04)00344-X

**158** Lewis, J. A. (2006). Direct ink writing of 3D functional materials. *Advanced Functional Materials*, 16(17), 2193–2204. doi:10.1002/adfm.200600434

**159** Hon, K., Li, L., & Hutchings, I. (2008). Direct writing technology – Advances and developments. *CIRP Annals*, 57(2), 601–620. doi:10.1016/j.cirp.2008.09.006

**160** Cesarano, J. (1998). A review of robocasting technology. *MRS Proceedings*, 542, 133. doi:10.1557/PROC-542-133

**161** Kuang, M., Wang, L., & Song, Y. (2014). Controllable printing droplets for high-resolution patterns. *Advanced Materials*, 26(40), 6950–6958. doi:10.1002/adma.201305416

**162** Lessing, J., Glavan, A. C., Walker, S. B., Keplinger, C., Lewis, J. A., & Whitesides, G. M. (2014). Inkjet printing of conductive inks with high lateral resolution on omniphobic R F paper for paper-based electronics and MEMS. *Advanced Materials*, 26(27), 4677–4682. doi:10.1002/adma.201401053

**163** Grouchko, M., Kamyshny, A., Mihailescu, C. F., Anghel, D. F., & Magdassi, S. (2011). Conductive inks with a built-in mechanism that enables sintering at room temperature. *ACS Nano*, 5(4), 3354–3359. doi:10.1021/nn2005848

**164** Sun, K., Wei, T.-S., Ahn, B. Y., Seo, J. Y., Dillon, S. J., & Lewis, J. A. (2013). 3D printing of interdigitated Li-ion microbattery architectures. *Advanced Materials*, 25(33), 4539–4543. doi:10.1002/adma.201301036

**165** Fu, K., Wang, Y., Yan, C., Yao, Y., Chen, Y., Dai, J., . . . Hu, L. (2016). Graphene oxide-based electrode inks for 3D-printed lithium-ion batteries. *Advanced Materials*, 28(13), 2587–2594. doi:10.1002/adma.201505391

**166** de Gans, B.-J., Duineveld, P. C., & Schubert, U. S. (2004). Inkjet printing of polymers: State of the art and future developments. *Advanced Materials*, 16(3), 203–213. doi:10.1002/adma.200300385

**167** de Gans, B.-J., Kazancioglu, E., Meyer, W., & Schubert, U. S. (2004). Ink-jet printing polymers and polymer libraries using micropipettes. *Macromolecular Rapid Communications*, 25(1), 292–296. doi:10.1002/marc.200300148

**168** Sculpteo. (2019). 3D printing speed. Retrieved from https://www.sculpteo.com/en/glossary/3d-printing-speed-definition/.

**169** Tofail, S. A., Koumoulos, E. P., Bandyopadhyay, A., Bose, S., O'Donoghue, L., & Charitidis, C. (2018). Additive manufacturing: Scientific and technological challenges, market uptake and opportunities. *Materials Today*, 21(1), 22–37. doi:10.1016/j.mattod.2017.07.001

**170** Vaezi, M., Chianrabutra, S., Mellor, B., & Yang, S. (2013). Multiple material additive manufacturing – Part 1: A review. *Virtual and Physical Prototyping*, 8(1), 19–50. doi:10.1080/17452759.2013.778175

**171** A. Bandyopadhyay, B. Heer, Additive manufacturing of multi-material structures, *Materials Science & Engineering R: Reports* 129 (April) (2018) 1–16. doi:10.1016/j.mser.2018.04.001.

**172** Gao, W., Singh, N., Song, L., Liu, Z., Reddy, A. L. M., Ci, L., . . . Ajayan, P. M. (2011). Direct laser writing of micro-supercapacitors on hydrated graphite oxide films. *Nature Nanotechnology*, 6(8), 496–500. doi:10.1038/nnano.2011.110

**173** Li, S., Wang, X., Xing, H., & Shen, C. (2013). Micro supercapacitors based on a 3D structure with symmetric graphene or activated carbon electrodes. *Journal of Micromechanics and Microengineering*, 23(11), 114013. doi:10.1088/0960-1317/23/11/114013

**174** Yan, W., Thai, M. L., Dutta, R., Li, X., Xing, W., & Penner, R. M. (2014). A lithographically patterned capacitor with horizontal nanowires of length 2.5 mm. *ACS Applied Materials & Interfaces*, 6(7), 5018–5025. doi:10.1021/am500051d

**175** Qi, D., Liu, Y., Liu, Z., Zhang, L., & Chen, X. (2017). Design of architectures and materials in in-plane micro-supercapacitors: Current status and future challenges. *Advanced Materials*, 29(5), 1602802. doi:10.1002/adma.201602802

# 10

# 3D-Printing for Solar Cells

*Marcel Di Vece[1], Lourens van Dijk[2], and Ruud E.I. Schropp[3]*

[1] *Interdisciplinary Centre for Nanostructured Materials and Interfaces (CIMaINa) and Physics Department "Aldo Pontremoli", University of Milan, Milan, Italy*
[2] *Soluxa B.V., Nijmegen, The Netherlands*
[3] *Department of Physics and Astronomy, University of the Western Cape, Belville, South Africa*

## 10.1 Introduction

Solar cell technology is important for many reasons, of which ecological sustainability is the most pregnant. The first commercial solar cells, based on silicon wafer technology, have been around for many decades and their technology has shown incremental progress. Currently, solar energy contributes to a bit more than 2% of the world energy demand (in some countries up to 7%) [1], which clearly indicates that the price per kWh should be further reduced in more locations worldwide to become competitive (with conventional electricity sources) and to obtain significantly higher market shares. Because the price of a solar cell is determined to a large extent by the module and installation costs, the efficiency of a solar cell determines its commercial success. This forms the basis of current ongoing solar cell research: increase of photovoltaic efficiency. Research aims at higher efficiencies with cheap materials and novel techniques [2, 3].

The efficiency of a solar cell has fundamental limits as determined by Shockley and Queisser (SQ) [4], who calculated a maximum efficiency using a single p–n junction of 32% for c-Si solar cells. With modern approaches such as tandem cells, the theoretical efficiency is 45%. One of the most important processes to limit solar cell efficiency is the recombination of electrons and holes inside the active layer and at the interfaces. The obvious advantage of lower cost and lower energy consumption during fabrication forms a strong motivation for fabricating thin film solar cells. It is clear that by reducing the thickness of the active layer in a solar cell,

*3D Printing for Energy Applications*, First Edition. Edited by Albert Tarancón and Vincenzo Esposito.

the optical absorption likely decreases as well. Therefore, the main aim in current thin film solar cell development lays in the improvement of absorption by choice of composition and with advanced light management. It is the light management of thin film solar cells were 3D-printing can have its strongest contribution.

Several parts of the solar cell and the module (the part containing the solar cell) are suitable for 3D-printing solutions [5]. For the module, parts of the frame and the sealing are obvious targets for improvement by 3D-printing. More promising and challenging aims to improve the solar cell by 3D-printing are related to the electric connections, light management, and composition/structure of the light absorbing layers. In this chapter, we will first discuss current promising developments in 3D-printing for PV, from interconnects to novel perovskite layer deposition. These examples demonstrate the current research activity and show possible future possibilities. In the last part of this chapter we describe how a light management system for solar cells using a light trap was investigated using 3D-printing.

## 10.2 Examples of 3D-Printing for PV

3D-printing technology may ultimately be used as part of the high-throughput large-scale roll-to-roll fabrication of thin-film PV cells on flexible substrates, such as transparent plastics and metallic foils. For example, 3D-printing of conductive or isolating lines in such thin film solar cells seems to be one of the most promising forms of 3D-printing technology in the PV sector [6]. The most relevant coating and printing techniques in the field of polymer solar cells, such as gravure coating, knife-over-edge and slot-die coating, are likely to provide opportunities for 3D-printing. Inkjet printing and (rotary) screen printing have been developed for complex patterns such as interconnects for monolithically series connected solar cells [7]. A 3D-printer assisted slot-die coater was used for scalable fabrication of printed perovskite solar cells with 11.6% efficiency and solar cell modules with 4.6% efficiency built on large-area ITO glass substrates $>48\,cm^2$ [8, 9]. Although rigid substrates were employed, such low (room) temperature fabrication technologies can easily be extended to flexible substrates making 3D-printing a potential lab-to-fab tool for solution-processed solar cells [10].

Arrays of ultra-thin semi-transparent solar microcells, which can reach an efficiency similar to conventional solar panels [11] are an alternative to flexible organic cells. This type of solar panel requires flexible interconnects, that is, flexible front electrodes. Because front electrodes are one of the major challenges for Si solar cells, 3D-printing has been extensively used to fabricate silver based electrodes [12–14]. Ahn et al. [15] were able to fabricate flexible, stretchable, and spanning silver microelectrodes using a three-axis micro-positioning stage coupled to a micro-nozzle (Figure 10.1). Similar bridging methods using the same 3D-printing technology were also shown for Sn-doped indium oxide (ITO), a

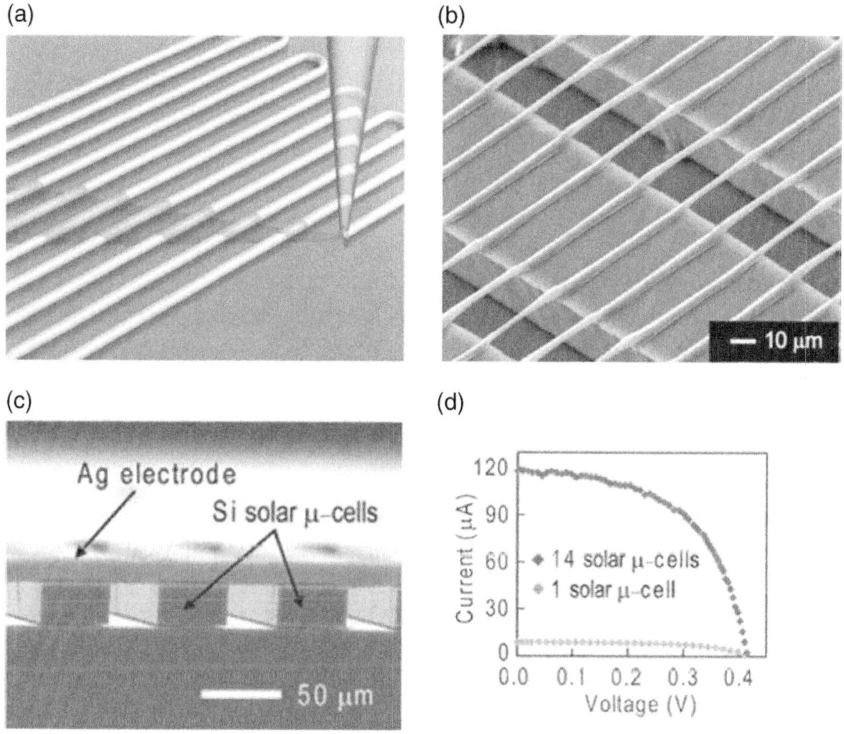


**Figure 10.1** (a) Schematic illustration of direct-write assembly. (b) micrograph of spanning ITO microelectrodes printed on Si ribbons [16]. *Source:* (a) and (b) From B. Y. Ahn, D. J. Lorang, E. B. Duoss, J. A. Lewis, Direct-write assembly of microperiodic planar and spanning ITO microelectrodes. *Chemical Communications* 2010, 46, 7118–7120. © 2010, Royal Society of Chemistry. (c) Optical image of a spanning silver microelectrode printed onto an unplanarized, silicon solar microcell array. (d) Current (*I*)–voltage (*V*) response of an individual silicon solar microcell and a 14-microcell array connected by silver microelectrodes [15]. *Source:* (c) and (d) From B. Y. Ahn, E. B. Duoss, M. J. Motala, X. Guo, S.-I. Park, Y. Xiong, J. Yoon, R. G. Nuzzo, J. A. Rogers, J. A. Lewis, Omnidirectional printing of flexible, stretchable, and spanning silver microelectrodes. *Science* 2009, 323(5921), 1590. © 2009, American Association for the Advancement of Science.

typical transparent conductor [16] (Figure 10.1). The omni-directionality of these flexible microelectrodes opens new possibilities for designing complex three-dimensional photovoltaic structures. Researchers at the Massachusetts Institute of Technology performed a computational study to optimize arbitrarily shaped three dimensional photovoltaic (3DPV) structures to maximize the energy production by using a genetic algorithm [17, 18]. The designed solar panels should be more efficient than flat solar panels by significantly extending the amount of solar radiation absorbed by the cells without sun tracking. Bernardi, Ferralis, Wan, Villalon, and Grossman [18] mounted commercially available Si solar cells onto

3D-printed plastic frames with optimized geometries demonstrating energy densities per projected area (kWh/$m^2$) higher than flat stationary panels by a factor between 2 and 20 and an enhancement factor in energy density of 1.5 to 4 times. A commercial realization by ZigZagSolar [19] is currently available, demonstrating its practical use. Beyond 3D-printed plastic frames, these simple or even more complex and efficient origami-like 3D geometries require the use of inexpensive and customized flexible cells and stretchable interconnects, which are currently under development by the 3D printing community. 3D-printing could therefore strongly contribute to the development of 3DPV.

Noncontact inkjet printing is a rapid and digital deposition technique which is combined with excellent control over the layer formation, which is interesting for printing perovskite solar cells. In the work of Mathies et al. [20] and Howard et al. [21], inkjet printing is used to deposit triple cation perovskite layers with 10% cesium in a mixed formamidinium/methylammonium lead iodide/bromide composite for solar cells with high temperature and moisture stability (Figure 10.2). A continuous power output at a constant voltage, resulting in a power conversion efficiency of 12.9%, demonstrates its feasibility. Since the stability of perovskite solar cells has been a considerable challenge, the extended resistance of the triple cation perovskite solar cells against heat and moisture from the ambient, demonstrates the suitability of the inkjet printing approach. This publication demonstrates that the fabrication of high-efficiency perovskite solar cells with good stability by using digital inkjet printing is a promising path toward next generation photovoltaic applications.

In the work of Andersen et al. [22], in-line printing and coating methods have been demonstrated to enable a high yield fabrication of fully roll-to-roll processed polymer tandem solar cell modules. An example and schematic image are shown in Figure 10.3. Different material sets and a fully printed and coated 14-layer flexible tandem solar cell stack have been demonstrated. The roll-to-roll methodologies involved were flexographic printing, rotary screen printing, slot-die coating, X-ray scattering, electrical testing, and UV-lamination. Their combination enabled the manufacture of completely functional devices with exceptionally high yields. The chosen technology can easily be transferred to industrial methods and the transfer to full roll-to-roll processing has been demonstrated.

Three-dimensional printing technology can also be employed to improve the photovoltaic and photo-thermal conversion efficiency of dye-sensitized solar cell (DSC) module. A 3D-printed concentrator is optically designed and improved the photovoltaic efficiency of the DSC module from 5.48 to 7.03% in the work by Huang et al. [23]. Additionally, the 3D-printed frame contained a microfluidic device serving as water cooling, with which the temperature of the DSC can be effectively controlled, which is beneficial for keeping a high photovoltaic

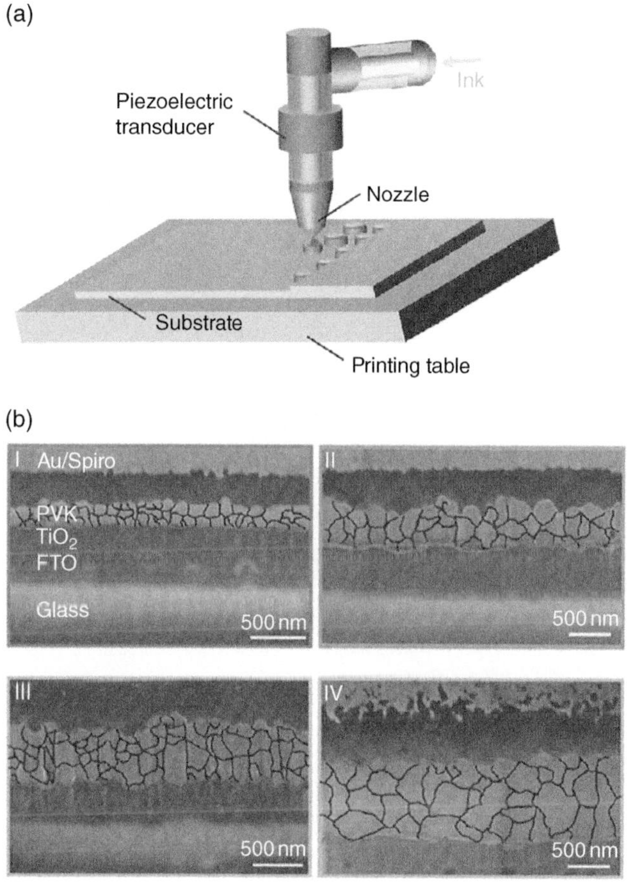


**Figure 10.2** Inkjet printing. (a) Schematic illustration of the inkjet printing process. (b) Scanning electron microscopy (SEM) cross-sectional images of inkjet-printed multi-cation absorbers with different thicknesses [21]. *Source:* From I. A. Howard, T. Abzieher, I. M. Hossain, H. Eggers, F. Schackmar, S. Ternes, B. S. Richards, U. Lemmer, U. W. Paetzold, Coated and printed perovskites for photovoltaic applications. *Advanced Materials* 2019, 0(0), 1806702. © 2019, John Wiley & Sons.

conversion efficiency in the DSC module (Figure 10.4). The 3D-printed microfluidic device can contribute to the photo-thermal conversion with an instantaneous photo-thermal efficiency of 42.1%. The integrated device produces a total photovoltaic and photo-thermal conversion efficiency of 49% at optimal working conditions. In solar tracking devices, in which a solar cell follows the direction of direct sunlight, a particular shape is often beneficial. 3D-printed solar tracking devices

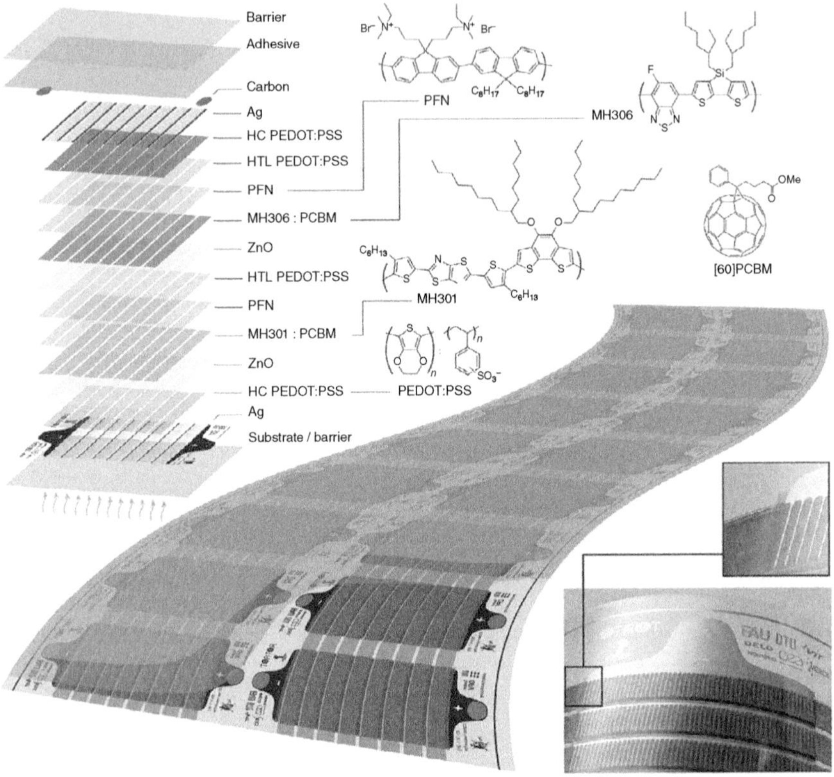


**Figure 10.3** The complete 14-layer tandem stack (upper left) along with structural formulae and names for the different materials involved (top right). The outline of the printed web is shown (middle) along with an actual photograph of a module (lower right). In the close-up photograph, the differently active materials are seen, representing the wide bandgap and low bandgap semiconductor junctions and the hole transport layer [22]. *Source:* From T. R. Andersen, H. F. Dam, M. Hösel, M. Helgesen, J. E. Carlé, T. T. Larsen-Olsen, S. A. Gevorgyan, J. W. Andreasen, J. Adams, N. Li, F. Machui, G. D. Spyropoulos, T. Ameri, N. Lemaître, M. Legros, A. Scheel, D. Gaiser, K. Kreul, S. Berny, O. R. Lozman, S. Nordman, M. Välimäki, M. Vilkman, Roar. R. Søndergaard, M. Jørgensen, C. J. Brabec, F. C. Krebs, Ambient atmosphere roll-to-roll manufacture of encapsulated large area, flexible organic tandem solar cell modules. *Energy & Environmental Science* 2014, 7(9), 2925–2933. © 2014, Royal Society of Chemistry.

are also being investigated, in which the geometry of the device has an important influence on the efficiency improvement [24]. The quality of 3D-printers is continuously improving as for example the ability to print different materials simultaneously and the increase in "resolution." 3D-printed optical components [25] may also contribute to more efficient solar cells.

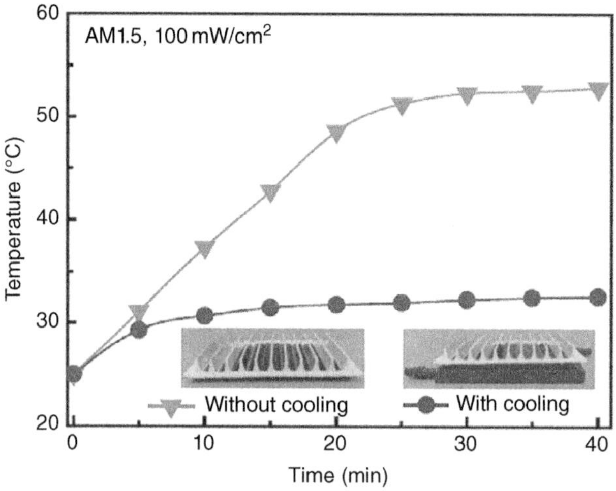


**Figure 10.4** Temperatures of DSC module without and with water cooling under AM 1.5 at 100 mW/cm$^2$ [23]. *Source:* From Q.-Z. Huang, Y.-Q. Zhu, J.-F. Shi, L.-L. Wang, L.-W. Zhong, G. Xu, Dye-sensitized solar cell module realized photovoltaic and photothermal highly efficient conversion via three-dimensional printing technology. *Chinese Physics B* 2017, 26(3), 038401. © 2017, Chinese Physical Society.

# 10.3 Geometric Light Management

## 10.3.1 Background

Light-management concepts that efficiently couple light into solar cells and enhance the absorption in optically thin absorber layers are essential to achieve high-power conversion efficiencies and low production costs [26–28]. To improve the energy conversion efficiency and reduce the material consumption of crystalline silicon (c-Si) cells, a continuation of the trend toward thinner wafers is imperative. However, this reduction in thickness results in a lower absorptance, which can be countered by light trapping schemes. It is therefore expected that light trapping will become increasingly more important for c-Si cells [29]. Light trapping is already important in thin-film solar cells like nanocrystalline silicon cells, as they do extensively rely on light management to increase the optical path length to realize a competitive efficiency. At the same time, there is considerable interest in methods that prevent incoming light from being reflected at the front surface due to the refractive index contrast, like structures similar to moth eyes [30]. One of the reasons why the efficiency of commercial modules is lower than the theoretical efficiency limit is the difficult trade-off between the optical and electrical cell properties. Internal light trapping schemes, like texturing of the surface of the absorber, improve the absorption, but at the same time deteriorate the electrical properties of the solar cell by inducing additional bulk and surface recombination centers [31–34]. Additionally, nano-textured metallic back reflectors and textured front contacts increase the parasitic absorptance [35–39]. It thus remains a challenge to realize internal light trapping schemes without affecting the electrical properties of the solar cell [40–42].

An external light trap is of interest for all solar cell technologies, because the light trap is placed as an add-on on top of the cell and retroreflects the light that is reflected as well as the light that is radiatively emitted by the solar cell. Although the theoretical concept of external light trapping has been wandering in literature and patents for several decades [43–49], this 3D-printed light trap is the first successful experimental demonstration. Figure 10.5 illustrates how sunlight can be externally trapped by "squeezing" the light through a small aperture in the light trap. The cross-sectional area of the hole is much smaller than that of the solar cell; thereby, the light trap retroreflects the light that would otherwise escape. Consequently, this external light trap reduces the optical losses and increases the light harvesting over the entire solar spectrum. The light trap consists of two parts:

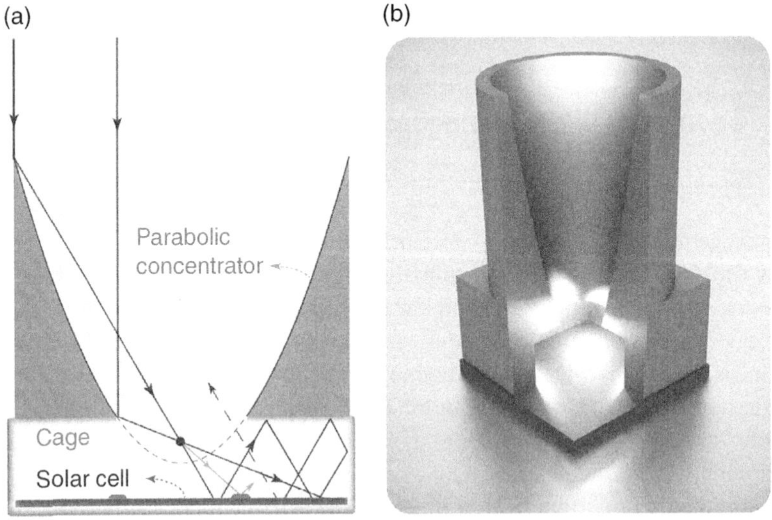


**Figure 10.5** (a) Schematic illustration of the external light trap. By concentration, light is funneled through a small aperture and is trapped. Also shown is the reflection from a metallic reflective front contact. (b) A rendered 3D-model of the light trap. The front quarter of the light trap is removed for visibility. A concentrator is placed on top of a square cage with the solar cell at the bottom. Most of the reflected light from the solar cell is recycled within the cage [70]. *Source:* Based on L. van Dijk, E. A. P. Marcus, A. J. Oostra, R. E. I. Schropp, M. Di Vece, 3D-printed concentrator arrays for external light trapping on thin film solar cells. *Solar Energy Materials & Solar Cells* 2015, 139, 19–26.

a parabolic concentrator and a cage; see Figure 10.5b for a 3D model. Most of the photons that are reflected by the front surface of the solar cell and those that passed the cell but were not absorbed are retroreflected by the external light trap.

It should be noted that this method of light trapping differs markedly from conventional concentrated photovoltaics. The external light trap reduces the need for internal light trapping schemes and antireflection coatings that interfere with the electrical performance of the cell [50–52]. An external light trap circumvents the electro-optical dilemmas, because it allows the cell designer to focus mainly on the electrical performance of the cell. The light trap targets three major optical loss mechanisms present in all solar cells: (a) reflection at the cell at its front surface, metal fingers, busbars, and the reflective area between the solar cells; (b) incomplete absorption, mainly at long wavelengths ($\lambda$); and (c) photons emitted by radiative recombination. For most solar cells, like c-Si, combating the losses due to reflection and poor absorption is most rewarding. Currently, the recycling of radiatively emitted photons is mainly interesting for gallium arsenide (GaAs) due to its relatively high external radiative efficiency [53–56]. Another benefit of the introduced concept is that it can help concentrated photovoltaics by reducing the heat load of the cell compared with conventional concentrated photovoltaics. Internal light trapping schemes theoretically result in a path length enhancement of maximally $2n^2$ the double cell pass [57]. Therefore, external light trapping is of special interest for low refractive index ($n < 2$) materials, such as organic and perovskite solar cells, because this method is independent of the refractive index. The top of the external light trap is a compound parabolic concentrator (CPC). CPCs are generally applied for concentrated photovoltaics and solar thermal applications. There are basically three different categories of CPCs: macroscopic-sized, millimeter-sized, and micrometer-sized concentrators. Macroscopic CPCs are commonly used in evacuated tube collectors and solar thermal applications [58–60]. Millimeter-sized CPCs were proposed to benefit from the relatively low Joule losses of small solar cells [61] and to realize a low-cost, flat plate-like concentrated photovoltaic module [62–64], while others fabricated micro-sized CPCs using two photon lithography [48, 49]. In this chapter, an important example of 3D-printed external light trap is discussed using a millimeter-scaled 3D-printed CPC placed on top of a 1.2-μm thick superstrate hydrogenated nanocrystalline silicon (nc-Si:H) cell. The usage of the light trap is certainly not restricted to nc-Si:H cells and no electro-optical compromises are needed, leaving more space to focus on the electrical performance of the solar cell.

### 10.3.2 Optical Model for External Light Trapping

Thermodynamics forbids the concept of a one-way mirror that would allow thermal heat to radiate unidirectionally from a cold to a hot body. Such a concept violates the second law of thermodynamics [65]. The external light trap, at first sight, also seems to violate this law because light can travel unhindered from the sun to the solar cell, while most radiation originating from the solar cell is retroreflected by the trap. However, the trap restricts the solid angle of light that is accepted, which enables an equivalent restriction of the escape area of the light confined in the trap [66]. Therefore, the angular étendue restriction by the CPC enables efficient retroreflection of the light that did not escape through the aperture. Because the étendue restriction of radiatively emitted and accepted photons is identical, the external light trap does not violate thermodynamics. For the external light trap, an absorption gain by light trapping is obtained upon placement of a cage between the CPC and the cell. The spacing of the cage enables sufficient divergence of the light beam, such that it does not escape through the aperture after the first reflection. The trap not only restricts the angular cone of acceptance, but also imposes an equivalent restriction for the trapped light: the light reflected by the solar cell outside this acceptance étendue is retroreflected. Because of this equivalency, the probability of photons to escape from the trap, *P*(escape), is in first order inversely proportional to the concentration factor (*C*).

$$\frac{1}{P(\text{escape})} = C = \frac{1}{\sin(\theta_{\text{acceptance}})^2} \tag{10.1}$$

where $\theta_{\text{acceptance}}$ is the half angle of the acceptance cone of the concentrator [66]. This reciprocity dictates that the restriction of the escape area within the light trap is inversely correlated to the angular acceptance restriction outside the trap, which is similar to rugate filters [54, 67, 68]. The restriction of the escape probability of photons increases the absorptance in the solar cell. In general, the intrinsic efficiency limit increases when the cell thickness is reduced while the absorptance stays constant [43]. The improved energy conversion efficiency limit is possible because thin solar cells have effectively less recombination and thereby a relatively high open circuit voltage. This is illustrated by c-Si solar cells. Because of the thickness-related Auger and Shockley–Read–Hall recombination, the maximum possible efficiency of ideally thin (3 μm) c-Si cells under maximum angular restriction is 33.4% and thus significantly higher than 29.5% of a thick solar cell without angular restriction [4, 43, 69].

The optical model that we used includes the losses originating from the CPC, cage, and cell. Initially, a fraction $T_c$ is transmitted by the CPC and penetrates into the solar cell. A fraction $T_C A_{sc}$ is absorbed. Most of the reflected light [$R_{sc} = T_C(1-A_{sc})$] does not escape, but is instead retroreflected by the top of the cage with reflectivity $R_{\text{cage}}$. For homogeneously distributed light a fraction $C-1$ escapes from the trap and $(1-C-1)R_{\text{cage}}$ is reflected backward by the cage. The cage absorbs a small fraction $(1 - C - 1)A_{cage}$. This process continues infinitely and the absorption accumulates. The total absorptance ($A_T$) is thus given by [70]:

$$A_T = T_C \cdot \frac{A_{SC}}{1 - R_{SC}\left(1 - C^{-1}\right)R_{cage}} \tag{10.2}$$

At a higher concentration factor, the escape probability is lower, and thereby, the total absorptance is higher. The quality of the CPC is more important for cells with a high absorptance as the optical loss in the CPC has to be compensated by light trapping. Low values of the external path length enhancement $\Pi_{ext}$ (2–10×) do already have a significant impact on the absorptance. For example, for $\Pi_{ext} = 10\times$, a cell with an absorptance of just 20% will be enhanced to 60% by the light trap. When $\Pi_{ext}$ increases, the total absorptance in the cell converges toward the transmittance of the CPC. It is of interest to compare internal and (concentrator based) external light trapping to identify the optimal method and to investigate how both trapping mechanisms can be combined. Internal light trapping is realized by scattering and total internal reflection within the cell. At low absorptivity, this results in a maximum theoretical path length enhancement factor of $\Pi_{int} = 2n^2$ times the path of a double cell pass [57, 71]. For silicon cells, this corresponds to a factor of $\sim 2 \cdot 3.5^2 = 24.5$. However, in practice, only broadband path length enhancement factors up to ~10× are realized [72]. For a perfect light trap, ($\Pi_{ext}$) is in the low absorbing limit equal to $C$. However, as a consequence of the non-zero absorptance values, $\Pi_{ext}$ ranges from 1 to 6×. From the experimental data, the effective path length enhancement factor can be derived using

$$\Pi_{ext} = \frac{\log\left(1 - A_T\right)}{\log\left(1 - A_{sc}\right)} \tag{10.3}$$

The external light trap thus results in a path length enhancement of $\Pi_{ext}$, which is equivalent to an effective thickness of $z_{eff} = \Pi_{ext} \cdot z$, with $z$ the solar cell thickness. Alternatively, one can consider the net effect of light trapping as an effective absorption coefficient ($\alpha$) enhancement from $\alpha$ to $\Pi_{ext} \cdot \alpha$. The total path length enhancement ($\Pi_{total}$) is determined by the product of the internal and external path length enhancement factors, $\Pi_{tot} = \Pi_{ext} \cdot \Pi_{int}$:

$$\Pi_{total} = \frac{2n^2}{\sin^2\left(\theta_{acceptance}\right)} \tag{10.4}$$

where $\theta_{acceptance}$ is the half angle of the acceptance cone [43]. As $\theta_{sun}$ is only ~0.266°, this translates to a remarkable maximum path length enhancement factor in the order of $2n^2 \times 46{,}000 \approx 10^6\times$ for direct sunlight. The theoretically maximum contribution to the total path length enhancement from external light trapping is thus considerably higher than that of internal light trapping.

### 10.3.3 Design and 3D-Printing of the External Light Trap

The CPC is designed in such a way that the angle of incidence at the cell at its front surface is below the Brewster angle of glass. The optimization of the concentration factor of the light trap is primarily a trade-off between the negative effects of design inaccuracies on the CPC transmission at high $C$ and the improved absorptance due to path length enhancement at high $C$. With state-of-the-art 3D print techniques and the CPC size used, it was found technically challenging to realize high transmittance for concentration factors of $C > 10$. This is mainly due to the need for higher accuracy with increase of $C$. Small deviations from the intended optical path will increasingly result in rejection of the light when the diameter of the exit aperture is reduced. Low concentration factors ($C \sim 2$–$10$) are already sufficient to make a significant absorptance change. Therefore, a CPC with a geometrical concentration factor of $C = 6$ was used, which experimentally showed both high transmittance and a substantial absorption enhancement.

One can engineer the light trap such that the path length enhancement for specular reflected light in the cage is even higher than the model based on homogeneously distributed of light would predict. The intensity ($I$) in the cell drops in first order exponentially according to the Beer–Lambert law, $I = I_0 \cdot e^{-\alpha l}$, with $l$ the distance traveled through the cell. When the trap is applied, the power escaping through the aperture is attenuated by the exponential intensity decay with the number of cell passes: $I_{escape} = I_0 \cdot A^{\#\text{cell passes}}$. The external light trap exploits this exponential decay, as the specularly reflected light cannot escape from the trap in the first recycle events because the light initially only propagates radially outward. This opportunity to postpone the first escape occurrence is an advantage over internal trapping schemes. This escape postponement cannot be obtained when the cell reflects the light diffusely. The largest path length enhancement is therefore expected for a flat, specular reflective cell, combined with a specular reflective cage. An additional advantage of a flat cell design is the superior electrical quality [40].

The light trap is 3D printed from acrylonitrile butadiene styrene (ABS) using an Ultimaker [73]. The use of ABS allows the prints to be vapor smoothened by exposure to a saturated acetone vapor; see Figure 10.6a. After smoothening, the cage and the CPC were silver coated, resulting in focusing of sunlight, as can be seen in Figure 10.6b. Figure 10.6c shows the square cage and the circular concentrator that are stacked to form the light trap. The opening area of both the CPC and the cage is 81 $mm^2$. A square cage was used to match the usual shape of the solar cell. Moreover, the square cage design is an interesting candidate for large area module integration. By arranging multiple cages into a square or hexagonal array, a large

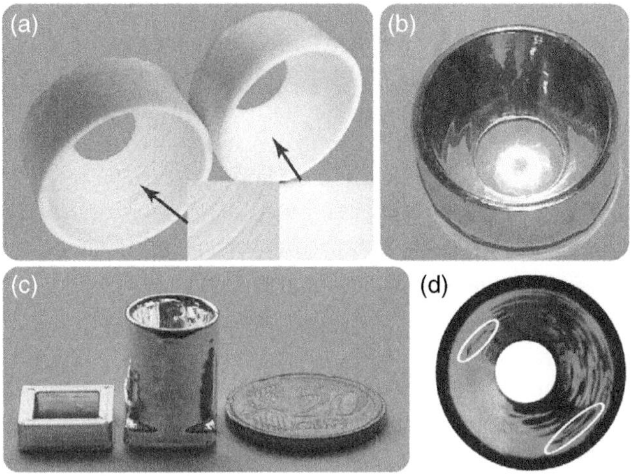


**Figure 10.6** (a) A 3D-printed compound parabolic concentrator (CPC) before and after chemical smoothening, the insets show the enlarged surface. (b) The silver-coated CPC with a focal ring of sunlight at the center. (c) The separated cage and CPC that can be combined to an external light trap. (d) View of the CPC from the entrance side, showing the reflection in the parabolic curve. Two wrinkles in the concentrator are encircled [70]. *Source:* From L. van Dijk, E. A. P. Marcus, A. J. Oostra, R. E. I. Schropp, M. Di Vece, 3D-printed concentrator arrays for external light trapping on thin film solar cells. *Solar Energy Materials & Solar Cells* 2015, 139, 19–26.

area can be efficiently covered by light traps [70]. Some undesired unevenness of the parabolic shape can be observed by eye from the small distortion in the projected image inside of the CPC; see Figure 10.6d. The formation of wrinkles might be further prevented by using a more accurate fused deposition modeling printer and by optimizing the acetone treatment. As an alternative, one could use a different 3D printer, such as a stereo-lithography printer (based on optical curing of a resin). The finished light trap was placed on top of a flat, square thin-film nc-Si:H solar cell prepared in superstrate configuration. The light trap changes the angle of incidence to non-normal angles. According to Snell's law, the path length thereby increases (relatively) more in the low refractive index ZnO than in the high index Si. This results in a relative increase of the parasitic losses. Depending on the exact geometry of the light trap, the illumination of the cell can become inhomogeneous, which may negatively affect the overall cell performance [74, 75].

### 10.3.4 Characterization

The square cage was placed on top of the square solar cell and used as a mask to determine the current density versus voltage (*J*–*V*) characteristics of the bare solar cell. Subsequently, the CPC (with equal opening area) is placed on top of the cage to determine the opto-electrical improvement by the external light trapping. Figure 10.7a shows the *J–V* characteristics at AM1.5G illumination of the bare (masked) solar cell and the cell with the external light trap (cage and concentrator). Because of the light trapping, the short circuit current density ($J_{sc}$) and the current at the maximum power point (Jmpp) improved by 14 and 15%, respectively. No significant change in the fill factor and open circuit voltage ($V_{oc}$) was measured. To understand the origin of this enhancement, the internal quantum efficiency (IQE), absorptance, and external quantum efficiency (EQE) were determined.

The IQE $\left(\mathrm{IQE} = \dfrac{\mathrm{EQE}}{A_{sc}}\right)$ of the bare cell was determined by measuring the absorptivity ($A_{sc} = 1 - R_{sc}$) and the EQE; see Figure 10.7b. For short wavelengths ($\lambda < 400$ nm), the absorptance of the bare cell is high (~80–90%). The relatively high parasitic absorptance in the ZnO front contact and the (not photovoltaically active) p-layer results in a low EQE and IQE. This implies that there is only a small potential for improvement (~10–20%$_{\mathrm{relative}}$) in this short wavelength regime. For wavelengths around 600 nm, the reflectance is 10–40%, while at the same time, the IQE is high. This provides significant room for improvement by external light trapping. For longer wavelengths ($\lambda > 700$ nm) there is 40–60% reflection. With further increase of the wavelength, the absorptance in the nc-Si reduces compared with the parasitic absorptance, and therefore, the IQE drops to zero. In the first order, the achievable absolute EQE improvement is given by $\Delta \mathrm{EQE}_{\uparrow} = R_{sc} \cdot \mathrm{IQE}$. Thus, the light trap mainly improves the response when both $R_{sc}$ and IQE are high, this is the case for ~$400 < \lambda <$ ~900 nm. The observed optical interference patterns are mainly due to the Fabry–Pérot modes in the 1.2 μm thick Si layer.

Figure 10.7b shows the EQE of the bare cell ($\mathrm{EQE}_{\mathrm{bare}}$), the EQE of the cell with trap ($\mathrm{EQE}_{\mathrm{trap}}$), and the EQE as calculated using the model ($\mathrm{EQE}_{\mathrm{model}}$). There is a significant EQE improvement from $\mathrm{EQE}_{\mathrm{bare}}$ to $\mathrm{EQE}_{\mathrm{trap}}$ over the full spectrum. At short wavelengths, the improvement is due to retroreflection of light that is reflected at the front interface of the cell. Thereby, the net reflectance is reduced somewhat, similar to that of an antireflection coating. For long wavelengths, there is also retroreflection of light that passed the solar cell but was not absorbed. The enhancement demonstrates that indeed the external light trap effectively prolongs the optical path length in the cell. The curve of $\mathrm{EQE}_{\mathrm{trap}}$ is smoother than that of $\mathrm{EQE}_{\mathrm{bare}}$, which can be explained by two effects. First, as the $\mathrm{EQE}_{\mathrm{trap}}$ converges toward the IQE, the interference fringes are reduced. Second, the trap offers a range of non-normal angles of incidence resulting in a range of different overlapping Fabry–Pérot resonances with a more constant average. A distribution of the path length in the cell with a small dispersion of just a few percent significantly smoothens the absorptance and EQE. From the introduced model (Equation 10.2, the achievable EQE was calculated using $\mathrm{EQE}_{\mathrm{bare}}$ and the IQE; see $\mathrm{EQE}_{\mathrm{model}}$ in Figure 10.7b. The

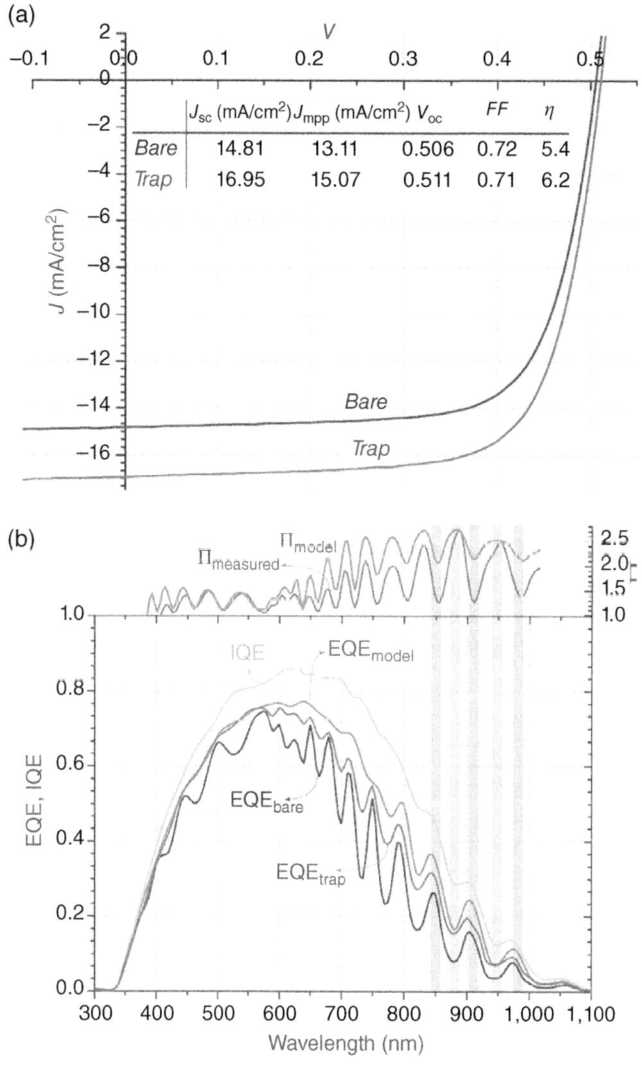


**Figure 10.7** (a) $J$–$V$ characteristics of the bare solar cell (with the cage used as a mask) and the solar cell with the external light trap. (b) Plot of the $EQE_{bare}$ and internal quantum efficiency (IQE) of the bare solar cell (without light trap). Also shown is the external quantum efficiency (EQE) of the solar cell with external light trap ($EQE_{trap}$) and the expected EQE of the cell with light trap based on the optical model ($EQE_{model}$). The external path length enhancement factor (top of figure) shows that there is an improved optical path length for all wavelengths. As expected, there is an anticorrelation of $\prod_{ext}$ with the interference pattern [70]. *Source:* Based on L. van Dijk, E. A. P. Marcus, A. J. Oostra, R. E. I. Schropp, M. Di Vece, 3D-printed concentrator arrays for external light trapping on thin film solar cells. *Solar Energy Materials & Solar Cells* 2015, 139, 19–26.

following parameter values were used: $T_c = 0.95$, $C = 6$, and $R_{cage} = 0.93$. The overall trend of $EQE_{model}$ corresponds well with the measured $EQE_{trap}$. There is excellent agreement at short wavelengths between the model and the experimental data. The observed decrease of the interference fringes of $EQE_{trap}$ corresponds with the trend of the model. Although the $EQE_{trap}$ approaches $EQE_{model}$, there is a discrepancy that is attributed to several losses that were not included in the model. (a) The glass between the light trap and the solar cell enables some light to escape sideways. (b) The printed cage and the CPC are not perfectly smooth, resulting in unwanted parasitic absorption and reflection in unintended directions. (c) Imperfect interconnections, the cell, the 3D-printed cage, and the CPC cause small cavities (around a few hundred μm) that give rise to a small escape loss (estimated to be a few percent). (d) The wavelength dependence of the parameters was not taken into account. Finally, as $IQE_{trap}$ cannot be determined experimentally, we assumed for the model that $IQE_{bare}$ is equal to $IQE_{trap}$. Based on the model, the $J_{sc}$ of the cell with light trap could ideally be 19.9 mA/cm$^2$. Therefore, by reducing the indicated loss mechanisms, one could gain another 2.9 mA/cm$^2$. The realized and expected external path length enhancements ($\prod_{experimental}$ and $\prod_{model}$) were calculated by evaluating Equation 10.3 on the implied absorptivity; see top of Figure 10.7b. The external path length enhancement ranges from 1.5 to 2.5× in the long wavelength regime, where light trapping is mostly needed. The pattern of $\prod_{experimental}$ agrees with the expected $\prod_{model}$. The path length enhancement is anticorrelated to the EQE, as can be seen from the vertical background bars in Figure 10.7b. This agrees with the optical model, which indicates that the achievable path length enhancement is roughly inversely related to the bare cell absorptance. Besides that, the before mentioned smoothening of $EQE_{trap}$ also results in an anticorrelation of the $\prod_{ext}$ and EQE data as the interference minima of the $EQE_{trap}$ relatively increase, while the maxima relatively decrease.

A high-quality external light trap with a moderate (e.g., ~10×) path length enhancement factor and low parasitic absorption losses result in an $EQE_{trap}$ just below the IQE. In the design process of solar cells that are to be combined with a light trap, one has to optimize the IQE. This can be carried out by preventing undesirable parasitic absorption, for example, in the back reflector [76].

The question now arises how external light trapping differs from conventional light concentration. Concentrators are commonly used to concentrate light directly on a solar cell to profit from reduced, relative expensive, materials usage, and the higher open circuit voltage. For a cell under concentrated light (without internal light trapping), light traverses the cell just back and forth, after which the non-absorbed light escapes through the front surface. However, for a cell with an external light trap, the light is forced to traverse the cell many times, which allows the use of thinner and un-textured cells, resulting in less bulk and surface recombination.

Several improvements for external light trapping can be envisaged. The (individual) external light trap can be scaled to a larger area by designing a CPC array as was shown recently [70]. Large trapping arrays can be made at low cost from ABS plastic (which we also used for the 3D-printing) by injection molding. It has been shown that the thickness of CPC arrays can be scaled down to the micrometer range, which benefits module integration [48, 49]. The need of sun tracking can be eliminated by matching the acceptance cone of the CPC to the path of the sun. Finally, the positive effects of an enhanced $V_{oc}$ due to light concentration can be combined with light trapping. For example, when a CPC with a concentration factor of 6 is combined with a solar cell with a smaller area of $A_{cell} = A_{CPC/2}$, there is both concentration (effectively 2×) and light trapping ($\prod_{ext} = 3\times$). This also enables mitigation of the high shading losses of the metal grid in concentrator cells of 10–20% [72]. Moreover, the reduced flux will help to mitigate heat problems at cell level compared with full concentration. The design of the cage is another aspect of the light trap that can be improved. It has been shown that an oblate hemi-ellipsoidal cavity with a specular reflective coating would be an interesting candidate [56, 66, 75, 77, 78]. In this design, the solar cell is located in between the focal circle of the ellipsoid, and thereby, all the light reflecting from the solar cell is reflected backward to the cell in just one reflection. This enables efficient recycling as it keeps the parasitic absorptance by the cage low. For isotropically distributed incident diffuse light, a fraction $1/C$ is transmitted by the parabolic concentrator [59]. Therefore, the use of low concentration factors ($C \sim 2$–10×) enables transmitting 10–50% of the diffuse light. There is an inevitable trade-off between loss of diffuse light and reduced escape probability (within the light trap) with increase of $C$. Further research should be performed to determine an optimal concentration factor. Alternatively, it is worthwhile to model and realize an earlier proposed concept in which both the diffuse and direct components are harvested by geometric separation [79].

Concluding, a combination of a 3D-printed parabolic concentrator and a light cage was used to demonstrate external light trapping for a thin-film nc-Si:H solar cell. A 15% enhancement of the energy conversion efficiency in this cell was demonstrated with respect to the same cell without an external light trap. The external light trap recycled both the unwanted reflection from the front side of the cell and enhanced the poor absorptance in the long wavelength regime. The measured absorptance enhancement at short wavelengths demonstrates the antireflection effect of the trap. An absorptance model for external light trapping was introduced, which showed good correspondence with the experimental data. As a result of light trapping, a path length enhancement factor up to 2.5 times was observed. Several optical loss mechanisms were found and described that can be targeted in future research to obtain better trapping than this initial result. For further research, it is recommended to focus on the accuracy and the related transmittance of the

CPC. The parasitic absorption losses of this first prototype of the CPC are more than compensated by the gain in the cell absorptance. The efficiency improvement is caused by an unprecedented broadband absorption enhancement by the external light trap [80–82], which is universally applicable to all solar cell technologies. Higher gains can be expected upon further optimization of the reflections in the trapping system. Reducing the cell thickness combined with the external light trap is a promising way for further efficiency improvement. The trap enables decoupling of the optical and electric optimization, which yields new perspectives for the design of solar cells. In the current solar cell market, revolutionary light trapping architectures are needed to obtain higher efficiencies. The demonstrated 3D-printed external light trap is an attractive candidate to accomplish this target.

In subsequent work by van Dijk, van de Groep, Di Vece, and Schropp [83], the 3D-printing of a proof of concept light trap design led to improved methods which are industrially interesting. A series of prototype millimeter-scale external metal light traps were milled and applied on an untextured c-Si photodiode, which is used as a model for future thin solar cells. This relatively simple and comprehensive light management solution is a promising candidate for highly efficient solar modules using thin c-Si solar cells. Furthermore, an external light trapping design is introduced [84] with as main concept the tackling of the reflection issues at the module level. In this module design, a lens array on the module cover glass funnels the incident sunlight through small apertures in a reflective coating at the backside of the cover glass. Highly efficient, colored, and even image displaying modules make this technology an interesting candidate for a new class of photovoltaic modules.

## 10.4 Conclusions

In this chapter we discussed the possibilities of using 3D-printing for solar cells with different approaches. 3D-printing methods are in use for the exterior or for electric contacting, while other 3D-printing methods aim at the deposition of the active solar cell layers, of which perovskites are currently very exciting. In the second part of this chapter, we showed an example of using a 3D-printed light trap to increase the solar cell conversion efficiency with light management. 3D-printing functions as an important tool for proof of concept research, while industrial production may follow with faster techniques. It is clear from these examples that 3D-printing opens important possibilities for research, rapid iterative prototyping, and mass producing solar cells with novel structures and materials. Since the efficiency of solar cells can and needs to be improved considerably, 3D-printing likely remains an important tool to achieve that goal, which offers research opportunities for the field.

## References


1 Schmela, M. (2019). *Global market outlook for solar power/2019–2023*. Brussels: Solar Power Europe. p 90.

2 Green, M. A. (2003). *Third generation photovoltaics: Advanced solar energy conversion*. Berlin: Springer.

3 Conibeer, G. (2007). Third-generation photovoltaics. *Materials Today*, 10(11), 42–50. doi:10.1016/S1369-7021(07)70278-X

4 Shockley, W., & Queisser, H. J. (1961). Detailed balance limit of efficiency of P-N junction solar cells. *Journal of Applied Physics*, 32(3), 510–519. doi:10.1063/1.1736034

5 Ruiz-Morales, J. C., Tarancón, A., Canales-Vázquez, J., Méndez-Ramos, J., Hernández-Afonso, L., Acosta-Mora, P., . . . Fernández-González, R. (2017). Three dimensional printing of components and functional devices for energy and environmental applications. *Energy & Environmental Science*, 10(4), 846–859. doi:10.1039/C6EE03526D

6 Krebs, F. C. (2009). Fabrication and processing of polymer solar cells: A review of printing and coating techniques. *Processing and Preparation of Polymer and Organic Solar Cells*, 93(4), 394–412. doi:10.1016/j.solmat.2008.10.004

7 Schweitzer, A., Geißendörfer, S., Siepmann, O., Sergeev, O., & Vehse, M. (2011). Monolithic series interconnection of flexible thin-film PV devices. Proceedings of the 26th European Photovoltaic Solar Energy Conference Proceedings, 1157–1160. doi:10.4229/EUPVSEC20152015-3CV.1.3.

8 Vak, D., Hwang, K., Faulks, A., Jung, Y.-S., Clark, N., Kim, D.-Y., . . . Watkins, S. E. (2015). 3D printer based slot-die coater as a lab-to-fab translation tool for solution-processed solar cells. *Advanced Energy Materials*, 5(4), 1401539. doi:10.1002/aenm.201401539

9 Hwang, K., Jung, Y.-S., Heo, Y.-J., Scholes, F. H., Watkins, S. E., Subbiah, J., . . . Vak, D. (2015). Toward large scale roll-to-roll production of fully printed perovskite solar cells. *Advanced Materials*, 27(7), 1241–1247. doi:10.1002/adma.201404598

10 Di Giacomo, F., Fakharuddin, A., Jose, R., & Brown, T. M. (2016). Progress, challenges and perspectives in flexible perovskite solar cells. *Energy & Environmental Science*, 9(10), 3007–3035. doi:10.1039/C6EE01137C

11 Yoon, J., Baca, A. J., Park, S.-I., Elvikis, P., Geddes, J. B., III, Li, L., . . . Rogers, J. A. (2008). Ultrathin silicon solar microcells for semitransparent, mechanically flexible and microconcentrator module designs. *Nature Materials*, 7, 907.

12 Gizachew, Y. T., Escoubas, L., Simon, J. J., Pasquinelli, M., Loiret, J., Leguen, P. Y., . . . Aguerre, J. P. (2011). Towards ink-jet printed fine line front side metallization of crystalline silicon solar cells. *Special Issue Thin Film Photovoltaic Applications TFPA*, 95, S70–S82. doi:10.1016/j.solmat.2010.12.031

13 Galagan, Y., Coenen, E. W. C., Sabik, S., Gorter, H. H., Barink, M., Veenstra, S. C., . . . Blom, P. W. M. (2012). Evaluation of ink-jet printed current collecting grids and busbars for ITO-free organic solar cells. *Solar Energy Materials & Solar Cells*, 104, 32–38. doi:10.1016/j.solmat.2012.04.039

14 Jiang, Y., Chen, Y., Zhang, M., Qiu, Y., Lin, Y., & Pan, F. (2016). 3D-printing Ag-line of front-electrodes with optimized size and interface to enhance performance of si solar cells. *RSC Advances*, 6, 51871–51876. doi:10.1039/C6RA08985B

15 Ahn, B. Y., Duoss, E. B., Motala, M. J., Guo, X., Park, S.-I., Xiong, Y., . . . Lewis, J. A. (2009). Omnidirectional printing of flexible, stretchable, and spanning silver microelectrodes. *Science*, 323(5921), 1590. doi:10.1126/science.1168375

16 Ahn, B. Y., Lorang, D. J., Duoss, E. B., & Lewis, J. A. (2010). Direct-write assembly of microperiodic planar and spanning ITO microelectrodes. *Chemical Communications*, 46, 7118–7120. doi:10.1039/C0CC01691H

17 Myers, B., Bernardi, M., & Grossman, J. C. (2010). Three-dimensional photovoltaics. *Applied Physics Letters*, 96(7), 071902. doi:10.1063/1.3308490

18 Bernardi, M., Ferralis, N., Wan, J. H., Villalon, R., & Grossman, J. C. (2012). Solar energy generation in three dimensions. *Energy & Environmental Science*, 5(5), 6880–6884. doi:10.1039/C2EE21170J

19 ZigZagSolar.

20 Mathies, F., Eggers, H., Richards, B. S., Hernandez-Sosa, G., Lemmer, U., & Paetzold, U. W. (2018). Inkjet-printed triple cation perovskite solar cells. *ACS Applied Energy Materials*, 1, 1834–1839. doi:10.1021/acsaem.8b00222

21 Howard, I. A., Abzieher, T., Hossain, I. M., Eggers, H., Schackmar, F., Ternes, S., . . . Paetzold, U. W. (2019). Coated and printed perovskites for photovoltaic applications. *Advanced Materials*, 0(0), 1806702. doi:10.1002/adma.201806702

22 Andersen, T. R., Dam, H. F., Hösel, M., Helgesen, M., Carlé, J. E., Larsen-Olsen, T. T., . . . Krebs, F. C. (2014). Ambient atmosphere roll-to-roll manufacture of encapsulated large area, flexible organic tandem solar cell modules. *Energy & Environmental Science*, 7(9), 2925–2933. doi:10.1039/C4EE01223B

23 Huang, Q.-Z., Zhu, Y.-Q., Shi, J.-F., Wang, L.-L., Zhong, L.-W., & Xu, G. (2017). Dye-sensitized solar cell module realized photovoltaic and photothermal highly efficient conversion via three-dimensional printing technology. *Chinese Physics B*, 26(3), 038401. doi:10.1088/1674-1056/26/3/038401

24 Apostoleris, H., Leland, J., Chiesa, M., & Stefancich, M. (2016, September 7). 3D-printed concentrators for tracking-integrated CPV modules. Proceedings of the SPIE 9955, Nonimaging Optics: Efficient Design for Illumination and Solar Concentration XIII—Commemorating the 50th Anniversary of Nonimaging Optics, 995507. doi:10.1117/12.2237786.

25 Luxexcel.

26 Mokkapati, S., & Catchpole, K. R. (2012). Nanophotonic light trapping in solar cells. *Journal of Applied Physics*, 112(10), 101101. doi:10.1063/1.4747795

**27** Polman, A., & Atwater, H. A. (2012). Photonic design principles for ultrahigh-efficiency photovoltaics. *Nature Materials*, 11, 174.

**28** Rau, U., & Kirchartz, T. (2014). On the thermodynamics of light trapping in solar cells. *Nature Materials*, 13, 103.

**29** Rau, U., Paetzold, U. W., & Kirchartz, T. (2014). Thermodynamics of light management in photovoltaic devices. *Physical Review B*, 90(3), 035211. doi:10.1103/PhysRevB.90.035211

**30** Sun, C.-H., Jiang, P., & Jiang, B. (2008). Broadband moth-eye antireflection coatings on silicon. *Applied Physics Letters*, 92(6), 061112. doi:10.1063/1.2870080

**31** Matsui, T., Tsukiji, M., Saika, H., Toyama, T., & Okamoto, H. (2002). Influence of substrate texture on microstructure and photovoltaic performances of thin film polycrystalline silicon solar cells. *19th International Conference on Amorphous and Microcrystalline Semiconductors*, 299–302, 1152–1156. doi:10.1016/S0022-3093(01)01083-3

**32** Schropp, R. E. I., Rath, J. K., & Li, H. (2009). Growth mechanism of nanocrystalline silicon at the phase transition and its application in thin film solar cells. *Proceedings of 4th International Asian Conference on Crystal Growth and Crystal Technology*, 311(3), 760–764. doi:10.1016/j.jcrysgro.2008.09.155

**33** Python, M., Dominé, D., Söderström, T., Meillaud, F., & Ballif, C. (2010). Microcrystalline silicon solar cells: Effect of substrate temperature on cracks and their role in post-oxidation. *Progress in Photovoltaics: Research and Applications*, 18(7), 491–499. doi:10.1002/pip.956

**34** Paetzold, U. W., Smeets, M., Meier, M., Bittkau, K., Merdzhanova, T., Smirnov, V., . . . Rau, U. (2014). Disorder improves nanophotonic light trapping in thin-film solar cells. *Applied Physics Letters*, 104(13), 131102. doi:10.1063/1.4869289

**35** Springer, J., Poruba, A., Müllerova, L., Vanecek, M., Kluth, O., & Rech, B. (2004). Absorption loss at nanorough silver back reflector of thin-film silicon solar cells. *Journal of Applied Physics*, 95(3), 1427–1429. doi:10.1063/1.1633652

**36** Haug, F.-J., Söderström, T., Cubero, O., Terrazzoni-Daudrix, V., & Ballif, C. (2008). Plasmonic absorption in textured silver back reflectors of thin film solar cells. *Journal of Applied Physics*, 104(6), 064509. doi:10.1063/1.2981194

**37** Moulin, E., Paetzold, U. W., Kirchhoff, J., Bauer, A., & Carius, R. (2012). Study of detached back reflector designs for thin-film silicon solar cells. *Physica Status Solidi RRL: Rapid Research Letters*, 6(2), 65–67. doi:10.1002/pssr.201105463

**38** Spinelli, P., & Polman, A. (2012). Prospects of near-field plasmonic absorption enhancement in semiconductor materials using embedded Ag nanoparticles. *Optics Express*, 20(S5), A641–A654. doi:10.1364/OE.20.00A641

**39** van Lare, M., Lenzmann, F., & Polman, A. (2013). Dielectric back scattering patterns for light trapping in thin-film SI solar cells. *Optics Express*, 21(18), 20738–20746. doi:10.1364/OE.21.020738

**40** Sai, H., Kanamori, Y., & Kondo, M. (2011). Flattened light-scattering substrate in thin film silicon solar cells for improved infrared response. *Applied Physics Letters*, 98(11), 113502. doi:10.1063/1.3565249

**41** Biswas, R., & Xu, C. (2011). Nano-crystalline silicon solar cell architecture with absorption at the classical 4n2 limit. *Optics Express*, 19(S4), A664–A672. doi:10.1364/OE.19.00A664

**42** Wang, K. X., Yu, Z., Liu, V., Cui, Y., & Fan, S. (2012). Absorption enhancement in ultrathin crystalline silicon solar cells with antireflection and light-trapping nanocone gratings. *Nano Letters*, 12(3), 1616–1619. doi:10.1021/nl204550q

**43** Campbell, P., & Green, M. A. (1986). The limiting efficiency of silicon solar cells under concentrated sunlight. *IEEE Transactions on Electron Devices*, 33(2), 234–239. doi:10.1109/T-ED.1986.22472

**44** Winston, R. (1974). Principles of solar concentrators of a novel design. *Solar Energy*, 16(2), 89–95. doi:10.1016/0038-092X(74)90004-8

**45** Gaspare, M. (1984). High yielding solar modulus for the transformation, on photovoltaic cells, of a substantially monochromatic radiation in electric energy.

**46** Goetzberger, A., Goldschmidt, J. C., Peters, M. (2011). Photovoltaic device and use thereof. *Patent No. WO2011098212.*

**47** Peumans, P., Bulović, V., & Forrest, S. R. (2000). Efficient photon harvesting at high optical intensities in ultrathin organic double-heterostructure photovoltaic diodes. *Applied Physics Letters*, 76(19), 2650–2652. doi:10.1063/1.126433

**48** Atwater, J. H., Spinelli, P., Kosten, E., Parsons, J., Van Lare, C., Van de Groep, J., . . . Atwater, H. A. (2011). Microphotonic parabolic light directors fabricated by two-photon lithography. *Applied Physics Letters*, 99(15), 151113. doi:10.1063/1.3648115

**49** Yan, W., Hossain, M. M., & Gu, M. (2013). High light-directing micrometer-sized parabolic mirror arrays. *Optics Letters*, 38(16), 3177–3180. doi:10.1364/OL.38.003177

**50** Spinelli, P., Macco, B., Verschuuren, M. A., Kessels, W. M. M., & Polman, A. (2013). Al2O3/TiO2 nano-pattern antireflection coating with ultralow surface recombination. *Applied Physics Letters*, 102(23), 233902. doi:10.1063/1.4810970

**51** Luque, A., & Hegedus, S. (2011). *Handbook of photovoltaic science and engineering*. Wiley.

**52** Zhao, J., Wang, A., Green, M. A., & Ferrazza, F. (1998). 19.8% efficient “honeycomb” textured multicrystalline and 24.4% monocrystalline silicon solar cells. *Applied Physics Letters*, 73(14), 1991–1993. doi:10.1063/1.122345

**53** Kosten, E. D., Atwater, J. H., Parsons, J., Polman, A., & Atwater, H. A. (2013). Highly efficient GaAs solar cells by limiting light emission angle. *Light: Science & Applications*, 2, e45.

**54** Kosten, E. D., Kayes, B. M., & Atwater, H. A. (2014). Experimental demonstration of enhanced photon recycling in angle-restricted GaAs solar cells. *Energy & Environmental Science*, 7(6), 1907–1912. doi:10.1039/C3EE43584A

**55** Green, M. A. (2012). Radiative efficiency of state-of-the-art photovoltaic cells. *Progress in Photovoltaics: Research and Applications*, 20(4), 472–476. doi:10.1002/pip.1147

**56** Braun, A., Katz, E. A., Feuermann, D., Kayes, B. M., & Gordon, J. M. (2013). Photovoltaic performance enhancement by external recycling of photon emission. *Energy & Environmental Science*, 6(5), 1499–1503. doi:10.1039/C3EE40377G

**57** Yablonovitch, E. (1982). Statistical ray optics. *Journal of the Optical Society of America*, 72(7), 899–907. doi:10.1364/JOSA.72.000899

**58** Winston, R., Miñano, J. C., & Benítez, P. (2005). *Nonimaging optics*, Electronics & Electrical. Burlington, MA.

**59** Rabl, A., O'Gallagher, J., & Winston, R. (1980). Design and test of non-evacuated solar collectors with compound parabolic concentrators. *Solar Energy*, 25(4), 335–351. doi:10.1016/0038-092X(80)90346-1

**60** Solargenix Energy.

**61** Paire, M., Lombez, L., Péré-Laperne, N., Collin, S., Pelouard, J.-L., Lincot, D., & Guillemoles, J.-F. (2011). Microscale solar cells for high concentration on polycrystalline Cu(In,Ga)Se2 thin films. *Applied Physics Letters*, 98(26), 264102. doi:10.1063/1.3604789

**62** Jared, B. H., Saavedra, M. P., Anderson, B. J., Goeke, R. S., Sweatt, W. C., Nielson, G. N., . . . Haney, M. W. (2014). Micro-concentrators for a microsystems-enabled photovoltaic system. *Optics Express*, 22(S2), A521–A527. doi:10.1364/OE.22.00A521

**63** Gu, T., Agrawal, G., Vessey, A., Sweatt, W. C., Jared, B. H., Luis Cruz-Campa, J., . . . Haney, M. W. (2015). Micro-concentrator module for microsystems-enabled photovoltaics: Optical performance characterization, modelling and analysis. IEEE 42nd Photovoltaic Specialists Conference PVSC, 1–5. doi:10.1109/PVSC.2015.7356183.

**64** Jutteau, S., Paire, M., Proise, F., Lombez, L., & Guillemoles, J. (2015). Micro solar concentrators: Design and fabrication for microcells arrays. IEEE 42nd Photovoltaic Specialists Conference PVSC, 1–4. doi:10.1109/PVSC.2015.7355767

**65** Würfel, P., & Würfel, U. (2009). *Physics of solar cells: From basic principles to advanced concepts*. Weinheim: Wiley.

**66** Luque, A., & Miñano, J. C. (1991). Optical aspects in photovoltaic energy conversion. *Solar Cells*, 31(3), 237–258. doi:10.1016/0379-6787(91)90026-L

**67** Ulbrich, C., Peters, M., Bläsi, B., Kirchartz, T., Gerber, A., & Rau, U. (2010). Enhanced light trapping in thin-film solar cells by a directionally selective filter. *Optics Express*, 18(S2), A133–A138. doi:10.1364/OE.18.00A133

**68** Höhn, O., Kraus, T., Bauhuis, G., Schwarz, U. T., & Bläsi, B. (2014). Maximal power output by solar cells with angular confinement. *Optics Express*, 22(S3), A715–A722. doi:10.1364/OE.22.00A715

**69** Kosten, E. D., Newman, B. K., Lloyd, J. V., Polman, A., & Atwater, H. A. (2015). Limiting light escape angle in silicon photovoltaics: Ideal and realistic cells. *IEEE Journal of Photovoltaics*, 5(1), 61–69. doi:10.1109/JPHOTOV.2014.2360566

**70** van Dijk, L., Marcus, E. A. P., Oostra, A. J., Schropp, R. E. I., & Di Vece, M. (2015). 3D-printed concentrator arrays for external light trapping on thin film solar cells. *Solar Energy Materials & Solar Cells*, 139, 19–26. doi:10.1016/j.solmat.2015.03.002

**71** Green, M. A. (2002). Lambertian light trapping in textured solar cells and light-emitting diodes: Analytical solutions. *Progress in Photovoltaics: Research and Applications*, 10(4), 235–241. doi:10.1002/pip.404

**72** Nelson, J. (2003). *The physics of solar cells*. World Scientific.

**73** Ultimaker Homepage.

**74** Luque, A., Sala, G., & Arboiro, J. C. (1998). Electric and thermal model for non-uniformly illuminated concentration cells. *Solar Energy Materials & Solar Cells*, 51(3), 269–290. doi:10.1016/S0927-0248(97)00228-6

**75** Gordon, J. M., Feuermann, D., & Mashaal, H. (2015). Micro-optical designs for angular confinement in solar cells. *Journal of Photonics for Energy*, 5(1), 1–10.

**76** Holman, Z. C., De Wolf, S., & Ballif, C. (2013). Improving metal reflectors by suppressing surface plasmon polaritons: A priori calculation of the internal reflectance of a solar cell. *Light: Science & Applications*, (2), e106.

**77** Weinstein, L., Kraemer, D., McEnaney, K., & Chen, G. (2014). Optical cavity for improved performance of solar receivers in solar-thermal systems. *Solar Energy*, 108, 69–79. doi:10.1016/j.solener.2014.06.023

**78** Weinstein, L. A., Hsu, W.-C., Yerci, S., Boriskina, S. V., & Chen, G. (2015). Enhanced absorption of thin-film photovoltaic cells using an optical cavity. *Journal of Optics*, 17(5), 055901. doi:10.1088/2040-8978/17/5/055901

**79** Tormen, M., Inganäs, O., Tvingstedt, K., & Zilio, S. D. (2010). Photovoltaic device with enhanced light harvesting. *U.S. Patent Application 12/601,798.*

**80** Tvingstedt, K., Zilio, S. D., Inganäs, O., & Tormen, M. (2008). Trapping light with micro lenses in thin film organic photovoltaic cells. *Optics Express*, 16(26), 21608–21615. doi:10.1364/OE.16.021608

**81** Peters, M., Ulbrich, C., Goldschmidt, J. C., Fernandez, J., Siefer, G., & Bläsi, B. (2011). Directionally selective light trapping in a germanium solar cell. *Optics Express*, 19(S2), A136–A145. doi:10.1364/OE.19.00A136

**82** Kim, S. J., Margulis, G. Y., Rim, S.-B., Brongersma, M. L., McGehee, M. D., & Peumans, P. (2013). Geometric light trapping with a V-trap for efficient organic solar cells. *Optics Express*, 21(S3), A305–A312. doi:10.1364/OE.21.00A305

**83** van Dijk, L., van de Groep, J., Di Vece, M., & Schropp, R. E. I. (2016). Exploration of external light trapping for photovoltaic modules. *Optics Express*, 24(14), A1158–A1175. doi:10.1364/OE.24.0A1158

**84** van Dijk, L., van de Groep, J., Veldhuizen, L. W., Di Vece, M., & Schropp, R. E. I. (2017). Concepts for external light trapping and its utilization in colored and image displaying photovoltaic modules. *Progress in Photovoltaics: Research and Applications*, 25(7), 553–568. doi:10.1002/pip.2863

# 11

# 3D Printing of Fuel Cells and Electrolyzers

*A. Hornés[1], A. Pesce[1], L. Hernández-Afonso[2], A. Morata[1], M. Torrell[1], and Albert Tarancón[1,3]*

[1] *Department of Advanced Materials for Energy, Catalonia Institute for Energy Research (IREC), Barcelona, Spain*
[2] *Department of Chemistry, University of La Laguna, Tenerife, Spain*
[3] *Catalan Institution for Research and Advanced Studies (ICREA), Barcelona, Spain*

## 11.1 Introduction

Fuel cell and electrolysis systems are highly efficient energy technologies based on hydrogen for clean power generation and chemical storage, respectively. Depending on the nature of the electrolyte, this technology can be mainly divided into (a) Solid Oxide Cells (SOCs), which are based on ceramic materials; (b) Polymer Exchange Membrane Cells (PEMCs), which are based on polymeric electrolytes; (c) Alkaline Cells (ACs), which are based on aqueous alkaline solution. Apart from the nature of the materials involved, the main differences between technologies are related to their level of maturity (being the alkaline cells the most well-established) and the maximum achievable efficiency (being the solid oxide cells the one with highest proved efficiencies due to their operation at high temperatures).

Since fuel cell and electrolysis technologies are based on complex systems consisting of multilayer electrochemical cells, coated interconnectors with flow channels and especial gas manifolds, 3D printing technologies can positively contribute from different perspectives. Regarding the core component, that is, the electrochemical cells, 3D printing technologies are promising tools to increase the active area of the electrodes, fabricate graded compositions and intermediate functional layers, and minimize the use of expensive catalysts by infiltration or impregnation using automatic methods. Moreover, the specific energy and power of this type of cells can be substantially increased by designing more complex architectures not

*3D Printing for Energy Applications*, First Edition. Edited by Albert Tarancón and Vincenzo Esposito.

realizable with current existing technologies. This can be especially relevant for materials with expensive and limited manufacturing possibilities such as ceramics. In relation to the interconnects, apart from the use of 3D printing technologies such as inkjet for depositing protective coatings that resist the harsh operation conditions, one can easily imagine a remarkable contribution of this near net-shape manufacturing techniques for the fabrication of plates with more efficient current collection, better gas/liquid flow distribution, and easy sealing geometries. Finally, the 3D printing of *ad hoc* designed manifolds, and similar encapsulation parts, will represent a big advance in the always complex coupling of this type of energy stacks to the rest of the Balance of Plant (BOP).

The next sections are devoted to cover these and other relevant aspects of the use of 3D printing technologies for fuel cell and electrolysis systems by reviewing the existing literature and prospecting future advances still in embryonic stage or even currently under conception. In this regard, special attention is dedicated to the ambitious idea of printing whole stacks in a single step or to consolidate the underpinnings of the use of 3D printing technologies for a future mass customization of this type of energy technologies.

## 11.2 3D Printing of Solid Oxide Cells Technology

Solid Oxide Cells (SOCs) are ceramic-based electrochemical devices able to generate power from hydrogen when operating in fuel cell mode (Solid Oxide Fuel Cells, SOFC) and hydrogen from water electrolysis in electrolyzer mode (Solid Oxide Electrolyzer Cells, SOECs). This power generation and chemical storage duality converts SOC technology in a key player for future zero-emission energy scenarios based on hydrogen. Compared to other type of fuel cells such as Proton-Exchange Membrane or Alkaline Fuel Cells (PEMFCs or AFCs), the high temperature operation of SOCs ($T$ = 650–850°C) allows reaching remarkable conversion efficiencies above 60% in fuel cell mode, which can be increased above 85% if coupled to a cogeneration unit in a Combined Heat and Power system (CHP), and even higher values (over 90%) in the conversion from water to hydrogen, (i.e., electrolysis mode). On the other hand, this high operating temperature typically involves more complex BOPs due to the required thermal management.

SOCs are multilayered devices made of functional ceramics, ceramic composites, and/or ceramic-metal parts (cermets). This ceramic nature strongly limits their geometry, which is typically restricted to planar and tubular architectures, due to the shaping restrictions of the currently existing ceramic manufacturing techniques (see Figure 11.1 where a planar SOFC stack and its components are shown). Reaching new shapes able to increase the power density while minimizing thermal stresses (occurring during the heating up and cooling down of the devices) have been a dream of the community from a long time ago. Moreover, since the microstructure of these ceramic components

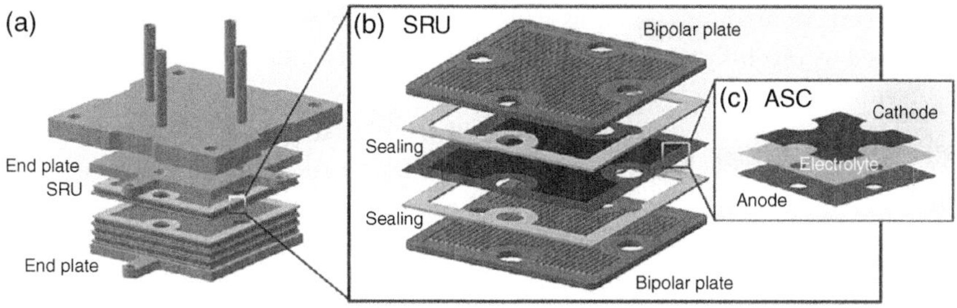


**Figure 11.1** (a) Scheme of a 5-cells SOFC stack, (b) components of a SOFC single-repeating unit (SRU), and (c) components of an anode-supported SOFC cell (ASC).

strongly affects the performance and durability of the cells, the precise design of the morphology of the layers and their interfaces has been a matter of study for decades. For these reasons, the recently offered new capabilities of 3D printing for implementing complex shapes and hierarchical features represents an ideal approach to positively contribute to this highly demanding application. An additional benefit is expected in the reduction of production costs by preventing the loss of valuable advanced materials moving from a subtractive manufacturing to an additive one and reducing the number of fabrication steps (shaping, thermal treatments, and assembly) [1].

### 11.2.1 Solid Oxide Fuel Cells

A SOFC cell consists of two electrodes, an anode and a cathode, which are separated by an oxide-ion conducting electrolyte. Electricity is produced by the direct fuel oxidation at the anode, while oxygen from air acts as oxidant in the cathode. The produced electrons flow through an external circuit, representing the most valuable power outcome, while ions flow through the ion conducting (but electronically insulating) electrolyte. As the involved reactions are exothermic, heat is also produced, which implies a self-sustained operation (providing an appropriate insulation) and usable heat.

Regarding microstructure, electrolyte must be dense to separate hydrogen and oxygen between the two chambers. In this way, oxide ions flow will be benefited. In contrast, electrodes should be porous to improve Triple Phase Boundary (TPB), which is the interphase where the gases get in contact with the electrode, and the electrodes meet the electrolyte, being, therefore, where electrochemical reactions take place.

Currently, yttria stabilized zirconia (YSZ) is the most commonly used electrolyte for SOFCs, due to its high stability in different atmospheres and its reasonable ionic conductivity at high temperatures. The ionic conductivity in YSZ mainly depends on yttria concentration, reaching a maximum value at 8 mol% (8YSZ). However, other compositions are commonly employed such as 3 mol% (3YSZ), which presents better mechanical properties but lower ionic conductivity [2].

Doped ceria compounds, such as gadolinia-doped ceria (GDC), have been also employed as electrolytes due to its high conductivity and lower reactivity than YSZ with typical cathode materials [3].

Regarding the electrodes, Pt was originally employed as an anode material, but nowadays, a wide range of cheaper materials have substituted it. Currently, the most commonly employed is the composite consisting of NiO and YSZ, which is *in situ* reduced to make a cermet of Ni-YSZ [4]. In this cermet, Ni plays the catalytic role while YSZ is a robust backbone that avoids agglomeration of the metal and increases the TPB region. On the cathode side, the states of art of materials are La-based perovskites, such as $La_{0.8}Sr_{0.2}MnO_3$ (LSM) and $La_{0.6}Sr_{0.4}Co_{0.2}Fe_{0.8}O_{2-\delta}$ (LSCF). LSM is typically used for high temperature applications, due to their limited catalytic activity at intermediate temperatures ($T < 800°C$), while LSCF is preferred for the lower temperature range even considering a high reactivity with YSZ that forces the use of diffusion barriers of CGO [5].

#### 11.2.1.1 SOFC Electrolyte

The electrolyte of SOFCs needs to hold high ionic conductivity with the purpose of minimizing the ohmic losses. It must be fully dense to separate the hydrogen and oxygen atmospheres while as thin as possible to reduce the ohmic contribution to the total resistance of the cell, which should be ideally below $0.5\,\Omega\,cm^2$ at the operating temperature. The former requirement restricts the number of additive manufacturing techniques that can be used for the fabrication of an ideal electrolyte, since full density is not easy to achieve by most of the available 3D printing techniques. Among others, 3D printing technologies including direct inkjet printing (IJP), selective laser sintering (SLS), robocasting, and stereolithography (SLA) have been used for this purpose so far.

Most of the reported trials to 3D print electrolytes have been carried out by employing direct inkjet printing. The reason is the great suitability of this technique to create very thin layers. A couple of pioneering works, published 10 years ago by Sukeshini, Cummins, Reitz, and Miller [6, 7], used inkjet printing to fabricate electrolytes, as well as electrodes. The cells exhibited maximum power densities of 430–460 $mW/cm^2$ at 850°C using hydrogen as fuel. More recently, Esposito et al. [8] deposited by IJP 1.2-μm-thick YSZ electrolytes on NiO-8YSZ tape casted anode supports. This, together with screen printed LSM-YSZ cathodes resulted in ASR values below $0.5\,\Omega\,cm^2$ at 750°C and open circuit voltages (OCV) of 1.15 V, which led to measure maximum power densities of 1,500 $mW/cm^2$ at 800°C. One year later, the same group presented YSZ structures obtained by hydrothermal synthesis of zirconia nano-colloids by the same technique. In this work, the possibility of implementing honeycomb patterns was demonstrated and, while not yet implemented in a measurable cell, the potential to enlarge the active area without thickness losses of the electrolyte was demonstrated [9]. Li, Shi, Ran, Su, and Shao [10] were also able to obtain YSZ electrolytes by using inkjet printing technology. They printed YSZ membranes of ~6 μm in thickness onto anode supports of NiO-YSZ. This thickness is obtained by four material deposition steps. Finally, they measured the performance of a full cell with a LSM cathode, producing a maximum power of density of 860 mW/cm$^2$ and 1.05 V of OCV at 800°C. Wang et al. focused on CGO as electrolyte material [11]. They developed stable CGO inks that were printed onto NiO-YSZ cermet substrates finally achieving 8 μm of thickness with 10 layer depositions. In the same direction, Mosiadz et al. [12] inkjetted CGO nanoparticles on highly textured Ni-5%W by using two techniques, namely, electromagnetic and piezoelectric inkjet printing. The piezoelectric actuator approach was able to produce layers free of cracks suitable for future implementation on SOFCs. El-Toni, Yamaguchi, Shimizu, Fujishiro, and Awano [13] used the same inkjet technique to print GDC electrolytes onto honeycomb LSM substrates, which were also fabricated with inkjet printing (20 printed layers). Similarly, Young et al. [14] used inkjet to print both components, the electrolyte and the anode.

In addition to the reduction of the thickness of the electrolyte, some works on this technology were addressed to improve the cell performance by structuration of the electrolyte/electrode interface [15, 16]. Following this strategy, Farandos, Kleiminger, Hankin, Ong, and Kelsall [17] and Farandos, Kleiminger, Li, Hankin, Kelsall [18] used for the first time IJP to produce 3D scaffolds of YSZ after previous simulation of the expected behavior of different patterns. In these works, they successfully printed YSZ micro-pillars and square lattices on NiO-YSZ anode substrates. Interestingly, Wang, Tomov, Kumar, and Glowacki [19] have dedicated efforts to avoid high-temperature and long-time densification of the electrolyte to reduce the risk of reactivity with the electrodes in future co-sintering approaches.

Beyond inkjet printing of thin layers, 3D printing is able to produce advantageous high-aspect ratio pieces with almost no shape restriction (free-form design). Using this approach, Masciandaro, Torrell, Leone, Tarancón [20] recently presented a seminal work focused on printing electrolytes using a commercial SLA printer. While SLA technology has long been subject of study for many years to produce zirconia parts [21], it was used for the first time for the fabrication of self-supported dense 3YSZ layers. Moreover, the high spatial resolution of this technique was exploited for structuring electrolytes with honeycomb conformation which were compared with a flat printed structure. Thickness values were 260 μm for the hexagonal cells with an active area of 1 mm$^2$ each, forming a network connected by 530 μm thick beams of 220 μm in width; the thickness for the flat cell was 340 μm with an active area of 1.54 cm$^2$. Complete fuel cells were made by painting electrode materials until complete cells were obtained with the structure: NiO-YSZ/YSZ(3D printed)/LSM-YSZ. Ionic conductivity values of

0.022 S/cm were obtained at 900°C in good agreement with previous works reported in the literature. Under fuel cell operation conditions, open circuit voltages of 1.14 V at 800°C were measured, perfectly matching the theoretical ones, indicating the gas-tightness of the 3D printed membrane. Power peak densities of 100 $mW/cm^2$ were obtained at 900°C for the flat cell, being compatible with the values expected for SOFCs based on 3YSZ electrolytes of the current thickness. In case of honeycomb cells, slightly higher values were found (115 $mW/m^2$) at the same temperature, attributed to an improvement of the active area. After this proof of concept, in the same group, Pesce et al. [22, 23] produced electrolytes in YSZ using new formulations with optimized yttria content. Corrugated cells were printed in order to increase the area approximately 60%, while keeping a homogeneous thickness over the electrolyte. After the functionalization with NiO-YSZ anode and LSM-YSZ cathode, the electrochemical characterization revealed an improvement compared to a planar design, printed as well, consistent with the improvement of the area, providing a maximum power density of 410 $mW/cm^2$ at 900°C for the proposed innovative design, compared to the 255 $mW/cm^2$ provided by the reference. Wei et al. [24] worked in the same direction, using DLP-stereo-lithography for the electrolyte production, manufacturing a 500 μm thick electrolytes based on 8YSZ and achieving 1.04 V of OCV and a maximum power density of 176 $mW/cm^2$ at 850°C.

Finally, it is worth mentioning that some activity in the fabrication of electrolyte materials has been recently reported using other techniques such as robocasting [25], Digital Light Processing (DLP) stereo-lithography [26], Ceramic On-Demand Extrusion (CODE) [27, 28], or Thermal Ink-Jet (TIJ) printing [29]. However, due to the previously mentioned specific features required for SOFC electrolytes, inkjet and SLA continue dominating this field of application.

#### 11.2.1.2 SOFC Electrodes

Due to the porous microstructure requirement for the electrodes, techniques such as: robocasting, laser sintering, or powder bed, are preferred for the fabrication of self-supported cathodes and anodes. Alternatively, inkjet is employed to deposit functional layers or infiltrate already existing porous scaffolds fabricated by other means. In these cases, 3D printing is used to carry out a functionalization more than a proper shaping step. This type of functionalization steps are considered a promising solution since latest results using functional layers and infiltrated composites in the field of SOFCs yielded excellent performances increasing the durability of the cells. Complementary, the current and future development of multi-material printing capabilities can be of major importance since graded compositions, especially in these functional layers, represent a reduction in the polarization resistances and, therefore, a great enhancement of the final device performance. This section summarizes the last advances including a comprehensive list of the technical parameters, cell designs, and performances recently reported in the literature for SOFC printed cells (Table 11.1).

***Anode*** As mentioned in previous sections, Ni-YSZ cermets are the standard material for SOFC anodes. In this particular case, although a certain control of the microstructure is preferred, porosity can be eventually reached after NiO reduction of dense NiO-YSZ layers. Therefore, it is interesting to study the possibility of controlling the microstructure of the layer by simply tuning 3D printing parameters. As an example, Yi, Effects, Hong, Kenneth, and Pei-chen [30] obtained dense parts of NiO-YSZ (60–40 wt.%) by SLS and SLM when using a $CO_2$ lasers while porous structures when employing a fiber laser. This tunability is interesting for the fabrication of anode supports, however, most of the existing works are focused on the deposition of anode functional layers on top of standard anode supports fabricated by other means (e.g., tape casting). In this other approach, the control of the porosity becomes less important while a gradation of the composition or a multi-material deposition become key factors. Regarding graded layers, Sukeshini, Meisenkothen, Gardner, and Reitz [31] reported functionally graded SOFC anode interlayers printed using Aerosol Jet Printing (AJP) achieving a power density of 235 mW/cm$^2$ at 850°C in anode-supported cells. Regarding the infiltration of existing porous scaffolds with complementary materials to increase the electrochemical performance, Wang et al. [32] detailed the successful use of inkjet printing for the infiltration of NiO-$Gd_{0.1}Ce_{0.9}O_{2-x}$ on NiO-8YSZ substrates obtaining a maximum power of density of 380 mW/cm$^2$ and an OCV of 0.9 V at 600°C in NiO-8YSZ/NiO-GDC/GDC/LSCF-GDC cells. More recently, Mitchell-Williams et al. [33] infiltrated CGO by IJP on a porous NiO-YSZ anode to reduce its polarization resistance while Dudek et al. fabricated multicomponent cermet anodes of Cu-Ni-YSZ for direct carbon fuel cells with solid oxide electrolyte (DC-SOFCs) [34, 35]. Combining Cu and Ni yielded an increase of 25% in the performance of the DC-SOFC cells.

Finally, other studies used inkjet technology to produce interlayers that improve the compatibility with the electrolyte, avoiding undesired reactions. This compatibility layers have been deposited over dense electrolytes [6, 7] or infiltrated on porous electrodes [31, 32]. As a particular example, Shimada, Ohba, Li, Hagiwara, and Ihara [36] impregnated an electrolyte-supported cell of ceria-doped and scandia-stabilized zirconia with $BaZr_{0.9}Y_{0.1}O_{3-\delta}$ (BZY) by IJP. BZY impregnation showed an enhanced performance of 790 mW/cm$^2$.

***Cathode*** Similar to anodes, most studies on 3D printing of cathodes were focused on the use of inkjet printing, covering the optimization of the necessary inks and the printing process. Since the cathode is typically the last layer in the fabrication

**Table 11.1** List of solid oxide fuel cells reported in the literature including at least one 3D printed layer.

| Support/printing method | Multilayer SOFC printed structure | OCV (V); MPD ($mW/cm^2$) | $T$ (°C) | References |
|---|---|---|---|---|
| Anode/inkjet | NiO-YSZ/**NiO-YSZ/YSZ/LSM-YSZ/LSM** | 1.1; 460 | 850 | [6] |
| Anode/inkjet | NiO-YSZ/**NiO-YSZ/YSZ**/LSM-YSZ/LSM | 1.1; 300 | 800 | [7] |
| Anode/inkjet | NiO-YSZ/**NiO-YSZ/YSZ/LSM-YSZ/LSM** | 1.1; 210 | 800 | [7] |
| Anode/inkjet | NiO-YSZ/**YSZ**/LSM-YSZ | 1.15; 1,500 | 800 | [8] |
| Anode/inkjet | Ni-YSZ/**YSZ**/LSM | 1.05; 860 | 800 | [10] |
| Anode interlayer/inkjet | **NiO/NiO-8YSZ/8YSZ**/LSM | 0.3; 1,750 | 750 | [14] |
| Anode/AJP | NiO-YSZ/**YSZ/LSM-YSZ/LSM** | 1.05; 400 | 850 | [48] |
| Anode/AJP | NiO-YSZ/**YSZ/GDC/LSCF** | 1.15; 600 | 800 | [48] |
| Electrolyte/SLA | NiO-YSZ/**3YSZ**/YSZ-LSM (flat) | 1.14; 100 | 900 | [20] |
| Electrolyte/SLA | NiO-YSZ/**3YSZ**/YSZ-LSM (honeycomb) | 1.14; 115 | 900 | [20] |
| Electrolyte/SLA | NiO-YSZ/**8YSZ**/YSZ-LSM (flat) | 1.1; 255 | 900 | [22] |
| Electrolyte/SLA | NiO-YSZ/**8YSZ**/YSZ-LSM (wave) | 1.1; 410 | 900 | [22] |
| Electrolyte/DLP | Ag-GDC/**8YSZ**/Ag-GDC | 1.04; 176 | 850 | [24] |
| Electrolyte/robocasting | NCAL/**SDC**/NCAL | 0.78; 448 | 550 | [25] |
| Electrolyte/AJP | NiO-YSZ/**NiO-8YSZ/8YSZ**/YSZ-LSM (infiltrated anode) | 1.2; 235 | 850 | [31] |
| Anode/inkjet | NiO-8YSZ/**NiO-GDC+GDC**/GDC/LSCF-GDC | 0.9; 380 | 600 | [32] |
| Electrolyte/inkjet | Ni-YSZ-**BZY**/$ScCeZrO_2$/LSM | ~1.1; 790 | 900 | [36] |
| Electrolyte/inkjet | **CGO**-NiO-8YSZ/8YSZ/LSCF | ~0.95; ~145 | 750 | [33] |
| Electrolyte/inkjet | **Ni-YSZ**/8YSZ/LSM-GDC/LSM | ~1.02; 96 | 800 | [34] |
| Electrolyte/inkjet | **Cu-Ni-YSZ**/8YSZ/LSM-GDC/LSM | ~1.02; 118 | 800 | [34] |

| | | | | |
|---|---|---|---|---|
| Electrolyte/inkjet | **Ni-8YSZ+Cu**/8YSZ/LSM-GDC/LSM | ~1.02; 80 | 780 | [35] |
| Anode/inkjet | NiO-YSZ/**8YSZ**/**LSM-YSZ**/**LSM** | 0.77; 690 | 788 | [47] |
| Anode/inkjet | NiO-GDC/GDC/**LSCF-GDC** | 0.43; 110 | 600 | [40] |
| Anode/inkjet | NiO-GDC/GDC/**LSCF-GDC**/LSCF-GDC | 0.94; 710 | 600 | [40] |
| Electrolyte/inkjet | NiO-GDC/3YSZ/LSCF/**LSC** | 1.04; 350 | 700 | [41] |
| Electrolyte/inkjet | NiO-GDC/3YSZ/LSCF/**GDC** | 1.04; 310 | 700 | [41] |
| Anode/inkjet | NiO-8YSZ/YSZ/SDC/**SSC-SDC** | 1.10; 950 | 750 | [46] |

Details on the support part of the cell, the printing method, the materials of the different layers, the OCV and maximum power density (MPD), and the operating temperature were included. The bold text indicates the layer manufactured by 3D printing.

flow of SOFCs, one have less restrictions regarding compatibility with the electrolyte or undesired densification during subsequent thermal treatments. Likely due to this advantage, there is an abundant collection of works on 3D printing of cathodes in the literature. Regarding the printing process itself, Chen et al. [37] were able to print LSCF using environmentally friendly aqueous suspension and Hill, Reitz, Rottmayer, and Huang [38] reached enough resolution to print arrays consisting of microdots of 60–90 μm in diameter. Still working on LSCF, Han et al. [39] optimized different parameters, such as deposition rate, viscosity, and porosity of the sintered layer, obtaining a maximum power density of 377 $mW/cm^2$ at 600°C. More specifically, Yashiro, Usui, and Kikuta [40] studied the advantages of using inkjet printing technology by comparing the performance of cells based on standalone air-brushed cathodes and cathodes combining inkjet printed and air-brushed layers. The material was again the state-of-the-art LSCF, this time in combination with GDC (LSCF-GDC), deposited on GDC dense electrolytes. The cells including the printed layers presented enhanced performances with maximum power densities c. 710 $mW/cm^2$ at 600°C (compared to 540 $mW/cm^2$ measured for the conventional cell). This work claims that the improvement is due to an induced higher number of Triple Phase Boundaries (TPBs) available for the electrochemical reaction.

As mentioned for anodes, the infiltration of existing scaffolds is a common strategy that involves inkjet printing. For instance, Venezia, Viviani, Presto, Kumar, and Tomov [41] infiltrated LSCF cathode layers with different materials such as $La_{0.6}Sr_{0.4}CoO_{3-\delta}$ (LSC) and GDC. The aim was to prove the beneficial effects of higher electrochemical activity of LSC compared with the pure LSCF, together with an improvement in the stability at the cathode–electrolyte interface due to the addition of GDC. Electrochemical measurements showed an improvement attributed to the implementation of these materials compared to a not infiltrated reference sample. The cell manufactured with the cathode infiltrated with LSC presented a power density of 350 $mW/cm^2$ at 700°C while the sample with GDC offered 310 $mW/cm^2$ at 700°C, both of them remarkably higher than the not-infiltrated sample which held 250 $mW/cm^2$ at 700°C. Similarly, Tomov, Mitchell-Williams, Gao, Kumar, and Glowacki [42] and Tomov et al. [43] used the inkjet printing to infiltrate LSCF/GDC cathodes with GDC and GDC-cobalt oxide nanoparticles reaching an improvement of the electrochemical properties and the stability of the cathode. Due to the excellent results obtained for infiltrated cells, 3D printing is also considered a promising candidate to scale up such type of functionalization processes giving them the required level of homogeneity, reproducibility, and velocity [44].

The optimization of the electrode porosity is one of the most important factors to increase the performance and durability of the cells. Different works were therefore devoted to study the capability of 3D printing to control this important parameter in SOFCs. Yu et al. [45] showed unique porous structures based on silver nanoparticles deposited by inkjet printing while Li, Chen, Shi, Tade, and Shao [46] introduced pore formers to produce cathode layers. Although the use of pore formers yielded in an improved performance of the cells fabricated by Li, Chen, Shi, Tade, and Shao [46] (maximum power of 950 vs. 550 $mW/cm^2$ at 750°C in NiO-8YSZ/YSZ/SDC/SSC-SDC configuration), even higher performances were obtained when using conventional techniques, which stresses the need for improving the current strategies to generate porosity by inkjet printing.

To end up with this section, it is important to remark that more complex multi-layered systems were also inkjet printed and tested by some authors. Farandos, Li, and Kelsall [47] used the inkjet printing technique to deposit LSM-YSZ/LSM bilayers over 3D printed YSZ electrolytes building up cells that reached a maximum power density of 690 $mW/cm^2$ at 788°C. Even one step beyond, in the already mentioned works of Sukeshini, Cummins, Reitz, and Miller [6, 7, 48], anode/electrolyte/cathode layers were all deposited by 3D printing giving rise to the first generation of fully printed cells (although supported on commercial anodes). With no doubt, this high value-added full-device concept becomes the most interesting approach when thinking about mass production of SOFCs by 3D printing since a single-step fabrication strategy can represent a competitive advantage compared to traditional manufacturing techniques.

### 11.2.2 Solid Oxide Electrolysis Cells

As previously mentioned, Solid Oxide Electrolysis Cells are electrochemical devices able to convert water into hydrogen and oxygen by injecting electricity. In other words, SOECs are SOFCs operated in reverse mode. Due to this analogy, the materials and configurations described above for SOFCs are also used in the state-of-art SOEC technology and, therefore, most of the developments on 3D printing of this type of fuel cells can be extrapolated to SOECs [49]. Despite these similarities and due to the relevance of the SOEC technology for future energy scenarios, it is worth mentioning some works specifically devoted to develop 3D printing for SOECs.

Farandos, Kleiminger, Hankin, Ong, and Kelsall [17] and Farandos, Kleiminger, Li, Hankin, Kelsall [18] proposed inkjet printing formulations for fabricating SOFC and SOEC functional layers reporting for the first time an electrolysis measurement upon 3D printing manufactured electrolytes. The cells were made by inkjet printing ~20-μm-thick 8YSZ layers over NiO-8YSZ substrates while painting LSM/YSZ and LSM as oxygen electrodes. OCV values of 0.84 V at 800°C in $CO_2$-CO/air atmospheres were obtained while a current density injection of −0.78 $A/cm^2$ was applicable at 1.5 V in SOEC mode. Further studies of the same group show results on SOEC cells with printed electrolyte and cathode. Dense layers of 8YSZ with 9 μm in thickness were obtained after inkjet printing and sintering them on top of NiO-8YSZ substrates. On the oxygen electrode side, LSM-YSZ interlayers combined with LSM cathodes (with a thickness of 9 and 20 μm, respectively) were deposited to complete the cells. The performance of such cells increased, in comparison with their previous study, obtaining an astonishing current density injection of $-3.33\,A/cm^2$ at 1.5 V at 788°C. These values clearly exceed the ones for cells produced by traditional manufacturing techniques proving the potential of inkjet printing for the fabrication of highly performing SOECs [47]. Apart from these seminal works based on inkjet printing, Pesce et al. [22] reported electrolysis measurements on electrolyte-supported cells fabricated by SLA. As previously mentioned, this work takes advantage of real high-aspect ratio 3D printed electrolytes to manufacture high-area corrugated YSZ cells. The increase of active area by design is directly correlated to the improvement of the performance also when tested in co-electrolysis mode. In this case, the corrugated cell allowed a current density injection of $-0.6\,A/cm^2$ at 1.3 V ($T = 900°C$), which represents a substantial increase when compared to the values of $-0.46\,A/cm^2$ reached for the planar counterpart (Figure 11.2).

### 11.2.3 SOC Stacks and Components

Apart from the fabrication of individual SOCs, which was extensively covered in the previous sections, 3D printing offers a promising alternative for the fabrication of other components of the stack or, eventually, the monolithic integration of different cells and functionalities into advanced systems. Conventional stacks consist of (a) tens of cells connected in series or in parallel by interconnectors (or bipolar plates), which are metal pieces that play the twofold role of electronic collection and gas flow distribution; (b) gas manifolds, which distribute the gases from the outside to the inside of the stack, and (c) current collectors.

Among other interesting advantages, 3D printing can be used to easily optimize the manifolds and flow distribution in the channels of the interconnectors or to improve the specific energy/power density by modifying the structural elements of the stacks. In these regards, Linne et al. [50] studied how to reduce the weight and homogenize the flow of SOEC stacks by 3D printing advanced manifolds to be used in spatial missions to Mars, where robustness, efficiency and power density per unit mass and volume are major drivers. Aligned with this enhancement of the structural and mechanical properties of the stacks is the fabrication of seal-free SOC architectures [51], which has been pursued for many years due to major issues observed when joining ceramic and metal pieces (usually by glass sealing) that

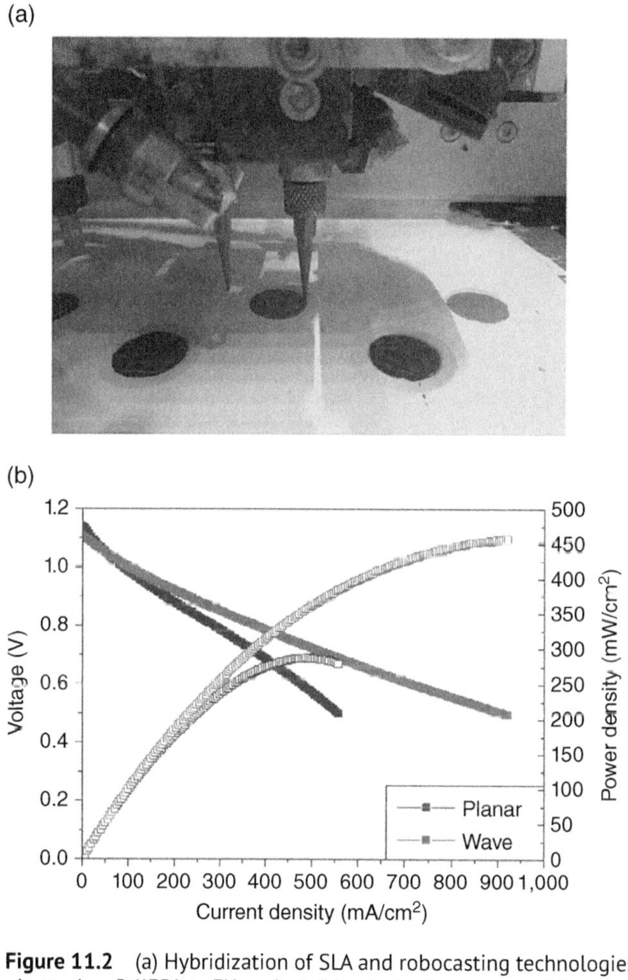


**Figure 11.2** (a) Hybridization of SLA and robocasting technologies within the pioneering Cell3Ditor EU project, (b) polarization curves of the planar and corrugated SLA printed electrolyte-supported SOFC cells (NiO-YSZ/YSZ/LSM) operated at 900°C under pure hydrogen and air atmospheres.

should resist strong thermal shocks and continuous cycling. 3D printing technologies can positively contribute to reach this monolithic approach by developing ambitious strategies to fully fabricate stacks using 3D printers. This is the case of the Cell3Ditor European project that targets the fabrication of full stacks with embedded functionality by combining SLA and robocasting into a hybrid machine [52, 53] (Figure 11.3). According to the Cell3Ditor project, the use of this technology will decrease the total cost of the fabrication of SOFC stacks in 63% (compared

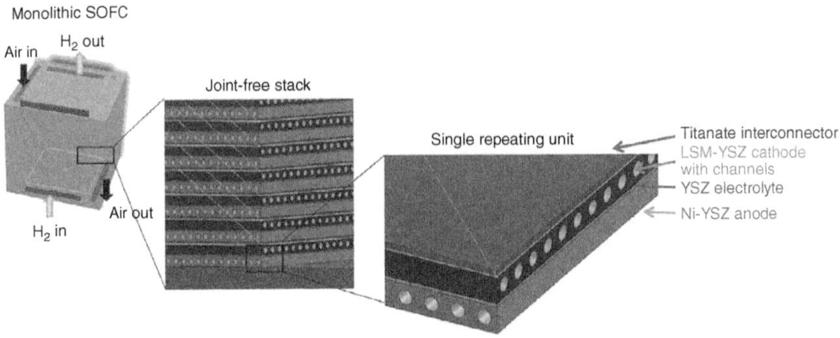


**Figure 11.3** Design of a fully 3D printed stack of Cell3Ditor project. From right to the left, single-repeating unit, joint-free stack and monolithic SOFC.

to conventional techniques) while reducing the amount of waste material above 80% and the energy consumption above 70% [54].

## 11.3 3D Printing of Polymer Exchange Membranes Cells Technology

PEM cells are a promising power generation or energy storage (in fuel cell mode or electrolyzer mode) technology which includes remarkable features such as: low operating temperature, high power density and easy scale-up, making them especially interesting for transport and mobile applications. The core of a PEM cell is the membrane-electrode assembly (MEA) that consists of the membrane, the catalysts layers (CLs, anode and cathode) and the gas diffusion layers (GDLs). Moreover, hardware components required to incorporate the MEA into a fuel cell contain: gaskets and bipolar plates (BPPs) (see Figure 11.4).

In these devices, the control of the microstructure of the different cell components plays a major role in their performance. In turn, hierarchical structured MEAs with functionally graded designs are capable of improving the cell performance by providing extended active area for reaction, interfacial transport and shorten diffusion paths while enhancing the mass transport of water, heat transfer, and electronic and protonic conductivity [55–57]. Moreover, smart design of the cell flow field plates contributes positively to water removal due to enhanced mass transport properties and to the uniform distribution of reactants to the CL which enables the utilization of ultra-low catalyst loading [57]. All this evidences the high impact of morphological aspects on the efficiency and performance of the devices, affecting ultimately their commercialization [58]. In this regard, one of the challenges involved in PEM systems commercialization is finding new

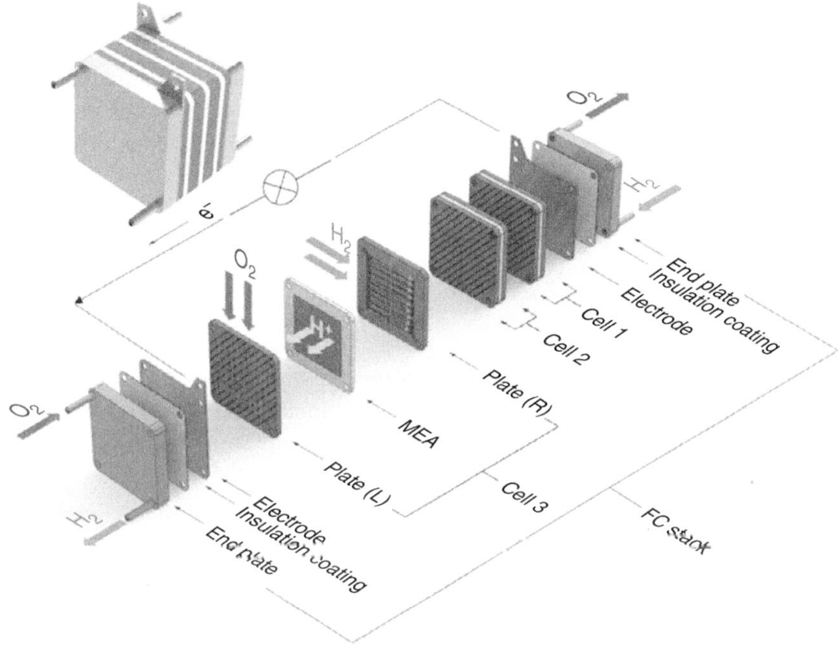


**Figure 11.4** Schematics of a PEM fuel cell stack operation and components where plates, R and L, contain their respective CL and GDL. *Source:* Sutharssan, T., Montalvao, D., Chen, Y. K., Wang, W.-C., Pisac, C., & Elemara, H. A review on prognostics and health monitoring of proton exchange membrane fuel cell. Renewable and Sustainable Energy Reviews 2017;75:440–450. © 2017, Elsevier.

fabrication methods which enable production of highly efficient and reproducible components at reduced costs.

### 11.3.1 Polymeric Exchange Membrane Fuel Cells

3D printing technologies enable the fabrication of PEM components in any desired 3D shape and dimensions, while preserving the multifunctional properties of the active materials, paving the way to optimized and more efficient designs for energy devices. In the subsequent sections, the different approaches followed for the fabrication of each of PEM fuel cell components by 3D printing techniques are unveiled. Afterward, the application of such technologies for the development of PEM electrolyzers will be described.

The type of materials currently used in PEM cells remain being the same since early 1990s [59]. For the electrolyte, perfluorosulfonic acid/tetrafluoroethylene copolymer (PFSA) membranes have been used for decades. These thin and flexible sheet membranes conduct electricity only when containing water, providing high proton conductivity as well as high chemical stability.

For the electrodes, platinum on carbon catalysts are used in both anode and cathode. Commercial electrodes usually contain c. 0.2–0.4 $mg/cm^2$ of platinum (Pt), generating a power density of 0.5–0.7 $W/cm^2$ [59]. Minimization of Pt content per kW is a critical issue of PEM technology due to the impact of its utilization in the final cost of the device. Water removal is another key factor in PEM systems performance, as oxygen diffusion coefficient in water is 5,700 times lower than in air at 60°C. In turn, gas diffusion media are added to the system in order to help with water management.

Finally, flow plates or bipolar plates are used for the homogeneous distribution of fuel gas and air, ensure electrical conduction between cells, remove heat from the active area, and prevent leakages. Common bipolar plates are made of high-density graphite, while currently metal or moldable graphite/polymer composites are used.

#### 11.3.1.1 PEMFC Electrolyte

Commercial PEM fuel cells operating at temperatures below 100°C mainly employ Nafion® membranes due to their suitable protonic conduction, low permeability to the reactant gas and electrons and, chemical and mechanical stability [60]. However, high production costs of these membranes, especially for small batches, have a negative impact on PEM fuel cells commercialization [61]. Recently, an original approach consisting in the direct deposition via inkjet printing of Nafion® dispersion onto the CLs of commercial gas diffusion electrodes have shown remarkable results [62]. In particular, the membrane showed a very low resistance and reduced charge transfer which led to a high power density (c. 4 $W/cm^2$) under pure $H_2/O_2$ operation. The relatively low thicknesses (8–25 μm) as well as the strongly decreased interfacial contact resistance between the membrane and the ionomer at the CLs were thought to be behind this improvement. It is noteworthy to mention the stable behavior shown by the system under completely dry conditions, keeping 1 $W/cm^2$, attributed to the internal humidification of the thin membrane via back-diffusion of the generated water. Subsequently, Breitwieser et al. [63] used this methodology to develop PEM fuel cells that incorporate GDEs with ultra-low platinum loading (0.102 and 0.029 $mg/cm^2$), attaining the highest platinum-utilization efficiency reported until that moment: 88 kW/g.

#### 11.3.1.2 PEMFC Catalysts Layer

Catalysts layer manufacturing has been of major importance in PEM cells development since their entry into mass-production industry, particularly due to the utilization of platinum as main component. The state-of-the-art commercial fabrication process presents two different routes [64, 65]: in the first method, an ink containing the catalyst is deposited onto the membrane and then combined with two GDLs while, in the second, the catalyst ink is coated onto the GDLs that are then hot-pressed with the membrane to form the MEA. Conventional techniques used for depositing catalyst layers include: hot-pressing [66], gravure coating [67], and screen printing [64, 68]. Even though great improvements have been done since the 1960s until now decreasing Pt loading by two orders of magnitude [69], catalysts still account for a 34% of the total cost of PEM stack [70]. In this regard, new methodologies based on additive manufacturing have been recently tested looking for an even greater reduction of Pt loading without any loss in performance and durability.

IJP is a promising alternative extensively used for fabrication of CLs especially due to its high cost-effectiveness, saving ink and reducing waste, extremely high resolution (with capacity to reach up to 1,200 dpi) [71], and superior control of Pt-loading and its distribution [65]. Among the early adopters [71, 72], special mention deserves the work of Taylor et al. where inkjet printed MEAs with ultra-low catalyst loading (0.020 mgPt/cm$^2$) were prepared. Outstanding ratio of obtained power by Pt utilization (17.6 kW/g of Pt) was attributed to TPB maximization associated to the decrease of CL thickness achieved by IJP. Moreover, graded catalyst deposition (a methodology that has demonstrated a better Pt utilization) [57, 73, 74] was attained by using IJP showing better performance than the uniformly distributed catalyst with the same loading. Another interesting analysis made by Saha et al. [75] where a comparison between the performance displayed by CLs prepared by different coating techniques, revealed a more homogeneous distribution of carbon support and ionomer, and a similar performance than a conventional decaled catalyst despite having 20% lower Pt loading. In a more recent work, where the effect of ionomer loading on thin, ultra-low Pt loading electrodes was assessed, IJP was selected as deposition technique [76]. In this study, a lack of correlation between the cell performance and Nafion loading was found, indicating a limiting effect of the interfacial oxygen transport which resistance dominates over bulk thin film transport resistance. A remarkable Pt utilization of 47.6 kW/g was found as well. Instead of depositing the catalyst layer, Wang and Nagao [77] attempted to deposit the Nafion® ionomer as transport media onto the CL by IJP in comparison with spraying. By using this methodology, a more intimate membrane/electrode interface was evidenced with much higher Pt utilization and a greater number of TPBs, showing a better performance for the inkjet-printed PEMFC.

#### 11.3.1.3 PEMFC Gas Diffusion Layer

The distribution of the reactive gases all over the membrane surface area is a critical factor governed by the GDL determining the power output of a PEM fuel cell [78]. A GDL is usually composed by two layers, the Macro-porous layer (MPL) and the Micro-porous layer formed both over the base of a carbon paper or carbon cloth substrates. The former is generated by applying a hydrophobic agent (i.e., PTFE) on the substrate while the latter, which helps in managing the water content of the MEA [79], is a far thinner layer made by coating the MPL with carbon black and additional PTFE. Recently, there have been several attempts for taking advantage of the benefits of 3D printing techniques for manufacturing GDLs. In particular, aiming to avoid issues derived from the use of carbon-based materials in the GDL [79–81], Jayakumar, Ramos, and Al-Jumaily [82] developed a carbon-free gas diffusion substrate by using selective laser sintering (SLS) based on a mixture of aluminum and polyamide (alumide). Although this study served for validating the feasibility of the approach proposed, additional post-processing by using graphene-based materials was necessary for increasing the electrical conductivity of the printed GDL. Moreover, alumide suffered electrochemical degradation under operating conditions. Based on the promising results obtained by SLS and its straightforwardness [83], the same authors developed a GDL substrate using a titanium/polyamide mixture [84]. Even if the microstructural features provided by SLS fabrication method were optimal and the higher conductivity of titanium powder, the electrical conductivity of the GDL was rather low, showing poor electrochemical performance and evidencing that further development is needed. In any case, the simplicity, cost-effectiveness, reduced lead time, and saving of material make SLS a good candidate.

#### 11.3.1.4 PEMFC Bipolar Plates and Flow Fields

Among the main parameters that control the performance of a PEM single-cell, water management plays a major role. Water is necessary to keep the membrane hydrated, being its level of hydration directly related to the conductivity displayed. In turn, with increasing the relative humidity (RH) of the membrane environment its water concentration raises, growing the volume fraction of the conducting phase within the membrane. However, an excessive amount of water can cause a flooding, filling the pores of the GDL and, consequently, obstructing the path for the reactants to reach the CL. Therefore, an efficient balance between the generation and removal of water is mandatory for achieving a good fuel cell performance. In this sense, the design of the flow fields is a key factor for avoiding water accumulation in the system and providing even gas distribution to the whole electrode surface area [85]. On the other hand, BPPs, which host the flow fields, cover essential functions as well: they provide mechanical support to the whole stack, electrical connection among cells, and aid with heat management [86]. Additionally, BPPs are a critical component to reduce the volume and the weight (especially relevant when thinking in transport applications) and, thus, the cost of the PEMFC stack. Considering the impact of the design in these elements, they have been one of the components where 3D printing techniques have been used more extensively.

A very preliminary approach was made by Chen et al. [87] where Fused Deposition Modelling (FDM) was used for manufacturing the flow fields of a miniature fuel cell. In comparison with Micro-Electro-Mechanical processes and CNC machining, FDM showed superior precision and lower production time. Moreover, power density of the FDM stack was comparable to that of state-of-the-art planar array stack.

Several authors have investigated the fabrication of graphite-based BPPs by means of indirect SLS. In comparison with conventional fabrication methodologies, such as, injection or compression molding, SLS allows building complex flow field patterns consuming less time and financial resources [88]. In turn, Guo and Leu [88] and Chen, Bourell, and Wood [89] fabricated composite graphite and phenolic resin BPPs with complex flow field designs. Unfortunately, SLS plates showed high porosity, and low mechanical and electrical properties, making necessary a post-processing stage. Liquid phenolic infiltration succeeds in this regard achieving a remarkable improvement of the electrical conductivity by c. 43% [89]. Another issue of SLS fabrication method is the high surface roughness of the pieces produced which induces poor contact between the MEA and BPP [90]. Laser re-melting process proposed by Yasa, Kruth, and Deckers [91] could be used to improving the surface smoothness. Meanwhile Guo and Leu [88, 92] in different works, developed BPPs with complex designs by SLS using a mixture of carbon-based materials which showed stable and comparable performance to conventional BPPs.

Recently, manufacturing of metallic-based BPPs has been assessed by several authors using different 3D printing techniques. Basically, they can be divided in two groups depending on the material chosen: titanium-based alloy or stainless steel. The material selection constrains the fabrication technique. In turn, Netwall, Gould, Rodgers, Nasello, and Swider-Lyons [93] and Gould et al. [94] have developed Direct Metal Laser Sintering (DMLS) process to make titanium-alloy BPPs with embedded flow channels. This methodology allows manufacturing of complex design flow fields and internal cooling channels with higher resolution than conventional machining processes, avoiding expensive tooling [95]. Previous studies [93] revealed the necessity, just after DMLS process, of a surface polishing posttreatment to reduce the contact resistance between the BPP and the GDL due to surface roughness as well as an additional coating in order to minimize surface resistance and improve the corrosion behavior under PEMFC environment. By using this methodology, Gould et al. [94] printed 21 $cm^2$ titanium-alloy BPPs with embedded flow channels which were afterwards assembled with the rest of the PEM components into both a single-cell and a 40-cell stacks and, subsequently, tested. While the single-cell stack performance was essentially similar to a standard stack, small differences attributed to higher contact resistance and higher hydrophobicity of the 3D printed BPP were found, the 40-cell stack displayed a

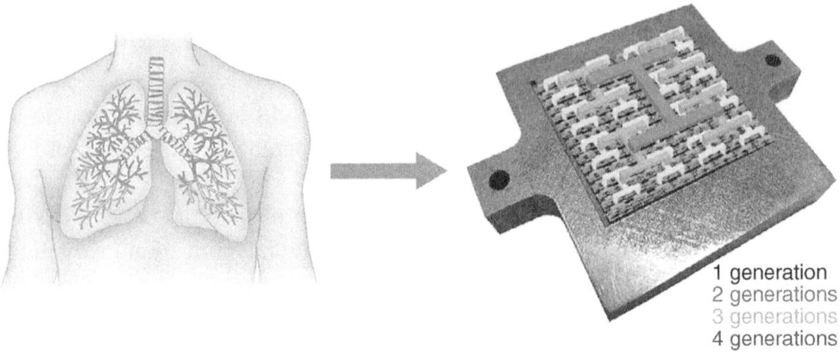


**Figure 11.5** Representation of the flow fields design for PEMFC inspired by the fractal characteristics of the lung [96]. *Source:* From Trogadas P, Cho JIS, Neville TP, Marquis J, Wu B, Brett DJL, & Coppens M O. A lung-inspired approach to scalable and robust fuel cell design. *Energy & Environmental Science* 2018;11(1):136–143. Licensed under CC BY 3.0.

20% lower current density than expected stemming from high contact resistances between the BPP and GDL caused by bending of the BPP during the fabrication.

In relation with BBPs made by stainless steel, special attention deserves the work made by Trogadas et al. [96] where flow fields with a fractal design mimicking the fractal geometry of the lung were 3D printed in stainless steel by DLMS (see Figure 11.5). In this work, a systematic approach using a formal mathematical description was used for optimizing the channel geometries (following the Murray's law [97]) and scaling them up. At small scale (c. 10 $cm^2$), 3D printed lung inspired flow fields with "N = 4 generations" (fractal with 4 generations of branching) outperform all other serpentine-based flow fields tested for comparison at high current densities and 50–75% of RH due to a more homogenous distribution of reactive gases across the catalyst layer. However, at higher RH the fractal 3D printed-based PEMFC suffered from flooding due to inadequate gas flow rate within the fractal field. This fact revealed the difficulty of fractal flow fields to purge water under high RH conditions. At larger scale (c. 25 $cm^2$), the same effect was observed, matching the performance of both systems: serpentine and fractal. In any case, commercial PEFC operating conditions never reach 100% RH [98] favoring the utilization of the proposed fractal flow fields.

Another approach for manufacturing stainless steel BPPs has been made by means of Selective Laser Melting (SLM) [99]. This study showed that the as-produced SLM samples exhibited lower endurance capacity under PEMFC operating conditions, although applied heat treatment improved the durability due a thicker passivation film. Finally, Piri [100] investigated the feasibility of Stereolithography (SLA) and Polyjet 3D printing as rapid prototyping techniques for testing novel flow field designs in BPPs. In this study SLA 3D printing was revealed as a promising screening tool for this purpose.

## 11.3.2 Polymer Exchange Membrane Electrolysis Cells

Similarly to what occurred for Solid Oxide Cells, PEM electrolyzers (PEMEC) present mainly the same components than their fuel cell counterparts. In contrast to PEMFCs, the diffusion layer must allow the entrance of water to reach the MEA, losing its hydrophobic character and being named liquid/gas diffusion layer (LGDL) due to that.

### 11.3.2.1 PEMEC Liquid Gas Diffusion Layer

In LGDLs, carbon-based materials, commonly used in PEMFCs, are not appropriate due to the high potential of the oxygen electrode [101]. For this reason, metallic LGDLs have attracted considerable attention for PEMEC. In this sense, Mo et al. [102] assessed the utilization of electron beam melting (EBM) 3D printing technique for the manufacturing of a titanium-based LGDL. Apart from the remarkable results obtained in terms of greater performance and efficiency when compared to conventional woven LGDLs, this work represents an important innovation and opens the possibilities to manufacture LGDLs with controlled microstructural features.

### 11.3.2.2 PEMEC Bipolar Plates and Flow Fields

As it happens for PEMFC, fabrication by 3D printing of BPPs and flow fields for PEMECs has been a quite active field in last years. The first example is the work presented by Chisholm, Kitson, Kirkaldy, Bloor, and Cronin [103] where FDM was used for printing flow plates from polypropylene that were afterward silver coated for conferring them electrical conductivity. Besides the promising electrochemical performance of a PEMEC built with these plates, this development opened the door to the fabrication of lightweight and cheaper electrolyzers. The work published by Hudkins, Wheeler, Peña, and Berlinguette [104] went in the same direction showing the fabrication in Acrylonitrile Butadiene Styrene (ABS) and Polylactic Acid (PLA) of 3D printed serpentine-based flow fields by FDM. Similarly, a coating process by Ni electroplating was applied, being the first demonstration of electroplating of any kind of metal on a 3D-printed plastic. Techno-economic assessment carried out for this study indicated that the proposed methodology could reduce by half the raw materials costs of the BPP, even more if stainless steel BPPs are replaced, facilitating future PEMEC mass deployment. Following the same trend, very recently Yang et al. [105] reported the fabrication by FDM of a non-conductive PLA bipolar plate only with the purpose of mass distribution and mechanical support. In this approach, current collection comes from a novel thin film LGDL fabricated by means of wet-etching. With this methodology, a reasonable low voltage was accomplished ($-1\,A/cm^2$ at 2.21 V) while degradation rate was kept relatively stable (c. 1 mV/hr) for 100 hr. From the economical point of view, this methodology decreases by one tenth the cost of production of the BPPs in comparison with graphite plates. Lately, Husaini, Kishida, and Affiqah Arzaee [106] manufactured a BPP of a photopolymer by digital light processing (DLP) 3D printing. Coating was necessary for attaining the required electrical conductivity. Overall, both conductivity and corrosion tests performed showed that the sample meets DOE standards for BPPs. On the other hand, Yang et al. [107] proposed a stainless steel plate with parallel flow field made by SLM for the first time. With such BPP, excellent performance for hydrogen production was shown (c. 1.78 V at $-2.0\,A/cm^2$) while the plate suffered very limited oxidation, demonstrating the benefits of this fabrication route.

#### 11.3.2.3 Fully Printed PEMEC

Only few groups have attempted to generate a fully 3D printed PEM electrolyzer; however, very promising results arise from their works. In this sense, Yang et al. [108] developed a PEMEC concept where the BPP, LGDL, gasket and current distributor were integrated into a single component that was fabricated by SLM of stainless steel powders. Significant improvement of the performance was observed due to virtually zero contact resistance of the integrated components. Moreover, stability test showed comparatively good results. Ambrosi and Pumera [109], trying to assess the catalytic behavior of several materials under operating conditions, have fabricated a PEMEC prototype using separately two different 3D printing techniques. In turn, SLM was used for printing the metal electrodes while FDM was employed for generating the polymeric gas/liquid handling components. An additional electrodeposition method was used for the application of catalysts onto the electrodes surface. Following the same idea, Lee, Taylor, Beirne, and Wallace [110] have fabricated a 3D printed PEMEC combining SLM for printing metallic electrodes with a tunable array of Ti-based conical structures, Polyjet 3D printing for the water-tight anodic and cathodic chambers and FDM for the locating base.

## 11.4 3D Printing of Bio-Fuel Cells Technology

Biofuel cells (BFCs), is a kind of PEM cell standing on catalytic reactions that provide electrons from a fuel and force it through an output circuit to a second electrode, while a flow of ions through an electrolyte balance the charges (See Figure 11.6 where an example of Microbial Electrolysis Cell is shown). The peculiarity of biofuel cells resides in the use of living organisms (microbial fuel cells)

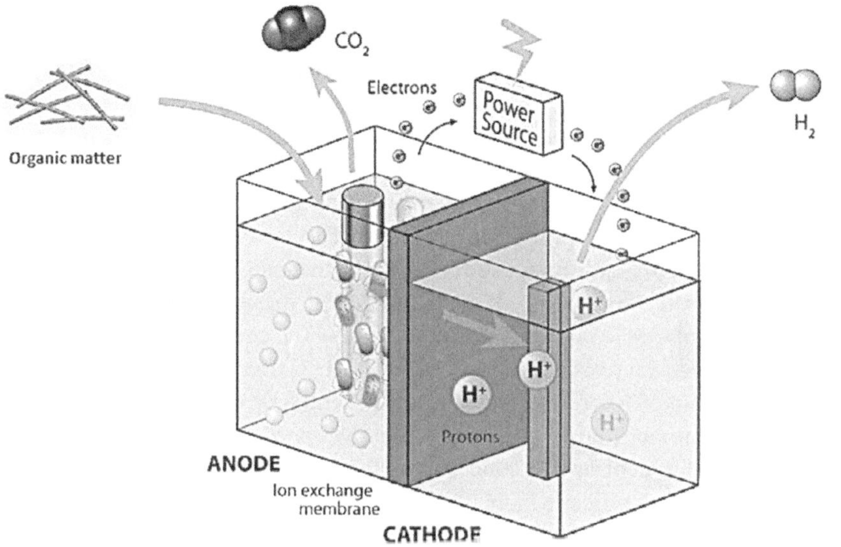


**Figure 11.6** A schematic of a microbial electrolysis cell (MEC) [114]. *Source:* From Deretsky Z. *A schematic of a microbial electrolysis cell (MEC)*. Wikimedia Commons/Public Domain; 2010. Retrieved from https://commons.wikimedia.org/wiki/File:Microbial_electrolysis_cell.png. Public Domain.

or enzymes from living cells (enzymatic fuel cells) to produce the electrochemical reactions, rather than metals or other inorganic catalysts [110, 111]. One of the main advantages of biofuel cells is the inherently cheap and environmentally friendly nature of their components. In contrast to the typically expensive or scarce catalyst, as nickel and platinum, which have to be mined, bacteria cells are very fast reproduced in the appropriate environment and enzymes are relatively straightforward to industrially fabricate [112]. Moreover, also the appropriate fuels to feed these devices, that is, alcohols and sugars, are very abundant and easy to extract from nature in a sustainable fashion, even typically coming from waste and effluent sources. Despite all these benefits, the use of microbial and enzymatic fuel cells has barely yet surpassed the boundaries of research laboratories, essentially due to a still modest provided power [113]. However, the technology is still considered highly promising thanks to the fruits of materials research and diversity of configurations tested in the last decades, which led to a remarkable increase of their performance.

As part of these improvement efforts, 3D manufacturing techniques have recently been adopted for the fabrication of biofuel cells [113–119]. The attractiveness of 3D printing technologies founds in the possibility of implementing appropriate structures at different scales. At the macroscopic level, bespoke configurations can provide highly compact designs with appropriate fluidics of the liquid feeds and exhausts, reduced ohmic losses, and overall biocompatible encapsulations [114, 115]. Papaharalabos, Greenman, Melhuish, and Ieropoulos [116] demonstrated the improvement of the performance due to the implementation of sophisticated architectures, examining the impact of using three different printable plastic feedstocks: medical-grade biocompatible Polycarbonate, ABS, and ceramic-filled photo curable resins. The access to complex structures has also shown to be useful for reducing the resistance of separator membrane elements, mostly because it allows overcoming the geometrical restrictions from typical planes to complex shapes, reducing the corresponding overpotentials and, hence, increasing the power density [118]. In this respect, early works from the group of Ieropoulos showed the potential of this approach using 3D printed membranes and non-(ion) selective porous polymeric materials. Further development of this concept and the implementation of a monolithic design led to a significant increase of performance, improving by a 40% the power output compared to a control sample made of commercially available components [120].

Additionally, 3D printing also offers the possibility of tuning the shape of the surfaces at the submillimeter scale. This offers very interesting opportunities mostly at the electrode components, which require (a) good electrical conductivity to provide efficient power transmission (minimizing ohmic loses), (b) high surface area for a maximum catalytic activity and a good contact with the environment, and (iii) biocompatibility that allows the preservation of microbes or enzymes [122]. The implementation of appropriate 3D printed microstructures can help enhancing surface area, favoring the release of reaction products (thus controlling the PH level), and implementing percolation paths to retain a good electronic conductivity. Following this concept, Calignano, Tommasi, Manfredi, and Chiolerio [119] used 3D printed anodes inspired the coral skeletal network structure, which led to an energy recovery close to 3 kWh/m$^3$/day. In a recent work from Bian et al. [121], this strategy demonstrated to improve the anode power outcome by up to a fourfold using DLP printed microstructures with periodicities ranging from 100 to 500 μm.

Finally, the flexibility of 3D printing also offers the possibility of tuning the surface chemistry by implementing tailored materials, which can provide the suitable finish for a good attachment of the bacteria or enzymes, ensuring appropriate interactions (hydrogen bonding, electrostatic or Van der Waals forces, hydrophilicity, etc.) [122]. Taking into account the last results in the field, 3D printing technologies are expected to play an important role in the development of future generations of biofuel cells.

## 11.5 Conclusions and Outlook

Fuel cell and electrolysis systems can clearly benefit from implementing 3D printing technologies in their manufacturing flow processes. Although the biggest impact is expected in the fabrication of the core components of these technologies, that is, the electrochemical cells, positive contributions are also envisaged concerning the interconnectors and other encapsulation components, such as manifolds or current collectors. Moreover, a more complex approach with fully printed high-aspect ratio compact designs or, eventually, the whole stack single-step fabrication with embedded functionality can represent an upcoming revolution in the field.

Among several reported advantages regarding an increase of the performance (by increasing the active area of the cells or improving the fuel flow distribution along the stack), already proved competitive advantages directly related to cost-effectiveness and low waste of expensive materials and energy make the 3D printing technology a promising candidate to replace currently existing manufacturing techniques for fuel cells and electrolysis (especially if a high level of reproducibility is achieved and mass customization is somehow required in the application scenario). In this regard, mobile applications like drones or variable-size markets such is the commercial segment, can lead this implementation.

Finally, it is worth mentioning the increasing interest in developing and standardizing at the industrial level multi-material 3D printing processes that allow fabricating complex devices like the here discussed fuel and electrolysis cells. 3D printing of such a high value-added products opens the door to transform an essentially prototyping technique into a fabrication tool.

## References


**1** Ruiz-Morales, J. C., Tarancón, A., Canales-Vázquez, J., Méndez-Ramos, J., Hernández-Afonso, L., Acosta-Mora, P., ... Fernández-González, R. (2017). Three dimensional printing of components and functional devices for energy and environmental applications. *Energy & Environmental Science*, 10(4), 846–859.

**2** Badwal, S. P. S. (1992). Zirconia-based solid electrolytes: Microstructure, stability and ionic conductivity. *Solid State Ionics*, 52(1–3), 23–32.

**3** Gao, Z., Mogni, L. V., Miller, E. C., Railsback, J. G., & Barnett, S. A. (2016). A perspective on low-temperature solid oxide fuel cells. *Energy & Environmental Science*, 9(5), 1602–1644.

**4** Edition, S., & Virginia, W. (2013). Fuel cell handbook. *Choice Reviews*, 26(11), 7.3–7.4.

5 Ruiz-Morales, J. C., Canales-Vázquez, J., Marrero-López, D., Peña-Martínez, J., Pérez-Coll, D., Núñez, P., ... Martín, C. S. (2008). *Pilas de combustible de óxidos sólidos (SOFC)* (2nd ed., pp. 198). Santa Cruz de Tenerife: Centro de la Cultura Popular Canaria (CCPC).

6 Sukeshini, A. M., Cummins, R., Reitz, T. L., & Miller, R. M. (2009 Dec). Inkjet printing of anode supported SOFC: Comparison of slurry pasted cathode and printed cathode. *Electrochemical and Solid-State Letters*, 12(12), B176.

7 Sukeshini, M. A., Cummins, R., Reitz, T. L., & Miller, R. M. (2009). Ink-jet printing: A versatile method for multilayer solid oxide fuel cells fabrication. *Journal of the American Ceramic Society*, 92(12), 2913–2919.

8 Esposito, V., Gadea, C., Hjelm, J., Marani, D., Hu, Q., Agersted, K., ... Højgaard Jensen, S. (2015, 2017). Fabrication of thin yttria-stabilized-zirconia dense electrolyte layers by inkjet printing for high performing solid oxide fuel cells. *Journal of Power Sources*, 273, 89–95.

9 Rosa, M., Gooden, P. N., Butterworth, S., Zielke, P., Kiebach, R., Xu, Y., ... Esposito, V. (2019). Zirconia nano-colloids transfer from continuous hydrothermal synthesis to inkjet printing. *Journal of the European Ceramic Society*, 39(1), 2–8.

10 Li, C., Shi, H., Ran, R., Su, C., & Shao, Z. (2013). Thermal inkjet printing of thin-film electrolytes and buffering layers for solid oxide fuel cells with improved performance. *International Journal of Hydrogen Energy*, 38(22), 9310–9319.

11 Wang, C., Hopkins, S. C., Tomov, R. I., Kumar, R. V., & Glowacki, B. A. (2012). Optimisation of CGO suspensions for inkjet-printed SOFC electrolytes. *Journal of the European Ceramic Society*, 32(10), 2317–2324.

12 Mosiadz, M., Tomov, R. I., Hopkins, S. C., Martin, G., Hardeman, D., Holzapfel, B., & Glowacki, B. A. (2010). Inkjet printing of Ce0.8Gd0.2O2 thin films on Ni-5%W flexible substrates. *Journal of Sol-Gel Science and Technology*, 54(2), 154–164.

13 El-Toni, A. M., Yamaguchi, T., Shimizu, S., Fujishiro, Y., & Awano, M. (2008). Development of a dense electrolyte thin film by the ink-jet printing technique for a porous LSM substrate. *Journal of the American Ceramic Society*, 91(1), 346–349.

14 Young, D., Sukeshini, A. M., Cummins, R., Xiao, H., Rottmayer, M., & Reitz, T. (2008). Ink-jet printing of electrolyte and anode functional layer for solid oxide fuel cells. *Journal of Power Sources*, 184(1), 191–196.

15 Bertei, A., Tariq, F., Yufit, V., Ruiz-Trejo, E., & Brandon, N. P. (2017). Guidelines for the rational design and engineering of 3D manufactured solid oxide fuel cell composite electrodes. *Journal of the Electrochemical Society*, 164(2), F89–F98.

16 Chueh, C.-C., Bertei, A., & Nicolella, C. (2019). Design guidelines for the manufacturing of the electrode-electrolyte interface of solid oxide fuel cells. *Journal of Power Sources*, 437, 226888.

17 Farandos, N., Kleiminger, L., Hankin, A., Ong, C. K., & Kelsall, G. H. (2015). 3D printing of functional layers for solid oxide fuel cells and electrolysers. *Meeting*

*Abstracts of ECS XIV International Symposium on Solid Oxide Fuel Cells, Glasgow 2015*, 03(1), 343–343.

**18** Farandos, N. M., Kleiminger, L., Li, T., Hankin, A., & Kelsall, G. H. (2016). Three-dimensional inkjet printed solid oxide electrochemical reactors. *Electrochimica Acta*, 213, 324–331.

**19** Wang, C., Tomov, R. I., Kumar, R. V., & Glowacki, B. A. (2011). Inkjet printing of gadolinium-doped ceria electrolyte on NiO-YSZ substrates for solid oxide fuel cell applications. *Journal of Materials Science*, 46(21), 6889–6896.

**20** Masciandaro, S., Torrell, M., Leone, P., & Tarancón, A. (2019). Three-dimensional printed yttria-stabilized zirconia self-supported electrolytes for solid oxide fuel cell applications. *Journal of the European Ceramic Society*, 39(1), 9–16.

**21** Kirihara, S. (2014). Creation of functional ceramics structures by using stereolithographic 3D printing. *Transactions of JWRI*, 43(1), 5–10.

**22** Pesce, A., Hornés, A., Núñez, M., Morata, A., Torrell, M., & Tarancón, A. (2020). Next-generation of 3D printed electrolytes for solid oxide cells with high power density. *Journal of Materials Chemistry A*. doi:10.1039/D0TA02803G

**23** Pesce, A., Hernández, C., Sánchez, I. S. M., Morata, A., Torrell, M., & Tarancón A. (2018). Improving performance of SOC through 3D printing of electrolytes. Proceedings of XIII European SOFC & SOE Forum. Chapter 09-Sessions B05, B08-6/153, Luzerne.

**24** Wei, L., Zhang, J., Yu, F., Zhang, W., Meng, X., Yang, N., & Liu, S. (2019). A novel fabrication of yttria-stabilized-zirconia dense electrolyte for solid oxide fuel cells by 3D printing technique. *International Journal of Hydrogen Energy*, 44(12), 6182–6191.

**25** Feng, Z., Liu, L., Li, L., Chen, J., & Liu, Y. (2019). 3D printed Sm-doped ceria composite electrolyte membrane for low temperature solid oxide fuel cells. *International Journal of Hydrogen Energy*, 44(26), 13843–13851.

**26** He, R., Liu, W., Wu, Z., An, D., Huang, M., Wu, H., ... Xie, Z. (2018). Fabrication of complex-shaped zirconia ceramic parts via a DLP- stereolithography-based 3D printing method. *Ceramics International*, 44(3), 3412–3416.

**27** Ghazanfari, A., Li, W., Leu, M. C., Watts, J. L., & Hilmas, G. E. (2017). Additive manufacturing and mechanical characterization of high density fully stabilized zirconia. *Ceramics International*, 43(8), 5847–6636.

**28** Li, W., Ghazanfari, A., McMillen, D., Leu, M. C., Hilmas, G. E., & Watts, J. (2018). Characterization of zirconia specimens fabricated by ceramic on-demand extrusion. *Ceramics International*, 44(11), 12245–12252.

**29** Özkol, E. (2013). Rheological characterization of aqueous 3Y-TZP inks optimized for direct thermal ink-jet printing of ceramic components. *Journal of the American Ceramic Society*, 96(4), 1124–1130.

**30** Yi, K. H., Pei-Chen, S., Chen-Nan, S., & Jun, W. (2016). Effects of laser processing on nickel oxide-yttria stabilized zirconia. Proceedings of the 2nd International Conference on Progress in Additive Manufacturing, 367–373.

**31** Sukeshini, M. A., Meisenkothen, F., Gardner, P., & Reitz, T. L. (2013). Aerosol jet printing of functionally graded SOFC anode interlayer and microstructural investigation by low voltage scanning electron microscopy. *Journal of Power Sources*, 224, 295–303.

**32** Wang, C., Tomov, R. I., Mitchell-Williams, T. B., Kumar, R. V., & Glowacki, B. A. (2017). Inkjet printing infiltration of Ni-Gd:CeO2 anodes for low temperature solid oxide fuel cells. *Journal of Applied Electrochemistry*, 47(11), 1227–1238.

**33** Mitchell-Williams, T. B., Tomov, R. I., Saadabadi, S. A., Krauz, M., Aravind, P. V., Glowacki, B. A., & Kumar, R. V. (2017). Infiltration of commercially available, anode supported SOFC's via inkjet printing. *Materials for Renewable and Sustainable Energy*, 6(2), 1–9.

**34** Dudek, M., Tomov, R. I., Wang, C., Glowacki, B. A., Tomczyk, P., Socha, R. P., & Mosialek, M. (2013). Feasibility of direct carbon solid oxide fuels cell (DC-SOFC) fabrication by inkjet printing technology. *Electrochimica Acta*, 105, 412–418.

**35** Tomov, R. I., Dudek, M., Hopkins, S. C., Krauz, M., Wang, H., Wang, C., ... Glowacki, B. A. (2013). Inkjet printing of direct carbon solid oxide fuel cell components. *ECS Transactions*, 57(1), 1359–1369.

**36** Shimada, H., Ohba, F., Li, X., Hagiwara, A., & Ihara, M. (2012). Electrochemical behaviors of nickel/yttria-stabilized zirconia anodes with distribution controlled yttrium-doped barium zirconate by ink-jet technique. *Journal of the Electrochemical Society*, 159(7), 360–367.

**37** Chen, Z., Ouyang, J., Liang, W., Yan, Z. C., Stadler, F., & Lao, C. (2018). Development and characterizations of novel aqueous-based LSCF suspensions for inkjet printing. *Ceramics International*, 44(11), 13381–13388. doi:10.1016/j.ceramint.2018.04.174

**38** Hill, T. Y., Reitz, T. L., Rottmayer, M. A., & Huang, H. (2015). Controlling inkjet fluid kinematics to achieve SOFC cathode micropatterns. *ECS Journal of Solid State Science and Technology*, 4(4), P3015–P3019.

**39** Han, G. D., Neoh, K. C., Bae, K., Choi, H. J., Park, S. W., Son, J. W., & Shim, J. H. (2016). Fabrication of lanthanum strontium cobalt ferrite (LSCF) cathodes for high performance solid oxide fuel cells using a low price commercial inkjet printer. *Journal of Power Sources*, 306, 503–509. doi:10.1016/j.jpowsour.2015.12.067

**40** Yashiro, N., Usui, T., & Kikuta, K. (2010). Application of a thin intermediate cathode layer prepared by inkjet printing for SOFCs. *Journal of the European Ceramic Society*, 30(10), 2093–2098. doi:10.1016/j.jeurceramsoc.2010.04.012

**41** Venezia, E., Viviani, M., Presto, S., Kumar, V., & Tomov, R. I. (2019). Inkjet printing functionalization of SOFC LSCF cathodes. *Nanomaterials*, 9(4), 1–16.

**42** Tomov, R. I., Mitchell-Williams, T., Gao, C., Kumar, R. V., & Glowacki, B. A. (2017). Performance optimization of LSCF/Gd:CeO2 composite cathodes via single-step inkjet printing infiltration. *Journal of Applied Electrochemistry*, 47(5), 641–651.

43 Tomov, R. I., Mitchel-Williams, T. B., Maher, R., Kerherve, G., Cohen, L., Payne, D. J., ... Glowacki, B. A. (2018). The synergistic effect of cobalt oxide and Gd-CeO2 dual infiltration in LSCF/CGO cathodes. *Journal of Materials Chemistry A*, 6(12), 5071–5081.

44 Da'As, E. H., Irvine, J. T. S., Traversa, E., & Boulfrad, S. (2013). Controllable impregnation via inkjet printing for the fabrication of solid oxide cell air electrodes. *ECS Transactions*, 57(1), 1851–1857.

45 Yu, C. C., Baek, J. D., Su, C. H., Fan, L., Wei, J., Liao, Y. C., & Su, P. C. (2016). Inkjet-printed porous silver thin film as a cathode for a low-temperature solid oxide fuel cell. *ACS Applied Materials & Interfaces*, 8(16), 10343–10349.

46 Li, C., Chen, H., Shi, H., Tade, M. O., & Shao, Z. (2015). Green fabrication of composite cathode with attractive performance for solid oxide fuel cells through facile inkjet printing. *Journal of Power Sources*, 273, 465–471.

47 Farandos, N. M., Li, T., & Kelsall, G. H. (2018). 3-D inkjet-printed solid oxide electrochemical reactors. II. LSM - YSZ electrodes. *Electrochimica Acta*, 270, 264–273.

48 Sukeshini, A. M., Gardner, P., Meisenkothen, F., Jenkins, T., Miller, R., Rottmayer, M., & Reitz, T. L. (2011). Aerosol jet printing and microstructure of SOFC electrolyte and cathode layers. ECS Meeting Abstracts, Volume MA2011-01, B7 - Solid Oxide Fuel Cells XII (SOFC-XII), 2151–2160.

49 Hauch, A., Ebbesen, S. D., Jensen, S. H., & Mogensen, M. (2008). Highly efficient high temperature electrolysis. *Journal of Materials Chemistry*, 18(20), 2331–2340.

50 Linne, D. L., Kuczmarski, M. A., Johnston, J. C., Farmer, S. C., Green, R. D., & Mital, S. K. (2018, September). Advanced manifolds for improved solid oxide electrolyzer performance. 2018 AIAA Sp Astronaut Forum and Exposition, Orlando, 1–6. 10.2514/6.2018-5174.

51 Barabasz, R., Dutton, E., Feldman, B, Lin, G., Mohanram, A., Narendar, Y., ..., Zandi, M. (2014). Saint-Gobain's all ceramic SOFC stack: Architecture and performance. Proceedings of the ASME 2014 12th International Conference on Fuel Cell Science, Engineering and Technology collocated with the ASME 2014 8th International Conference on Energy Sustainability, 6. Boston, MA: ASME. doi:10.1115/FuelCell2014-6715.

52 (2016). Cell3Ditor project on 3D printing tech for SOFC stacks. *Fuel Cells Bulletin*, 2016(10), 12. doi:10.1016/S1464-2859(16)30291-7

53 Cell3Ditor. (2016). Cell3Ditor. Retrived from http://www.cell3ditor.eu/

54 Tarancón, A., Morata, A., & Hornes, A. (2018). Mid-term review report Cell3Ditor (H2020-FCH JU, GA#700266).

55 Li, Y., Fu, Z. Y., & Su, B. L. (2012). Hierarchically structured porous materials for energy conversion and storage. *Advanced Functional Materials*, 22, 4634–4667.

**56** Xing, L., Shi, W., Su, H., Xu, Q., Das, P. K., Mao, B., ... Scott, K. (2019). Membrane electrode assemblies for PEM fuel cells: A review of functional graded design and optimization. *Energy*, 177, 445–464.

**57** Trogadas, P., Nigra, M. M., & Coppens, M. O. (2016). Nature-inspired optimization of hierarchical porous media for catalytic and separation processes. *New Journal of Chemistry*, 40(5), 4016–4026.

**58** Wang, Y., Chen, K. S., Mishler, J., Cho, S. C., & Adroher, X. C. (2011). A review of polymer electrolyte membrane fuel cells: Technology, applications, and needs on fundamental research. *Applied Energy*, 88(4), 981–1007.

**59** de Bruijn, F. A., Makkus, R. C., Mallant, R. K. A. M., & Janssen, G. J. M. (2007). Materials for state-of-the-art pem fuel cells, and their suitability for operation above 100°C. In T. S. Zhao, K.-D. Kreuer, & T. B. T.-A. Van Nguyen (Eds.), *Advances in fuel cell* (pp. 235–336). Elsevier Science.

**60** Napporn, T. W., Karpenko-Jereb, L., & Pichler, B. (2018). Polymer electrolyte fuel cells. In V. Hacker & S. Mitsushima (Eds.), *Fuel cells and hydrogen* (pp. 63–89). Amsterdam: Elsevier.

**61** James, B. D., Huya-Kouadio, J. M., Houchins, C., & DeSantis, D. A. (2010). Mass production cost estimation of direct H2 PEM fuel cell systems for transportation applications: 2016 update. Final Report to the Department of Enegy by Strategic Analysis Inc.

**62** Klingele, M., Breitwieser, M., Zengerle, R., & Thiele, S. (2015). Direct deposition of proton exchange membranes enabling high performance hydrogen fuel cells. *Journal of Materials Chemistry A*, 3(21), 11239–11245.

**63** Breitwieser, M., Klingele, M., Britton, B., Holdcroft, S., Zengerle, R., & Thiele, S. (2015). Improved Pt-utilization efficiency of low Pt-loading PEM fuel cell electrodes using direct membrane deposition. *Electrochemistry Communications*, 60, 168–171. Retrieved from http://www.sciencedirect.com/science/article/pii/S1388248115002507

**64** Hill, M. R. (2013). *Preparation of catalyst coated membranes using screen printing*. University of Cape Town.

**65** Santangelo, P., Cannio, M., & Romagnoli, M. (2019). Review of catalyst-deposition techniques for PEMFC electrodes. *Tecnica Italiana-Italian Journal of Engineering Science*, 63(1), 65–72.

**66** Therdthianwong, A., Manomayidthikarn, P., & Therdthianwong, S. (2007). Investigation of membrane electrode assembly (MEA) hot-pressing parameters for proton exchange membrane fuel cell. *Energy*, 32(12), 2401–2411.

**67** Kapur, N. (2003). A parametric study of direct gravure coating. *Chemical Engineering Science*, 58(13), 2875–2882.

**68** Tang, H., Wang, S., Jiang, S. P., & Pan, M. (2007). A comparative study of CCM and hot-pressed MEAs for PEM fuel cells. *Journal of Power Sources*, 170(1), 140–144.

**69** Larminie, J., & Dicks, A. (2003). *Fuel cell systems explained* (pp. 406). Wiley.

**70** Andress, D., Das, S., Joseck, F., & Dean Nguyen, T. (2012). Status of advanced light-duty transportation technologies in the US. *Energy Policy*, 41, 348–364.

**71** Krebs, F. C. (2009). Fabrication and processing of polymer solar cells: A review of printing and coating techniques. *Solar Energy Materials & Solar Cells*, 93(4), 394–412.

**72** Taylor, A. D., Kim, E. Y., Humes, V. P., Kizuka, J., & Thompson, L. T. (2007). Inkjet printing of carbon supported platinum 3-D catalyst layers for use in fuel cells. *Journal of Power Sources*, 171(1), 101–106.

**73** Towne, S., Viswanathan, V., Holbery, J., & Rieke, P. (2007). Fabrication of polymer electrolyte membrane fuel cell MEAs utilizing inkjet print technology. *Journal of Power Sources*, 171(2), 575–584.

**74** Paganin, V. A., Ticianelli, E. A., & Gonzalez, E. R. (1996). Development and electrochemical studies of gas diffusion electrodes for polymer electrolyte fuel cells. *Journal of Applied Electrochemistry*, 26(3), 297–304.

**75** Saha, M. S., Tam, M., Berejnov, V., Susac, D., McDermid, S., Hitchcock, A. P., & Stumper, J. (2013). Characterization and performance of catalyst layers prepared by inkjet printing technology. *ECS Transactions*, 58(1), 797–806.

**76** Shukla, S., Domican, K., Karan, K., Bhattacharjee, S., & Secanell, M. (2015). Analysis of low platinum loading thin polymer electrolyte fuel cell electrodes prepared by inkjet printing. *Electrochimica Acta*, 156, 289–300.

**77** Wang, Z., & Nagao, Y. (2014). Effects of Nafion impregnation using inkjet printing for membrane electrode assemblies in polymer electrolyte membrane fuel cells. *Electrochimica Acta*, 129, 343–347.

**78** Williams, M. V., Begg, E., Bonville, L., Kunz, H. R., & Fenton, J. M. (2004). Characterization of gas diffusion layers for PEMFC. *Journal of the Electrochemical Society*, 151(8), A1173–A1180.

**79** Passalacqua, E., Squadrito, G., Lufrano, F., Patti, A., & Giorgi, L. (2001). Effects of the diffusion layer characteristics on the performance of polymer electrolyte fuel cell electrodes. *Journal of Applied Electrochemistry*, 31(4), 449–454.

**80** Antolini, E. (2003). Formation, microstructural characteristics and stability of carbon supported platinum catalysts for low temperature fuel cells. *Journal of Materials Science*, 38(14), 2995–3005.

**81** Kangasniemi, K. H., Condit, D. A., & Jarvi, T. D. (2004). Characterization of vulcan electrochemically oxidized under simulated PEM fuel cell conditions. *Journal of the Electrochemical Society*, 151(4), E125–E132.

**82** Jayakumar, A., Ramos, M., & Al-Jumaily, A. (2016). A novel 3D printing technique to synthesise gas diffusion layer for PEM fuel cell application. Proceedings of the ASME 2016 International Mechanical Engineering Congress and Exposition. IMECE2016. 6B, 1–5. Phoenix, US: American Society of Mechanical Engineers.

83 Schweiss, R., Meiser, C., Damjanovic, T., Galbiati, I., and Haak, N. (2016, February). *SIGRACET® Gas diffusion layers for PEM fuel cells, electrolyzers and batteries* [White Paper]. Meitingen, Germany. Retrieved from https://www.researchgate.net/profile/Ruediger_Schweiss/publication/295859224_SIGRACETR_Gas_Diffusion_Layers_for_PEM_Fuel_Cells_Electrolyzers_and_Batteries_White_Paper/links/56cece8308aeb52500c5416a.pdf

84 Jayakumar, A., Singamneni, S., Ramos, M., Al-Jumaily, A., & Pethaiah, S. (2017). Manufacturing the gas diffusion layer for PEM fuel cell using a novel 3D printing technique and critical assessment of the challenges encountered. *Materials (Basel)*, 10(7), 796–805.

85 Li, X., & Sabir, I. (2005). Review of bipolar plates in PEM fuel cells: Flow-field designs. *International Journal of Hydrogen Energy*, 30(4), 359–371.

86 Taherian, R. (2014). A review of composite and metallic bipolar plates in proton exchange membrane fuel cell: Materials, fabrication, and material selection. *Journal of Power Sources*, 265, 370–390.

87 Chen, C.-Y., Lai, W.-H., Weng, B.-J., Chuang, H.-J., Hsieh, C.-Y., & Kung, C.-C. (2008). Planar array stack design aided by rapid prototyping in development of air-breathing PEMFC. *Journal of Power Sources*, 179(1), 147–154.

88 Guo, N., & Leu, M. C. (2012). Effect of different graphite materials on the electrical conductivity and flexural strength of bipolar plates fabricated using selective laser sintering. *International Journal of Hydrogen Energy*, 37(4), 3558–3566.

89 Chen, S., Bourell, D. L., & Wood, K. L. (2005). Improvement of electrical conductivity of SLS PEM fuel cell bipolar plates. Proceedings of the 16th Solid Freeform Fabrication Symposium, SFF, Austin, TX.

90 Chen, S. (2006). *Fabrication of PEM fuel cell bipolar plate by indirect selective laser sintering*. Austin: University of Texas.

91 Yasa, E., Kruth, J.-P., & Deckers, J. (2011). Manufacturing by combining selective laser melting and selective laser erosion/laser re-melting. *CIRP Annals*, 60(1), 263–266.

92 Guo, N., & Leu, M. C. (2013). Performance investigation of polymer electrolyte membrane fuel cells using graphite composite plates fabricated by selective laser sintering. *Journal of Fuel Cell Science and Technology*, 11(1), 011003.

93 Netwall, C. J., Gould, B. D., Rodgers, J. A., Nasello, N. J., & Swider-Lyons, K. E. (2013). Decreasing contact resistance in proton-exchange membrane fuel cells with metal bipolar plates. *Journal of Power Sources*, 227, 137–144.

94 Gould, B. D., Rodgers, J. A., Schuette, M., Bethune, K., Louis, S., Rocheleau, R., & Swider-Lyons, K. (2015). Performance and limitations of 3D-printed bipolar plates in fuel cells. *ECS Journal of Solid State Science and Technology*, 4(4), P3063–P3068.

95 Khaing, M., Fuh, J. Y., & Lu, L. (2001). Direct metal laser sintering for rapid tooling: Processing and characterisation of EOS parts. *Journal of Materials Processing Technology*, 113(1–3), 269–272.

**96** Trogadas, P., Cho, J. I. S., Neville, T. P., Marquis, J., Wu, B., Brett, D. J. L., & Coppens, M. O. (2018). A lung-inspired approach to scalable and robust fuel cell design. *Energy & Environmental Science*, 11(1), 136–143.

**97** Murray, C. D. (1926). The physiological principle of minimum work: I. The vascular system and the cost of blood volume. *Proceedings of the National Academy of Sciences of the United States of America*, 12(3), 207–214.

**98** Yoshida, T., & Kojima, K. (2015). Toyota MIRAI fuel cell vehicle and progress toward a future hydrogen society. *Interface Magazine*, 24(2), 45–49.

**99** Kong, D., Ni, X., Dong, C., Zhang, L., Man, C., Yao, J., ... Li, X. (2018). Heat treatment effect on the microstructure and corrosion behavior of 316L stainless steel fabricated by selective laser melting for proton exchange membrane fuel cells. *Electrochimica Acta*, 276, 293–303.

**100** Piri, H. (2017). *Flow visualization in 3D printed PEM fuel cell bipolar plates.* University of British Columbia. Retrieved from https://open.library.ubc.ca/cIRcle/collections/ubctheses/24/items/1.0348828

**101** Mo, J., Steen, S. M., Zhang, F.-Y., Toops, T. J., Brady, M. P., & Green, J. B. (2015). Electrochemical investigation of stainless steel corrosion in a proton exchange membrane electrolyzer cell. *International Journal of Hydrogen Energy*, 40(36), 12506–12511.

**102** Mo, J., Dehoff, R. R., Peter, W. H., Toops, T. J., Green, J. B., & Zhang, F.-Y. (2016). Additive manufacturing of liquid/gas diffusion layers for low-cost and high-efficiency hydrogen production. *International Journal of Hydrogen Energy*, 41(4), 3128–3135.

**103** Chisholm, G., Kitson, P. J., Kirkaldy, N. D., Bloor, L. G., & Cronin, L. (2014). 3D printed flow plates for the electrolysis of water: An economic and adaptable approach to device manufacture. *Energy & Environmental Science*, 7(9), 3026–3032.

**104** Hudkins, J. R., Wheeler, D. G., Peña, B., & Berlinguette, C. P. (2016). Rapid prototyping of electrolyzer flow field plates. *Energy & Environmental Science*, 9(11), 3417–3423.

**105** Yang, G., Yu, S., Kang, Z., Dohrmann, Y., Bender, G., Pivovar, B. S., ... Zhanga, F. Y. (2019). A novel PEMEC with 3D printed non-conductive bipolar plate for low-cost hydrogen production from water electrolysis. *Energy Conversion and Management*, 182, 108–116.

**106** Husaini, T., Kishida, R., & Affiqah Arzaee, N. (2019). Digital light processing 3-dimensional printer to manufacture electrolyzer bipolar plate. *IOP Conference Series: Earth and Environmental Science*, 268(1), 012039.

**107** Yang, G., Mo, J., Kang, Z., List, F. A., Green, J. B., Babu, S. S., & Zhang, F. Y. (2017). Additive manufactured bipolar plate for high-efficiency hydrogen production in proton exchange membrane electrolyzer cells. *International Journal of Hydrogen Energy*, 42(21), 14734–14740.

**108** Yang, G., Mo, J., Kang, Z., Dohrmann, Y., List, F. A., Green, J. B., ... Zhang, F. Y. (2018). Fully printed and integrated electrolyzer cells with additive manufacturing for high-efficiency water splitting. *Applied Energy*, 215, 202–210.

**109** Ambrosi, A., & Pumera, M. (2018). Multimaterial 3D-printed water electrolyzer with earth-abundant electrodeposited catalysts. *ACS Sustainable Chemistry & Engineering*, 6(12), 16968–16975.

**110** Lee, C., Taylor, A. C., Beirne, S., & Wallace, G. G. (2019). A 3D-printed electrochemical water splitting cell. *Advanced Materials Technologies*, 4(10), 1900433.

**111** Rabaey, K., & Verstraete, W. (2005). Microbial fuel cells: Novel biotechnology for energy generation. *Trends in Biotechnology*, 23(6), 291–298.

**112** Santoro, C., Arbizzani, C., Erable, B., & Ieropoulos, I. (2017). Microbial fuel cells: From fundamentals to applications. A review. *Journal of Power Sources*, 356, 225–244.

**113** Logan, B. E., & Regan, J. M. (2006). Microbial fuel cells—Challenges and applications. *Environmental Science & Technology*, 40(17), 5172–5180.

**114** Deretsky, Z. (2010). *A schematic of a microbial electrolysis cell (MEC)*. Wikimedia Commons/Public Domain. Retrieved from https://commons.wikimedia.org/wiki/File:Microbial_electrolysis_cell.png

**115** Yang, Y., Liu, T., Tao, K., & Chang, H. (2018). Generating electricity on chips: Microfluidic biofuel cells in perspective. *Industrial and Engineering Chemistry Research*, 57(8), 2746–2758.

**116** Papaharalabos, G., Greenman, J., Melhuish, C., & Ieropoulos, I. (2015). A novel small scale microbial fuel cell design for increased electricity generation and waste water treatment. *International Journal of Hydrogen Energy*, 40(11), 4263–4268.

**117** Pasternak, G., Greenman, J., & Ieropoulos, I. (2016). Comprehensive study on ceramic membranes for low-cost microbial fuel cells. *ChemSusChem*, 9(1), 88–96.

**118** Philamore, H., Rossiter, J., Walters, P., Winfield, J., & Ieropoulos, I. (2015). Cast and 3D printed ion exchange membranes for monolithic microbial fuel cell fabrication. *Journal of Power Sources*, 289, 91–99.

**119** Calignano, F., Tommasi, T., Manfredi, D., & Chiolerio, A. (2015). Additive manufacturing of a microbial fuel cell—A detailed study. *Scientific Reports*, 5, 17373.

**120** You, J., Preen, R. J., Bull, L., Greenman, J., & Ieropoulos, I. (2017). 3D printed components of microbial fuel cells: Towards monolithic microbial fuel cell fabrication using additive layer manufacturing. *Sustainable Energy Technologies and Assessments*, 19, 94–101.

**121** Bian, B., Shi, D., Cai, X., Hu, M., Guo, Q., Zhang, C., ... Yang, J. (2018). 3D printed porous carbon anode for enhanced power generation in microbial fuel cell. *Nano Energy*, 44, 174–180.

**122** Guo, K., Prévoteau, A., Patil, S. A., & Rabaey, K. (2015). Engineering electrodes for microbial electrocatalysis. *Current Opinion in Biotechnology*, 33, 149–156.

# 12

# DED for Repair and Manufacture of Turbomachinery Components

*S. Linnenbrink[1], M. Alkhayat[1], N. Pirch[1], A. Gasser[1,2], and H. Schleifenbaum[1,3]*

[1] *Fraunhofer Institute for Laser Technology (ILT), Aachen, Germany*
[2] *Chair for Laser Technology (LLT), Aachen, Germany*
[3] *Digital Additive Production (DAP), Aachen, Germany*

## 12.1 Introduction

Direct energy deposition (DED) is well known for its suitability for repair applications. Nowadays also DED based fabrication of new components becomes more and more profitable and is in the focus of many current R&D projects. Both application scenarios, the repair case and the new component fabrication, require highly adapted DED process chains and data consistency throughout the whole process chain. The data consistency is precondition to fulfill demands such as shortest lead time in individualized production surroundings. This is enabled by improved technology approaches along the additive manufacturing (AM) process chain, for example adaptive machining methods, advanced path planning tools, and hybrid machine concepts.

In this publication the main process steps of two exemplary DED process chains for turbomachinery use cases (repair of turbine blades and new fabrication of turbocharger nozzle rings) are described. The repair of turbine blades as well as the new fabrication of turbocharger nozzle rings consist of mainly five process steps. Key challenge for repair of turbine blades is the acquisition of the actual geometry and based on this the adaption of the deposition weld strategies. The new fabrication of turbocharger nozzle rings is based on a hybrid production approach in order to reduce the number of re-clamping operations and shorten the processing

*3D Printing for Energy Applications*, First Edition. Edited by Albert Tarancón and Vincenzo Esposito.

time. The hybrid production approach allows an on-demand production and in this way contributes to a reduction of storage costs for nozzle ring spare parts.

In the field of turbomachinery maintenance and new component production additive manufacturing technologies, such as direct energy deposition (DED), contribute to a flexible and profitable production of small batch sizes. In direct comparison to powder bed processes such as laser powder bed fusion (LPBF), the DED process stands out through its inherent characteristics, which allow direct processing on freeform surfaces on large components (Figure 12.1) [1]. These inherent characteristics are key enablers for use cases such as repair of turbine blades and for hybrid production approaches. The two exemplary DED process chains are described in the following (Figure 12.2). Process chain A is applied for the repair, and process chain B for the manufacturing of components.

Components in stationary gas turbines or jet engines are exposed to high thermomechanical loads and a corrosive atmosphere. Highly advanced materials and coatings are developed and used for components to increase the efficiency of turbines. For these high-valued components, the importance of repair and overhaul of service exposed parts increases immensely [2–4]. Especially turbine blades are subject to various damages after a certain period of operation [5–7], and the geometries of the used parts differ from the geometry provided by original CAD data. In order to obtain the actual geometry data of the component and to determine the volume to be recovered, reverse engineering is a common procedure [8, 9].

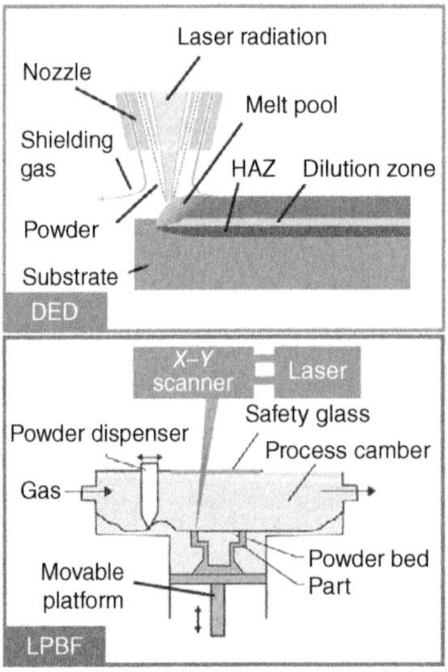


| Characteristics | Technology DED | LPBF |
|---|---|---|
| Attainable density | ≥ 99.5 % | ≥ 99.5 % |
| Component dimensions | Limited by path of machine tool | Limited by size of process chamber |
| Complexity of component | No hollow structures, limited overhangs | Hollow structures, overhangs |
| Gradient layers | Yes | No (limited) |
| Processing on | Almost all 3D-surfaces | Flat surfaces |
| Fields of application | Coating and repair, additive manufacturing | Additive manufacturing (repair) |

**Figure 12.1** Characteristics of DED and LPBF processes [1]. *Source:* Based on Alkhayat, M. & Gasser, A. (2016, April 27–29). LMD and SLM in additive manufacturing – A technology comparison. *AKL – International Laser Technology Congress.* Symposium conducted at the meeting of Fraunhofer ILT, Aachen. © 2016, Fraunhofer Institute for Laser Technology ILT.

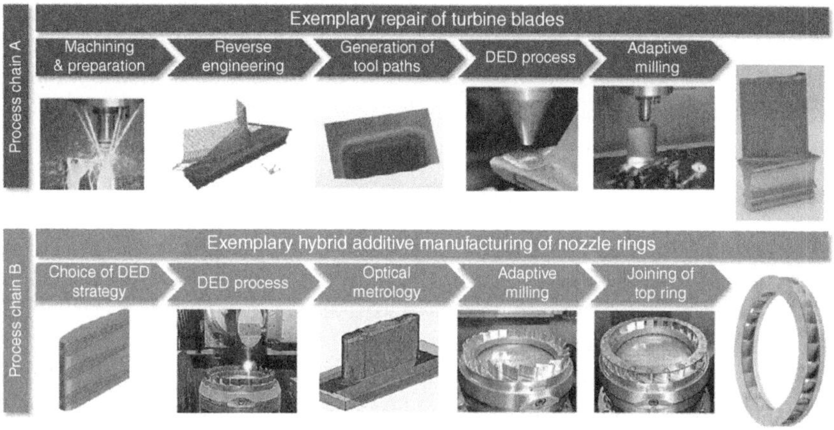


**Figure 12.2** Exemplary process chains (top: repair of turbine blades, bottom: fabrication of new nozzle rings).

The repair process chain (process chain A, Figure 12.2, above) is shown for a gas turbine blade platform.

The hybrid additive manufacturing of nozzle rings (process chain B, Figure 12.2, below) requires the design of an adapted and integrated hybrid production cell. This is precondition to exploit the full DED potential while minimizing and compensating the influence of inherent process characteristics (e.g., surface roughness and need for post-processing). Within the development of process chain B additional hardware is developed, integrated into the base machine and qualified.

## 12.2 DED Based Repair of Turbomachinery Components

Powder based DED is a suitable and proven additive manufacturing technology for the repair and overhaul of metal parts (e.g., from turbomachinery or aviation) [3, 10]. Due to typical technological advantages as low heat input and high accuracy, repair of gas turbine components such as turbine blade tips is already current practice [3, 4].

Due to the variety of damages that typically occur, the geometry of the components after operation differs from the geometry provided by the nominal CAD data [6, 7]. Therefore the nominal CAD data cannot be used directly for the tool path planning for the repair of the component [9]. Experience has shown that the damages of the components differ in such a way that the repair of the blades can only be performed in small series, or even individual repairs are necessary.

The damaged areas have to be machined before the DED process. This means, that a smooth, clean, crack, and oxide free surface must be achieved.

A minimally invasive approach is recommended to ensure that repairs are as economical as possible, and to affect the structure of the component as little as possible. Therefore, the prepared surfaces are digitized, the original CAD model is transferred to the undamaged area of the model by a best fit, and the missing volumes are obtained by reverse engineering [8, 9].

### 12.2.1 DED Process

During the DED process a melt pool is created on the surface of a workpiece by means of laser radiation. A powder nozzle is used to inject powder additive material into the melt pool simultaneously (Figure 12.1) [11]. In order to protect the melt pool from oxidation, an inert gas is applied both as a conveying gas and as protective gas. Argon, helium, or nitrogen is typically used for this purpose [11]. In general, single dots, tracks, layers, and volumes can be deposited by DED. Layers are produced by depositing overlapping tracks, and volumes consist of layers on top of each other [11]. Different types of powder nozzles are available for forming the powder-gas jet [9, 11]. The experiments presented are performed with a coaxial powder nozzle [12].

### 12.2.2 Work Environment

In general, a variety of machine tools and lasers can be combined for the procedure described here. In this example an *X–Y–Z*-machine tool in cross table design and an Nd:YAG laser are used. The laser radiation is guided by an optical fiber with a diameter of 600 μm. Both, collimation and focusing lens have a focal length of 200 mm. To avoid collision during processing between the turbine blade and the powder nozzle, the turbine blade is mounted in a slightly tilted position.

A powdery nickel-based superalloy is used as additive material.

For the acquisition of the actual point cloud of the worn out areas, a line scanner is used.

### 12.2.3 Process Chain for the Repair of Turbine Blades

The exemplary process chain for the repair of a turbine blade platform (process chain A) roughly consists of five steps (Figure 12.2, above).

#### 12.2.3.1 Step 1: "Machining & Preparation"

Main objective of the repair process is to obtain near net shape deposits without defects such as cracks, pores or lack of fusion, and also impurities as oxides have to be avoided. Depending on their size and frequency, all these undesired features can impair the mechanical properties and thus lead to premature failure of the component. Therefore it is essential to prepare the affected areas appropriately by milling, EDM, or grinding (Figure 12.3).

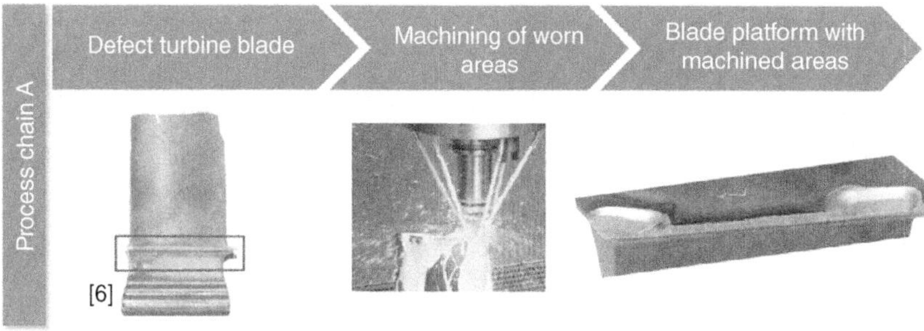


**Figure 12.3** Process chain A: defect areas (left, [6]) must be machined (middle) to create a smooth surface without defects or oxides and with a suitable geometry for the DED process. *Source:* From Błachnio, J. & Pawlak, W. I. (2011). Damageability of gas turbine blades – Evaluation of exhaust gas temperature in front of the turbine using a non-linear observer. *Advances in gas turbine technology*, Ernesto Benini (Ed.) InTech, ISBN: 978-953-307-611-9, doi:10.5772/29546. Retrieved from http://www.intechopen.com/books/advances-in-gas-turbine-technology/damageability-of-gas-turbine-bladesevaluation-of-exhaust-gas-temperature-in-front-of-the-turbine-us.

The final prepared surface has to be clean, smooth, and free from defects or oxides. The final machined surface is typically a 3D free-form surface. To avoid lack of fusion it is also important that the laser radiation can reach any point on the surface and that a suitable angle of incidence between the laser radiation and surface is achieved. The machined area is kept as small as possible. In this specific case, two areas on the platform of a turbine blade were damaged and therefore prepared as described for the DED process (Figure 12.3, right).

#### 12.2.3.2 Step 2: "Reverse Engineering"

Typically, the geometry of service exposed turbine blades do not match with the geometry provided by the original CAD data any more. In this case a reverse engineering procedure offers the possibility to obtain the actual geometry and determine the volume to be deposited.

First of all the actual geometrical data of the damaged platform is captured by optical metrology (Figure 12.4, left). For this purpose a line scanner is integrated into the machine setup and the laser line is adjusted with a resolution of 512 dots in a line of 23–41 mm, depending on the local distance between measured area and line scanner. The line scanner is connected via encoder interface with the control unit of the machine tool, and the current position of the scanner is saved. The scanner has an internal 2D coordinate system, which is aligned in parallel to the $X$–$Z$ plane of the machine coordinates. Consequently the turbine blade is moved in $Y$-direction during the measurement procedure. The distance between two adjacent line profiles was set to approximately 80 μm.

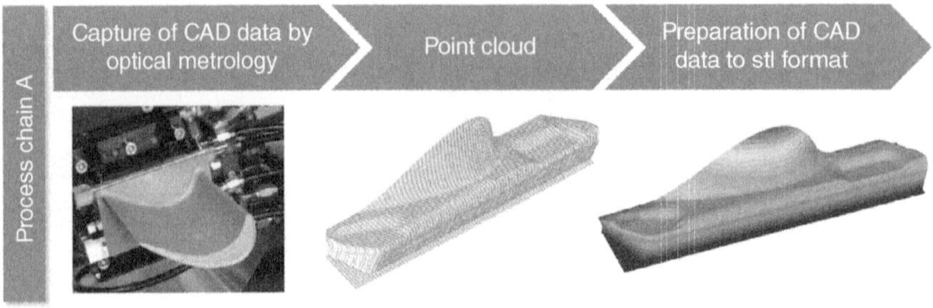


**Figure 12.4** Defect areas are digitized by optical metrology (left). The resulting point cloud (middle) has to be prepared and transferred to stl format (right).

The measured coordinates of each line profile are combined with the associated scanner position in the machine tool, and thus the position of each point in machine coordinates is determined. The result is a point cloud of the defect area (Figure 12.4, middle). From the points of the point cloud a triangulated mesh is derived and stored as an stl file (Figure 12.4, right). The data are further prepared to reduce noise and obtain a smooth surface.

In parallel, a master geometry of the area to be repaired is generated. For this purpose, a master area is separated from the original CAD model (Figure 12.5, left) which contains the damaged area and sufficient overlap with an undamaged area on the turbine blade with which a best fit can be carried out.

For the best fit, suitable points of the master geometry and the actual geometry are fitted by using the ICP (Iterative Closest Point) algorithm [13]. With this best fitted data a variance comparison is carried out, and the missing volume is determined (Figure 12.6, left and middle). After DED a machining process has

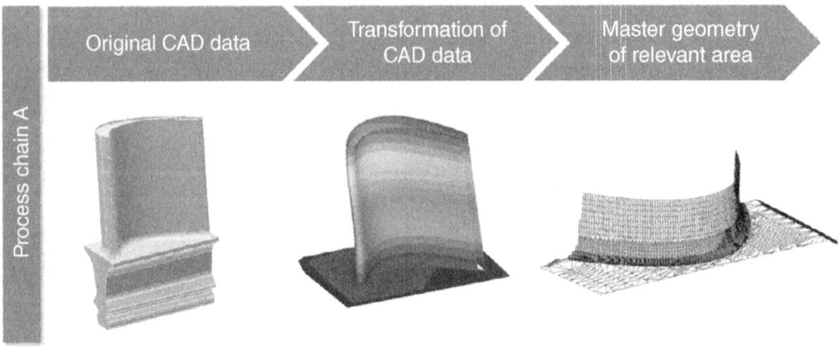


**Figure 12.5** In parallel the original CAD data (left) have to be prepared (middle), and the relevant area extracted as master geometry (right).

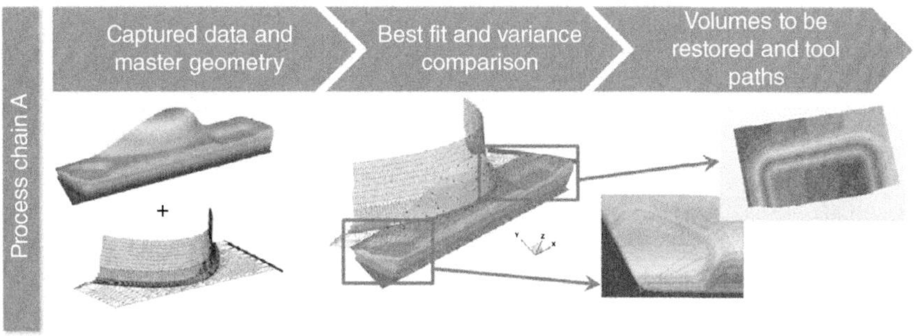


**Figure 12.6** With the stl file of captured data and master (left) best fit and variance comparison is performed (middle) to determine the volume to be restored. Tool paths are generated on the basis of the determined volume (right).

to be carried out, and for this a certain oversize of the deposit is necessary. Therefore, a suitable oversize is added to the volume determined by the variance comparison.

#### 12.2.3.3 Step 3: "Generation of Tool Paths"

For the generation of the tool paths an in-house CAM software especially designed for the DED process is used [14]. First the stl file of the measured CAD data is loaded into the CAM software. The area to be deposited is set by a boundary, and this boundary is filled with equidistant tool paths that lay directly on the 3D surface of the stl file. The distance between two tool paths is about half of the desired track width to obtain an overlap of adjacent tracks. For the other layers the tool paths of the first layer are duplicated once for each additional layer, and shifted in *Z* direction. To obtain an NC program a post processor exclusively programmed for the relevant machine tool is applied.

#### 12.2.3.4 Step 4: "DED Process"

The DED process is performed in the same setup as the optical metrology (Figure 12.7, left). The deposition of the volume to be repaired is carried out layerwise. Every layer consists of overlapping tracks which are welded in meanders. To achieve an improved robustness and preventing lack of fusion, the feed direction is rotated by 90° from layer to layer. In order to protect the melt pool from the surrounding atmosphere, an additional argon shielding gas jet is conducted coaxially to the laser radiation through the center of the powder nozzle onto the melt pool.

The surface of the deposited volume is silver shiny, which means that the melt pool is successfully protected from oxidation, and an oversize of some tenths of a mm is achieved (Figure 12.7, second from left).

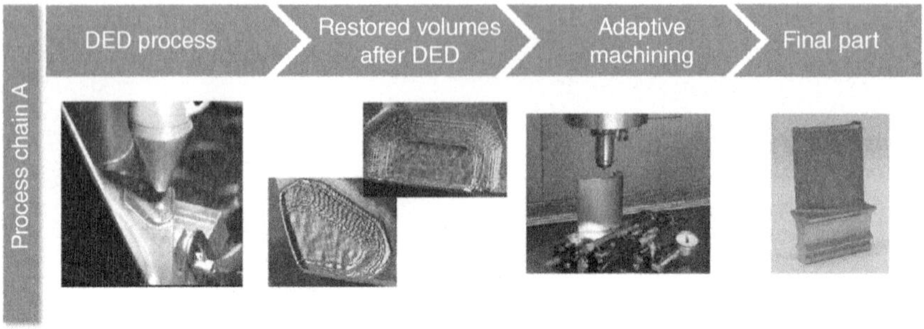


**Figure 12.7** Volumes to be restored are deposited by using a coaxial powder nozzle (left). After deposition small overhang and shiny surfaces are achieved (second from the left). After deposition the deposited parts have to be machined adaptively (second from the right). After Machining (right) the turbine blade is ready for heat treatment and recoating.

#### 12.2.3.5 Step 5: "Adaptive Machining"

The DED process delivers a wave-like surface after deposition. Therefore, the surface has to be machined after DED. A separate milling machine is used for this purpose (Figure 12.7, second from right).

Due to the run condition of the components, nominal CAD data cannot be used for post machining, too. Hence, the machine tool is equipped with a software module for adaptive post-machining [15].

With this software the tool paths can be automatically adapted to the actual geometry of the turbine blade.

To achieve this, the position of a suitable number of points on the non-welded surrounding on the platform is measured. With the coordinates of the measured points the desired final geometry of the platform can be calculated. The challenge is to achieve a smooth transition between welded and non-affected area and thus avoid notch effects which could lead to a failure of the component [15] (Figure 12.7, right).

## 12.3 DED Based Hybrid Manufacturing of New Components

The exemplary process chain (process chain B) for the fabrication of new nozzle rings, which is presented here, is investigated and qualified as part of the EU project HyProCell. Main goal is on one hand to make full use of DED advantages (such as its capability of transforming digital designs directly into physical products) while overcoming currently existing disadvantages of laser-based additive manufacturing processes by designing a fully integrated hybrid production cell on the other hand. In particular worth to be mentioned are the necessary post-processing operations, which are usually not fully integrated so that human intervention is needed to overcome technology gaps.

### 12.3.1 Hybrid Additive Manufacturing

A comprehensive classification of hybrid additive manufacturing (AM) is currently not provided in literature and according to [16] a distinction between additive and hybrid manufacturing methods is arbitrary and case-dependent. DED based hybrid manufacturing of new components is mainly motivated either by an increase of productivity (e.g., shortening of processing time) or by an increase of component value (e.g., increase of operating temperature). In general the implementation of DED based hybrid manufacturing can be achieved through three approaches (Figure 12.8) which are described in the following:

The first approach of DED based hybrid manufacturing is the combination of the DED technology together with an additional production technology in one machine. By the combination of at least two production technologies a hybrid process technology is created. In the common case hybrid process technologies include an additive and a subtractive manufacturing process. The integration of metal powder application (MPA) technology into a five-axis machining center is an example of this approach. Powder particles are accelerated to supersonic speed

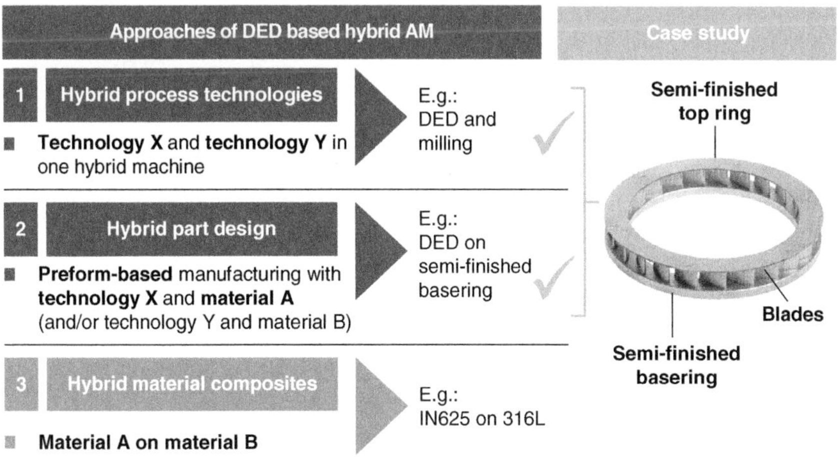


**Figure 12.8** Approaches of DED based hybrid AM [17]. *Source:* Based on Alkhayat, M. & Ulrichsohn, B. (2017, September 27–28). Hybrid additive manufacturing – An outlook on prospective manufacturing approaches. *3D Valley Conference.* Symposium conducted at the meeting of TEMA Technologie Marketing AG, Aachen.

and applied to a substrate. This has the advantage that the powder particles are not melted and only a small amount of energy is transferred to the component. In addition, different materials can be bonded together. By using a water-soluble carrier material, it is possible to integrate cooling channels within the component. The post-processing of the applied volume is carried out by the milling unit of the machining center [18].

A further possibility of combining additive and subtractive manufacturing processes is additive manufacturing using ultrasound (UAM, Ultrasonic Additive Manufacturing) as energy source for joining. Metal foils are joined by ultrasonic metal welding (USW) and the excess material is removed by CNC post-processing. This process offers advantages with regard to the range of joining materials with thermally and mechanically very different properties. In addition, functional modules can be integrated between the individual foils in a metal matrix [19].

As an example of the combination of a powder bed-based and a subtractive manufacturing process [20] can be cited. The LPBF process is combined with a high-speed milling process within one machine. This enables a high-precision construction of 3D geometries without intermediate steps. In comparison to the examples given so far, the following example represents a serial combination of two additive manufacturing processes. The advantages of the LPBF process, such as the freedom of geometric design possibilities, are be combined with the advantages of the DED process, such as the possibility of building up on almost any surface and the possibility of powder change during the process [21].

The so-called Controlled Metal Build Up (CMB) combines wire-based laser deposition welding as an additional module to an High Speed Cutting (HSC) milling machine for hybrid additive manufacturing processes. Using CMB, a volume is welded on semifinished components and reworked, usually for repair or modification of tools and dies [22].

In the present case study the combination of DED and multi axis milling in one hybrid production cell enables to forego additional re-clamping operations and time-consuming calibration processes, while running through the process chain.

The second approach of DED based hybrid manufacturing is based on the combination of materials in order to gain application-optimized component properties. The advantages of hybrid material composites with regard to an extended range of material properties when used for manufacture of injection molds are listed in [23]. In these investigations it proves to be advantageous that the tools produced can obtain excellent heat conduction properties as well as a high resistance to wear through the use of different materials. These advantages are also described by [24, 25] with focus on the bonding of high-performance fiber composites with classical structural materials such as metals to increase the mechanical properties of lightweight components. With the aim of reducing thermally induced stresses, the advantages of using hybrid additive processes are described in [17]. Stresses of turbomachinery components can be caused by temperature gradients in the hot gas section of engines and can be avoided by using two materials with different coefficients of thermal expansion. Due to the different coefficients of thermal expansion, a uniform expansion of the material combinations can be achieved and thermally induced stresses can be avoided. Another example of hybrid material combinations is the use of flexible coating processes using micro-DED. In the electrical industry, individualized gold contact layers are applied to substrate plates. By using the micro-DED, the material consumption of conductive materials can be significantly reduced [26].

The approach of hybrid material composites will not be further described in this case study.

The third approach of DED based hybrid manufacturing requires a hybrid part design, which allows the use of preforms or prefabricated semifinished goods. In most of the applications a redesign is necessary to reach a maximum of benefit. An example of this group is the modification of components in automobile construction in [27]. The aim of these investigations is the functional integration of mechanical and surface properties for component modification in small series. In order to achieve these modifications, reinforcement brackets are welded onto forged aluminum components by DED. Another example of hybrid component design are the approaches presented in Bremer [28] for the development of an economic process chain for the hybrid additive manufacture of large components. Flexible industrial robot systems are used, for example, for the assembly of functional elements on engine housings, to enable both wire and powder-based laser deposition welding with various filler materials. For the presented case study DED will be used to generate the vanes on a base ring and to bond the prefabricated top ring on top of the vanes.

### 12.3.2 Turbocharger Nozzle Ring

The exemplary component in the presented process chain B is a turbocharger nozzle ring (Figure 12.9). The nozzle ring is used for turbochargers of diesel and gas engines in the 400–3,300 kW power range. The nozzle ring is an important element for the efficiency of a turbocharger and it directs the exhaust gas flow to the turbine. Nozzle rings are available in different diameters and in a wide range of different angular vane positions in order to enable the turbocharger to be optimally designed with regard to the application and engine size. The variation in diameter and angular vane positions results in >1,000 variants of nozzle rings at one manufacturer. Such a large number of variants require extensive spare parts management. Various manufacturing processes are available for the conventional production of the nozzle rings, such as investment casting or the manufacture based on welded construction.

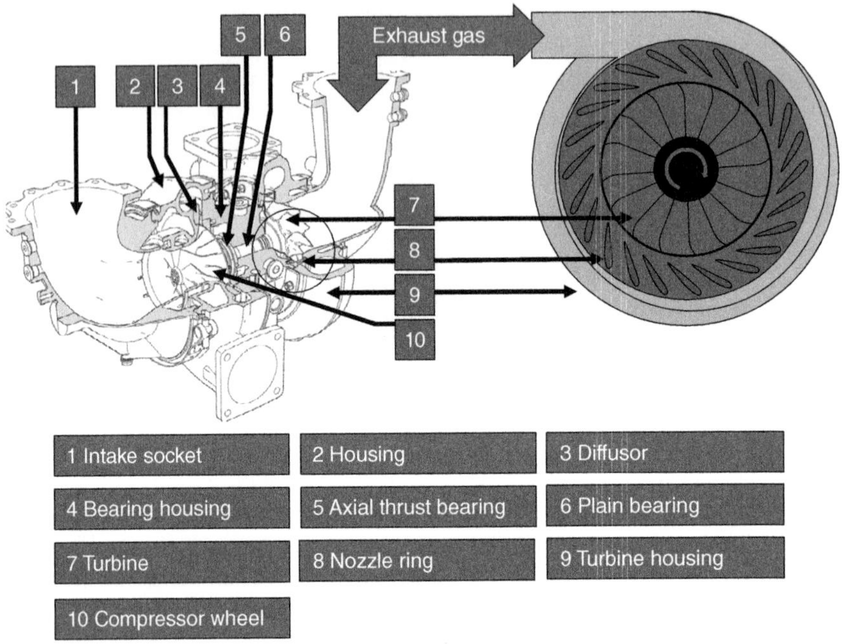


**Figure 12.9** Position of the nozzle ring inside of the turbocharger and further elements of a turbocharger.

The following hybrid approach shall provide a possibility to decrease the complexity of the spare part management and thus the cost.

The nozzle ring is divided into three areas within this work:

The base ring is used as a base plate, as a so-called preform for later additive assembly. In use, the base ring serves to fix and seal the turbine side of the turbocharger (Figure 12.8, right).

The elements of the nozzle ring that serve as guide vanes are defined as vanes in this publication and are generated via DED.

The top ring is joined onto the blades, also using DED. As with the base ring, a semifinished product is used as the material for the top ring. Like the base ring, the top ring serves to seal the turbine side of the turbocharger.

When being in operation, demanding conditions arise for turbocharger components. Due to the direct position of the nozzle ring within the 580–680 °C hot exhaust gas, its operating temperature lies within this range. This leads to a special requirement for the temperature resistance of the nozzle ring. The nozzle ring is made of austenitic, stainless steel X2CrNiMo17-12-2 (1.4404), which is suitable for operating temperatures up to 550 °C. Suitable hot gas corrosion resistance and scaling resistance are also required in this environment. Due to the large impurities in the fuel used on big vessels, in particular sulfur, an aggressive exhaust gas mixture is produced during combustion which settles on the turbocharger components. These deposits can lead to mechanical contact between the turbine and the nozzle ring or the turbo housing. As expected, a mechanical contact would result in increased wear of the components. As a result, a loss of efficiency is to be expected. This loss of efficiency can be remedied by replacing the components.

### 12.3.3 Hybrid Production Cell

The core part of the DED based machine tool for hybrid manufacturing of nozzle rings is capable of turning and milling processes, and is equipped with additional hardware (Figure 12.10).

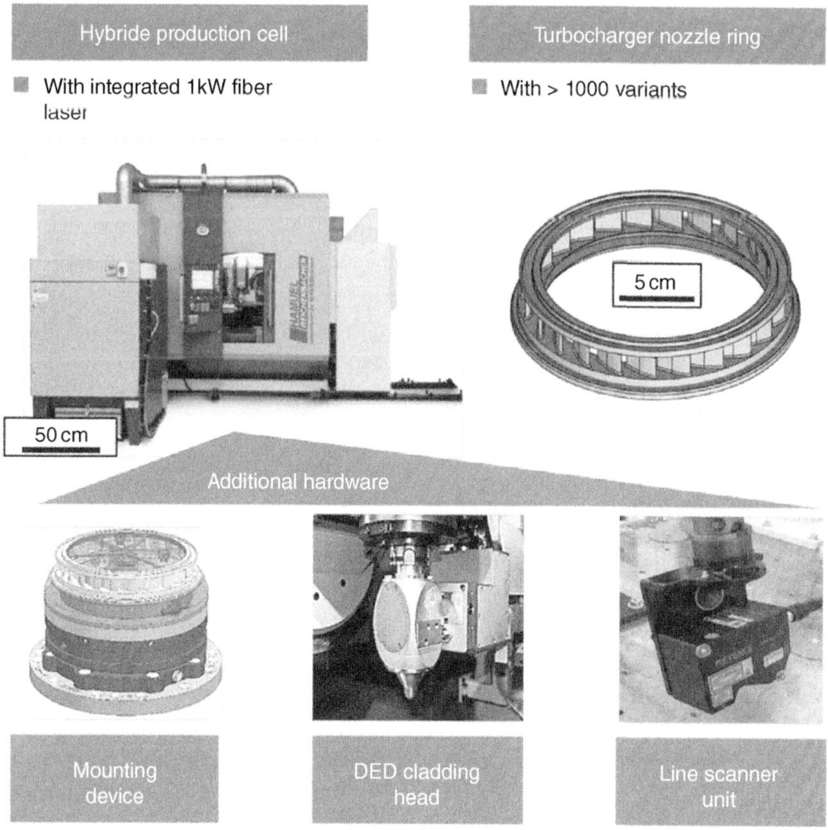


**Figure 12.10** Overview of components and additional hardware within the hybrid production cell.

The machine tool is particularly designed for machining of turbine and compressor blades, blisks or radial compressors. It is designed for high dynamic performance and a maximum of accuracy. The spindle unit is equipped with a hollow shank taper interface, called HSK for short. This HSK interface is suitable to mount conventional milling tools as well as the DED unit or the scanner system for metrology purposes.

The machine tool is additionally equipped with a DED head, a laser beam source (fiber laser), and a powder feeder. The DED head is treated as an additional tool and can be clamped at the main spindle via an HSK interface. The DED head can be stored in the tool magazine if not being used. The DED head is suitable to process powder of the grain size of 45–105 μm and can operate at a continuous laser power <500 W. The laser beam source and the powder feeder are controlled by the basic NC control of the main machine.

Since the production of nozzle rings requires adaptive processing steps such as adaptive milling of DED generated blades, a high precision fixture is added to the machining area. The fixture is sufficient for positioning of the base ring during five axis milling operations as well as during DED processes. In order to reach a maximum of flexibility and to enable a batchwise manufacturing process, the fixture is designed to be mounted on a pallet system. Fixture and pallet system together form the mounting unit. The mounting unit is operated by compressed air.

In order to develop and run integrated multiprocessing operations on the machine tool, additional measuring technology is precondition. The machine tool is also equipped with a line scanning measuring system. The measuring unit itself consists of three components, a line scanner, the controller of the scanner, and a computer. These three devices are connected to the CNC control of the machine. The line scanner of the measuring system is also handled as a regular tool and can be mounted at the HSK interface of the milling unit.

### 12.3.4 Process Chain for Hybrid Additive Manufacturing of Nozzle Rings

The exemplary process chain for hybrid additive manufacturing of nozzle rings (process chain B) consists of mainly five process steps.

#### 12.3.4.1 Step 1: "Choice of DED Strategy"

The selection of a suitable DED strategy is important to achieve an appropriate ratio of productivity and geometrical accuracy. The chosen laser beam diameter is a key parameter and has significant influence on the productivity and geometrical influence. In order to achieve the most accurate shape and geometry of the vanes while keeping an adequate productivity, a preliminary study is carried out (Figure 12.11). During this preliminary study the laser beam diameter is varied

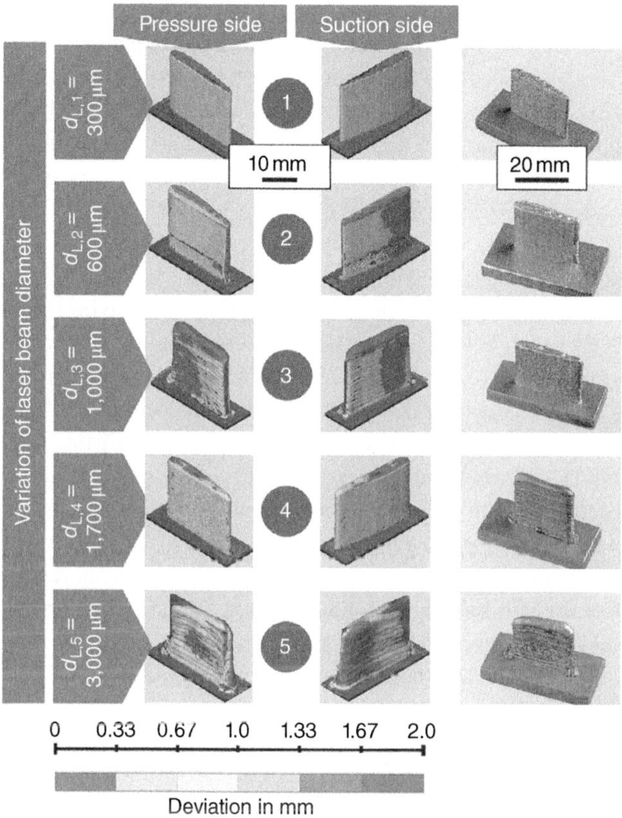


**Figure 12.11** Influence of laser beam diameter on geometrical accuracy.

from 0.3 to 3 mm, by adapting the laser power and the powder mass flow. The geometrical accuracy for several vanes is evaluated via 3D scans. An increase of the beam diameter leads to a reduction of geometrical accuracy. The blades deposited with a beam diameter of 0.3 mm are of an even and homogeneous surface with a maximum oversize of <0.67 mm. On the other side welding the blades with a beam diameter of 3 mm leads to an increased oversize up to <2 mm. The beam diameter of 1 mm is a compromise and is adopted for the chosen DED strategy for process step 2 (Section 12.3.4.2).

#### 12.3.4.2 Step 2: "DED Process"

Process step 2 takes place in the hybrid production cell. The DED strategy, which was identified during process step 1 (12.3.4.1), is used for the full buildup of the vanes. In the beginning of process step 2 the base plate is inserted into the fixture.

In the machine the tool is automatically changed to the DED cladding head and a layer-wise buildup of the vanes is started. A laser power of approximately 450 W and a powder mass flow of approximately 2.5 g/min are set by the NC of the hybrid production cell. The DED process itself takes about 7 min/vane. Each vane is built with a certain oversize to ensure a robust milling operation during process step 4 (Section 12.3.4.4). The actual geometry of each vane after the DED process is determined during process step 3 (Section 12.3.4.3).

#### 12.3.4.3 Step 3: "Optical Metrology"

In the beginning of process step 3, the whole machining area of the hybrid production cell gets cleaned by the coolant, so that the measurement of vanes won't be influenced by residual material (e.g., powder). Afterward the laser line scanner is attached to HSK interface. To obtain a three-dimensional representation of the part, usually several measurements from different positions must be carried out. The resulting point clouds are combined, and a polygon mesh is derived. To be able to combine scans captured with different sensor orientations, the kinematics of the machine has to be represented. This functionality is part of the software used to support adaptive manufacturing processes. Based on the laser scan result, the DED program and/or the milling program can be adjusted in order to increase accuracy, shorten process time, and compensate specific inaccuracies. In this case study especially the path definition of the milling program is required.

#### 12.3.4.4 Step 4: "Adaptive Milling"

Based on the data input of process step 3 (Section 12.3.4.3), the milling tracks are adjusted and the milling operation is started. Additional re-clamping or repositioning operations of the semifinished part aren't necessary. The hybrid production cell automatically changes for the defined tools and runs through roughing and finishing operations until the target geometry is achieved. With regard to the fifth process step (Section 12.3.4.5), the vanes are designed to have pins on the top created by milling (Figure 12.12). The pins are used as positioning aid for the top ring on the one hand, and also as filler material during the joining process.

#### 12.3.4.5 Step 5: "Joining of Top Ring"

The top ring is a semifinished product, which is placed on top of the vanes at the beginning of process step 5. When placing the top ring on top of the vanes, it is brought into a defined position by the six pins on top of the vanes (Figure 12.12). In addition to the positioning of the top ring, the pin is elementary for the joining process. The joining process is designed to work without additional filler material. The material volume of the pin is heated up by the laser radiation with a laser power of approximately 650 W. During the joining process the DED cladding head is moved in circular paths in order to reach an even bonding between vanes and top ring. After this step the nozzle ring is ready for application (Figure 12.2, Process Chain B, right).

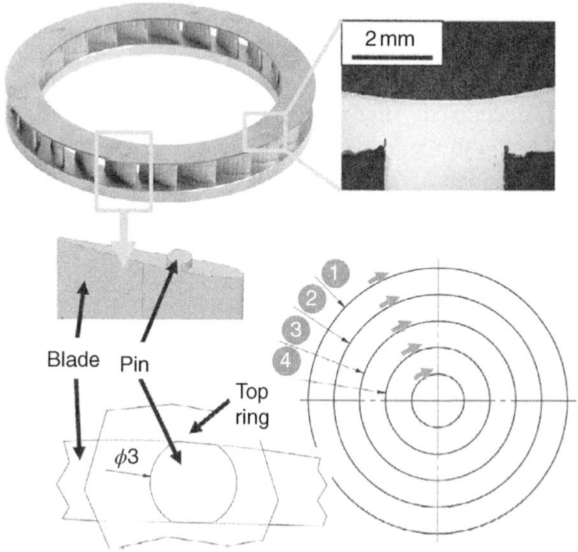


**Figure 12.12** Procedure of joining the top ring on top of the vanes.

## 12.4 Summary

Direct Energy Deposition (DED) is a suitable process for repair and manufacture of high valued metal parts. Typical process chains for DED-based repair and new component fabrication require adapted processes. Two successful exemplary process chains are presented.

The first example is the DED-based repair of a turbine blade platform. The challenge in the repair case is the variety of damages that may occur. In extreme cases individual tool paths have to be generated for each turbine blade. For turbine blades that come out of service, the original CAD data are no longer valid. Therefore, the actual geometry of each part has to be determined and taken into account.

The volumes to be deposited are determined by reverse engineering, and an oversize is added, which is needed for the machining after DED.

The tool paths are generated on the surface of the actual CAD data, are duplicated, shifted and trimmed with the volumes determined before.

After DED the deposited areas have a suitable oversize, and silver shiny surfaces. For the adaptive machining process a separate CNC milling machine tool is used. After these process steps the repaired turbine blade is ready for heat treatment and recoating.

The second example is the DED based hybrid AM of turbocharger nozzle rings. Nozzle rings are manufactured conventionally in a very large variety at one manufacturer (>1,000 variants). The hybrid production cell (machining, DED-based manufacturing, measuring) offers the possibility of individual production of numerous variants on one machine. Starting point for the production of the turbocharger nozzle ring is CAD data. For the design of the DED process, a suitable DED strategy for the vanes is developed. After measurement of the vanes in as-built condition, the structure of the blades is machined by adaptive milling. Finally the top ring is then laser joined.

## Acknowledgments

The results presented were developed in the following public funded projects:

Innovation Cluster "Turpro-Integrative production technology for energy-efficient turbomachinery" funded by the state government of North Rhine Westphalia and the Fraunhofer Gesellschaft für angewandte Forschung e.V. For this we want to express our sincere gratitude.

EU project "HyProCell" (founding code 723538). Special thanks to all project partners, especially IK4-Lortek, ABB Turbo Systems AG, HAMUEL Maschinenbau GmbH & Co. KG, BCT GmbH, EROWA AG, Autodesk Inc.

## References

1 Alkhayat, M. & Gasser, A. (2016, April 27–29). LMD and SLM in additive manufacturing – A technology comparison. *AKL – International Laser Technology Congress*. Symposium conducted at the meeting of Fraunhofer ILT, Aachen.

2 Heumer, W. (2017). Überholen statt neu bauen: Luftfahrt: Die Instandhaltung entwickelt sich für die Hersteller von Flugzeugtriebwerken zunehmend zum kräftigen Wachstumsmotor, Fokus: Flugzeuginstandhaltung. VDI Nachrichten (24).

3 Arjakine, N., Bruck, J., Grüger, B., Seeger, D. M., & Wilkenhoener, R. (2008, June 9–13). Advanced weld repair of gas turbine hot section components. *Proceedings of GT2008 ASME Turbo Expo 2008: Power for Land, Sea and Air* (pp. 559–564). Symposium conducted at the meeting of American Society of Mechanical Engineers (ASME), Berlin, Germany. doi:10.1115/GT2008-51534.

4 Wittig, N. (2018, June 11–15) Digitalization: Laser metal deposition – The future of spare parts and repairs for industrial steam turbines. *ASME Turbo Expo 2018: Turbomachinery Technical Conference and Exposition*. American Society of Mechanical Engineers Digital Collection. Symposium conducted at the meeting of American Society of Mechanical Engineers (ASME), Oslo, Norway. GT2018-75066, V008T29A002; 7 pages. doi:10.1115/GT2018-75066.

5 Bi, G., & Gasser, A. (2011). Restoration of nickel-base turbine blade knife-edges with controlled laser aided additive manufacturing. *Physics Procedia*, 12, 402–409.

6 Błachnio, J., & Pawlak, W. I. (2011). Damageability of gas turbine blades – Evaluation of exhaust gas temperature in front of the turbine using a non-linear observer. In E. Benini (Ed.), *Advances in gas turbine technology*. InTech, ISBN: 978-953-307-611-9. doi:10.5772/29546 Retrieved from http://www.intechopen.com/books/advances-in-gas-turbine-technology/damageability-of-gas-turbine-bladesevaluation-of-exhaust-gas-temperature-in-front-of-the-turbine-us

7 Dewangan, R., Patel, J., Dubey, J., Kumar Sen, P., & Kumar Bohidar, S. (2015). Gas turbines blades – A critical review of failure on first and second stages. *International Journal of Mechanical Engineering and Robotics Research*, 4(1), 216–223.

8 Görgl, R. & Brandstätter, E. (2017, September 5). Laser materials processing of complex components: from reverse engineering via automated beam path generation to short process development cycles. In D. Schuoecker, R. Majer, & J. Brunnbauer (Eds.), *XXI International Symposium on High Power Laser Systems and Applications 2016* (p. 102541D). Gmunden, Austria: SPIE.

9 Gao, J., Chen, X., Yilmaz, O., & Gindy, N. (2008). An integrated adaptive repair solution for complex aerospace components through geometry reconstruction. *The International Journal of Advanced Manufacturing Technology*, 36(11), 1170–1179.

10 Kelbassa, I. (2006). *Qualifizieren des Laserstrahl-Auftragschweißens von BLISKs aus Nickel- und Titanbasislegierungen* (Dissertation). RWTH Aachen University. Retrieved from http://publications.rwth-aachen.de/record/60728

11 Jambor, T. (2012). *Funktionalisierung von Bauteiloberflächen durch Mikro Laserauftragschweißen* (Dissertation). RWTH Aachen University. Retrieved from http://publications.rwth-aachen.de/record/229125

12 Poprawe, R. (2011). *Tailored light 2* (pp. 209–212). Berlin, Heidelberg: Springer.

13 Rusinkiewicz, S. & Levoy, M. (2001, May 28–June 1). Efficient variants of the ICP algorithm. *Third International Conference on 3-D Digital Imaging and Modeling: Proceedings* (pp. 145–152), Quebec City, Canada. Los Alamitos, CA: IEEE Computer Society.

14 Flemmer, J., Pirch, N., & Drinck, F. (2015, June 8–12). LMDCAM2: Software tool for near-net repair, cladding and built-up by laser metal deposition. *Proceedings of the ASME 2015 International Manufacturing Science and Engineering Conference*. Symposium conducted at the meeting of American Society of Mechanical Engineers (ASME), Charlotte, North Carolina. MSEC2015-9243, V001T02A015; 5 pages. doi:10.1115/MSEC2015-9243.

15 Bremer, C. & Kosche, T. (2018, September 25–27). Efficient post-machining of additively manufactured components. *3D Valley Conference*. Symposium conducted at the meeting of TEMA Technologie Marketing AG, Aachen.

16 Gebhardt, A., & Hötter, J.-S. (2016). *Additive manufacturing: 3D printing for prototyping and manufacturing*. München: Carl Hanser Verlag GmbH Co. KG. ISBN: 978-1-56990-664-4

**17** Alkhayat, M. & Ulrichsohn, B. (2017, September 27–28). Hybrid additive manufacturing – An outlook on prospective manufacturing approaches. *3D Valley Conference*. Symposium conducted at the meeting of TEMA Technologie Marketing AG, Aachen.

**18** Merklein, M., Junker, D., Schaub, A., & Neubauer, F. (2016). Hybrid additive manufacturing technologies – An analysis regarding potentials and applications. *LANE 2016 9th International Conference on Photonic Technologies, Physics Procedia*, 83, 549–559.

**19** Friel, R., & Harris, R. Ultrasonic additive manufacturing – A hybrid production process for novel functional products. *CIPR Procedia*, 6, 35–40.

**20** ALESCO GmbH. (2019, August 29). Retreived from https://www.lumex.alesco-gmbh.de

**21** Graf, B., Schuch, M., Kersting, R., Gumenyuk, A., & Rethmerer, M. (2015, June 22–25). Additive process chain using selective laser melting and laser metal deposition. *Lasers in Manufacturing Conference*. Symposium conducted at the meeting of German Scientific Laser Society (WLT e.V.), Munich.

**22** Freyer, C., & Klocke, F. (2001, May 28–30). Controlled metal build up (CMB): Fast manufacture of high strength tools of steel. In R. Meyer & H. Kühnle (Eds.), *International User's Conference on Rapid Prototyping & Rapid Tooling & Rapid Manufacturing, Amsterdam*. Retrieved from http://publica.fraunhofer.de/dokumente/B-70410.html

**23** Becker, D. & Meiners, W. (2011). Tools made out of combined materials. Retrieved from https://www.ilt.fraunhofer.de/content/dam/ilt/en/documents/product_and_services/laser_material_processing/HO_Tools_Made_out_of_Combined_Materials_2011.pdf

**24** Penter, L. (2019, August 29). M3D - Generative Fertigung von Multi-Material-Leichtbaustrukturen und –Werkzeugsystemen. Collaborative Project "InnoTeam". Retreived from https://tu-dresden.de/ing/maschinenwesen/iwm/lwm/forschung/forschungsgebiete/maschinen-und-prozessanalyse/MM3D

**25** Nezami, F. (2019, August 29). AdditiveCARBON – Hybridisierung von 3D-Druck mit Carbonfasern. Retreived from http://cikoni.com/de/3d-druck_additive_fertigung_kohlefaser_carbon.

**26** Belting, M. & Weisheit, A. (2014). Micro-laser deposition with gold paste. *Fraunhofer ILT Annual Report* 2014 (p. 78). Retrieved from https://www.ilt.fraunhofer.de/en/media-center/annual-report/archive.html

**27** Rittinghaus, S. & Weisheit, A. (2014). Additive laser metal deposition for 3D-component modification in automobile construction. *Fraunhofer ILT Annual Report* 2014 (p. 77). Retrieved from https://www.ilt.fraunhofer.de/en/media-center/annual-report/archive.html

**28** Bremer, J. (2019, August 29). Process and system technology for hybrid production of large components with laser material deposition (LMD). Collaborative Project "ProLMD". Retreived from https://www.ilt.fraunhofer.de/en/projects-technology-studies/cp_current/cp-prolmd.html

# 13

# Thermoelectrics

*Fredrick Kim, Seungjun Choo, and Jae Sung Son*

*School of Materials Science and Engineering, Ulsan National Institute of Science and Technology (UNIST), Ulsan, Republic of Korea*

## 13.1 Introduction

World energy consumption has rapidly grown thanks to a robust global economy and increased heating and cooling demand in developing countries. As fossil fuels still account for more than 80% of total global energy consumption, this increase in consumption has given rise to critical concerns over the depletion of hydrocarbon resources and the environmental pollution associated with their use; addressing these concerns will require renewable energy sources for a sustainable energy supply. To this end, various solar, wind, and biomass power generation technologies have been developed as renewable energy sources [1]. Of these, thermoelectric (TE) energy conversion can provide a unique solution to generate electricity from heat, more than 60% of which is dissipated to the environment from nature, industry, and transportation [2]. TE technology can convert heat directly into electricity, or vice versa, without any environmental pollution. Moreover, it can serve a wide range of temperatures—from room temperature to ~900 °C—without any moving mechanical parts [3, 4]; therefore, TE technology can be applied to low-grade heat, such as body heat [5], as well as high-grade heat such as that generated from atomic energy [6].

The output power of TE power generators (TEGs) depends on the inherent efficiency of the materials used as well as the amount of heat energy transferred from the source to devices. The efficiency of TE materials, the dimensionless figure-of-merit ZT, is defined by the equation $ZT = S^2\sigma T/\kappa$, where $S$, $\sigma$, $T$, and $\kappa$ are the Seebeck coefficient, electrical conductivity, absolute temperature, and thermal

*3D Printing for Energy Applications*, First Edition. Edited by Albert Tarancón and Vincenzo Esposito.

conductivity, respectively [7]. At present, the workhorse TE materials are semiconductors made of inorganic compounds such as chalcogenides, skutterudites, and half-Heusler alloys, which are preferred for use in TEGs at different operating temperatures (Figure 13.1a) [8]. Typically, the TE modules containing these alloys are fabricated by multiple-step material synthesis processes, such as zone-melting (ZM) [9], hot pressing [10], and hot extrusion [11]. This is followed by a series of assembly processes such as dicing to produce wafers and cuboid legs, plating, and soldering (Figure 13.1b) [2]. However, one potential issue with this procedure is its high production cost that is the result of complicated and energy-intensive processing and an unavoidable loss of material during the dicing processes. More importantly, the limited geometry of converting TE legs to cuboids by conventional procedures alone restricts the geometry of the resulting TE modules to planar structures only; this critically limits the widespread applications of TE technology [2].

Three-dimensional printing technologies can revolutionize several aspects of TE module manufacturing. First, the process can reduce the production cost considerably as it reduces the energy input needed to synthesize TE materials, simplify the assembly process, and realize the full usage of TE materials without loss. Second, geometrical control of TE legs and modules are facilitated by 3D printing technologies, thereby substantially extending the flexibility of device engineering. In this chapter, we review the recent progress in the development of 3D printing technologies for TE materials and devices. We will discuss various processes of 3D printing—such as the extrusion-based process, fused deposition modeling (FDM), stereolithography apparatus (SLA), and selective laser sintering (SLS)—available to synthesize and control the geometry of TE materials, including several examples of new types of TEGs created with 3D printing technologies. Finally, the outlook of this field regarding the issues and solutions facing the commercialization of TE power generation technology are presented.

## 13.2 Additive Manufacturing Techniques of Thermoelectric Materials

### 13.2.1 Extrusion-Based Additive Manufacturing Process

The extrusion-based additive manufacturing process is the most basic form of 3D printing, which relies solely on the rheological property of the ink. These processes include dispenser printing and robocasting. In dispenser printing, ink is dispensed from the micrometer nozzle directly to the substrate by pneumatic control. Since there is no complex constituent required for this printing process, the processing parameter itself is relatively simple compared to other manufacturing techniques. The processing parameter includes the diameter of the nozzle, the distance between the nozzle and the substrate, the extrusion pressure, and the printing speed. The

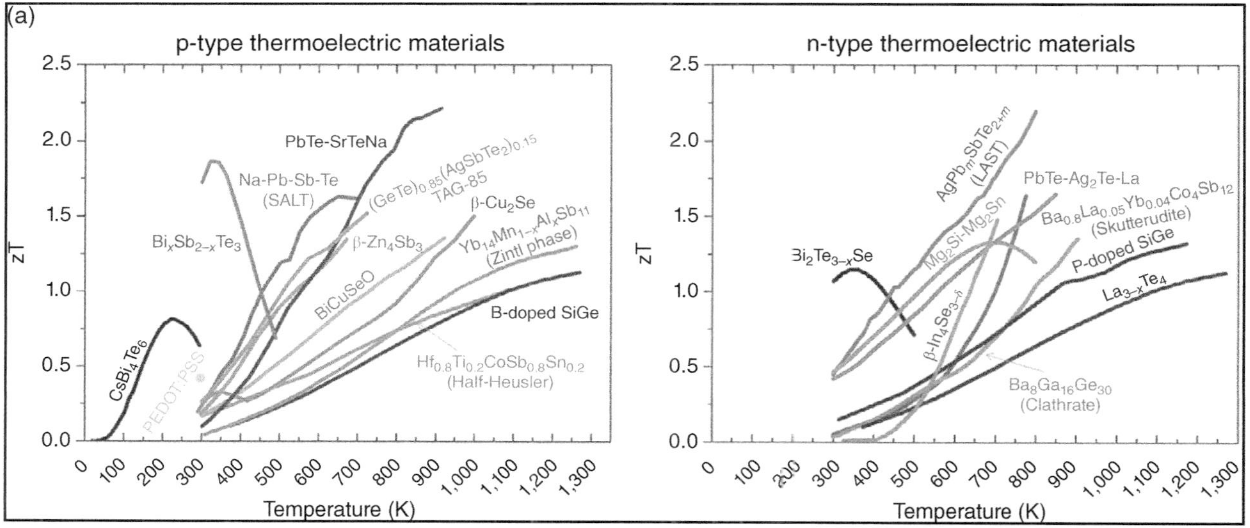


**Figure 13.1** (a) ZT for common TE materials as a function of working temperature [8]. *Source:* From M. Rull-Bravo, A. Moure, J. F. Fernández, M. Martín-González, Skutterudites as thermoelectric materials: Revisited. *RSC Advances* **2015,** *5,* 41653. © 2015, Royal Society of Chemistry. (b) Comparison of the fabrication process of a TEG by conventional ingot-based and 3D printing process [2]. *Source:* From S. Jo, S. Choo, F. Kim, S. H. Heo, J. S. Son, Ink processing for thermoelectric materials and power-generating devices. *Advanced Materials* **2019,** *31*, e1804930. © 2019, John Wiley & Sons.

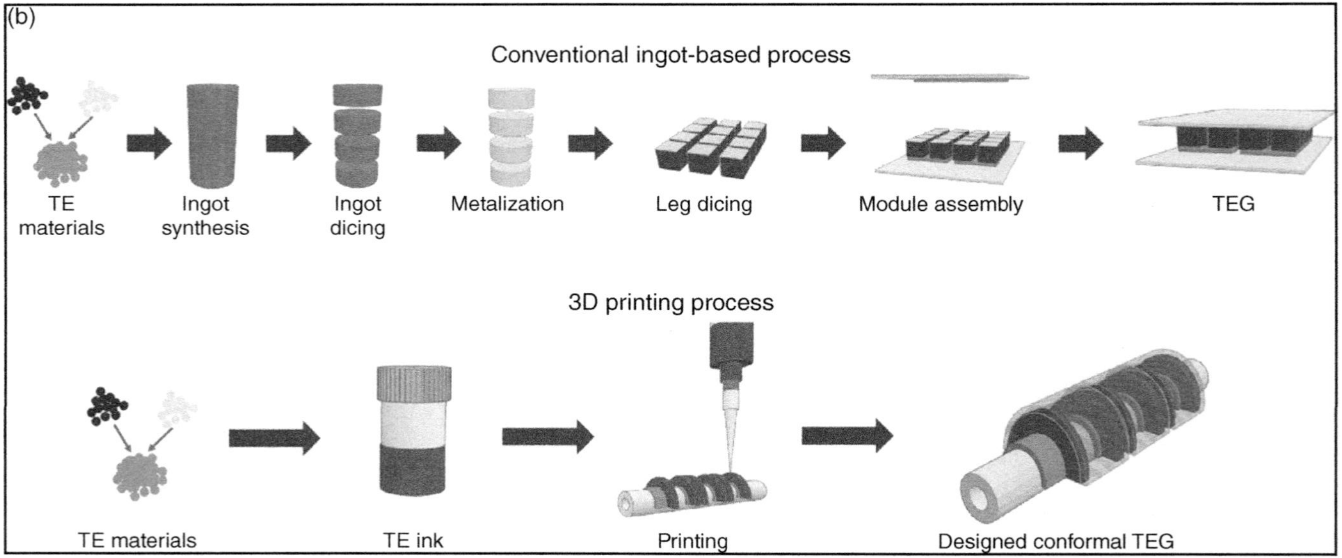


**Figure 13.1** (Continued)

design of colloidal inks has a considerable influence on the quality of the printed material. For the dispensing printing method, the ink must exhibit non-Newtonian and viscoelastic behavior [12, 13]. These rheological behaviors consist of plastic deformation upon critical yield stress and rapid elastic recovery to retain the shape of the ink after deposition [14]. The inks must be tailored to simultaneously have shear-thinning behavior, which facilitates a reliable extrusion flow through the nozzle to prevent clogging as well as high elasticity to avoid the collapse of the 3D structure [15]. Moreover, since shape deformation may occur due to gravity and phase separation among particles, these colloidal inks should also have a stabilizing agent to prevent potential agglomeration and sedimentation of the solid particles [16].

TE inks are often paired with a polymer-based organic binder to facilitate viscoelastic behavior [17]. The TE performance of these composite inks with a polymer matrix is strictly limited due to low power factors. TE ink with an organic binder suffers from low electrical properties due to its intrinsically insulation [5, 18, 19]. For the highest thermopower and the quality of the resulting material, a high colloid volume fraction of TE particles is required to minimize the volume shrinkage and voids created upon decomposition of the binders. Dispenser printing of flexible thick-film TEG using TE composite inks was introduced by Madan, Wang, Chen, Wright, and Evans [20]. A flexible p-type single leg TEG was fabricated using the dispenser printing technique. $Bi_{0.5}Sb_{1.5}Te_{3.0}$ particles, which yielded one of the highest thermopower results at room temperature, was mechanically alloyed and wet grinded to prevent agglomeration at the nozzle by reducing the average particle size to 10 μm. Commercial epoxy resin was chosen as the polymer matrix of the ink to produce thermoelectric composite slurries. With the optimum printing parameter, they printed a p-type single leg TEG with sixty high-aspect-ratio thermoelectric legs on a polyimide substrate. The dispenser-printed BST composite film was cured at 250 °C for 12 hr to remove excess epoxy resin. However, the SEM image of the composite film confirms the existence of epoxy binders among BST grains, which may have contributed to a poor electrical property of the composite film. The electrical conductivity was 12 S/cm, which is two orders of magnitude lower than that of the commercial bulk value, resulting in a ZT value of 0.2. Coupled with a high internal device resistance of 800 Ω, the maximum power output was 20.5 μW for $\Delta T$ of 20 K. Although the viscoelastic property of the TE ink is established by the addition of organic binders, the thermopower was retarded due to low mechanical and electrical connections. Furthermore, due to the intricate design requirement of TE ink, a fabrication of 3D structures was not realized.

As a solution to these issues related to organic binders, Kim et al. [21] developed viscoelastic TE inks with inorganic binders. The electrically insulating polymer matrix was replaced by molecular chalcogenidometallate-based (ChaM) ions, which are anions containing metal atoms ligated with chalcogens. The ChaM ions are synthesized by dissolving metal chalcogenide in dimensional reducing agents, such as 1,2-ethylenediamine and a 1,2-ethanedithiol mixture in the presence of excess chalcogens [2, 22, 23]. The octahedral framework of 3D $MX_6$ (M: metal atoms, X: anion) is dismantled upon the reaction with the $A_aX$ ionic compound. These compounds exhibit excellent solubility to polar solvents with a dielectric constant (ε, F $m^-$) ranging from 10 to 50, such as ethylenediamine (ε = 13) and glycerol (ε = 43). Furthermore, structure transformations occurred for molecular ChaM ions to the crystalline semiconducting metal chalcogenide phase upon mild heat treatment [24]. Webber and Brutchey [25] reported the successful synthesis of nine $V_2VI_3$ metal chalcogenide solutions using an amine/thiol mixture. These dissolved compounds were recrystallized after annealing at 350 °C and created crystalline thin films.

A previous study reported that ChaM ions can also function as surface ligands of nanoparticles, which stabilizes the dispersion of particles in polar solvents. Kovalenko et al. [26] developed a successful synthesis route for the colloidal nanocrystals of Pb and Bi chalcogenide capped with $Sb_2Te_3$-based ChaM. $Sb_2Te_3$ ChaM ions strongly improved interparticle coupling and stabilized the particles in a solution via electrostatic interactions. These unique qualities of molecular ChaM ions observed at the nanoscale have been expanded to microscale TE particles to realize the 3D printing of high-performance TE material.

The authors of this chapter prepared p- and n-type $Bi_2Te_3$-based TE powders with stoichiometric compositions of $Bi_{0.4}Sb_{1.6}Te_3$ and $Bi_2Sb_{2.7}Se_{0.3}$, respectively, using a high-energy ball milling process. The average TE particles' size was reduced by sieving to <45 μm. Soluble $Sb_2Te_3$ ChaM ions were synthesized to be compositionally compatible to the BST matrix. Sb and Te powders were completely dissolved in a cosolvent of ethanethiol and ethylenediamine. The solution was precipitated by excess acetonitrile and vacuum dried to produce ChaM ion powder. Appropriate amounts of ChaM ions and TE particles were dispersed in glycerol to create a viscous TE ink (Figure 13.2a).

A series of rheological analyses have been conducted to observe the effect of ChaM ions on the $Bi_2Te_3$-based ink [27]. Figure 13.2b demonstrates the zeta potential curve of ChaM ions. A peak was observed at −23 mV, which indicates the negatively charged surface of the ChaM ions. The negatively charged ChaM ions on the surface of the TE particles function as a dispersing and stabilizing agent. The charged surface creates electrostatically stable dispersion due to repulsive forces between particles (Figure 13.2c). The improved colloidal stability was clearly visible in the phase separation test in Figure 13.2d and e.

Reliable flows through a few needles of hundreds of micrometers are possible by structural deformation under stress. The dynamic viscosity test suggested that TE inks behaved like Bingham body fluid, indicating their deformation upon stress. A loss tangent ($\tan\delta$) value over 1 indicates liquid-like behavior, while a $\tan\delta$ value under 1 represents solid-like behavior. Furthermore, its structural integrity after deposition was assessed in terms of yield stress ($\tau_y$) and recoverable elastic strain ($S_R$). Increased $\tau_y$ and $S_R$ implied high elasticity achieved by electrostatic binding of ChaM ions. These rheological analyses prove the enhancement of viscoelasticity and phase stability with the addition of ChaM ions.

The precise control of rheological properties, coupled with an optimal ink composition, enabled the 3D printing of shape-conformal TE materials. Structures of various shapes were printed (Figure 13.2f) and subsequently heat-treated. During heat treatment, $Sb_2Te_3$ ChaM ions promoted the sintering process without the introduction of secondary phases (Figure 13.2g). The structural transformation from molecular ChaM ions to crystalline phases led to the formation of a mechanically robust product.

The improvement in mechanical properties of the material also resulted in an outstanding TE properties. The electrical conductivities of the n- and p-type samples were 500 and 550 S/cm, respectively (Figure 13.2h). The maximum Seebeck coefficients of the n- and p-type printed samples were −145 and 199 μV/K at 175 and 200 °C, respectively (Figure 13.2h). The thermal conductivity was recorded at a range of 0.50 and 0.63 W/mK, which was significantly lower than the hot-pressed BST materials (Figure 13.2i). The resulting materials had pores of multiple sizes ranging from the nano- to macroscales. The macroscale pores contributed to the ultra-low thermal conductivity by scattering phonons effectively. The electrical carrier transport behavior was not altered due to the relatively short mean free path of the electron and hole carriers. The low thermal conductivity resulted in ZT values of 0.6 and 0.9 at 170 and 125 °C for n- and p-type TE materials, respectively; this is comparable to the results obtained by hot-pressed bulk materials (Figure 13.2j).

Excellent TE properties coupled with an unlimited degree of geometrical freedom enabled the fabrication of a cylindrical TEG that conformed to the shape of the heat source (Figure 13.2k). p- and n-type 3D-printed TE half rings were mounted on an alumina pipe and connected with copper electrodes. The output power and voltage of the TEG were measured under a flow of hot water through the alumina pipe. The maximum output voltage of 27.0 mV and maximum power of 1.62 mW were achieved at a temperature difference of 39 °C. The output characteristics, which are low compared to the high TE properties, are due to the high contact resistance between the electrode and TE material.

The binders are essential in modulating the rheological properties and stability of thermoelectric ink for 3D printing. However, an appropriate mixture of solvents using nanoparticles can also regulate the rheological property of the ink. Saeidi-Javash, Kuang, Dun, and Zhang [28] reported direct 3D conformal printing on curved substrates using colloidal $Bi_2Te_{2.7}Se_{0.3}$ nanoplate ink and the rapid photonic sintering method (Figure 13.2l). $Bi_2Te_{2.7}Se_{0.3}$ nanoplates were dispersed in an appropriate ratio of glycerol, ethanol, and ethylene glycol in a reasonable viscosity range [28]. Nanoplates were dispersed in a stable manner without the

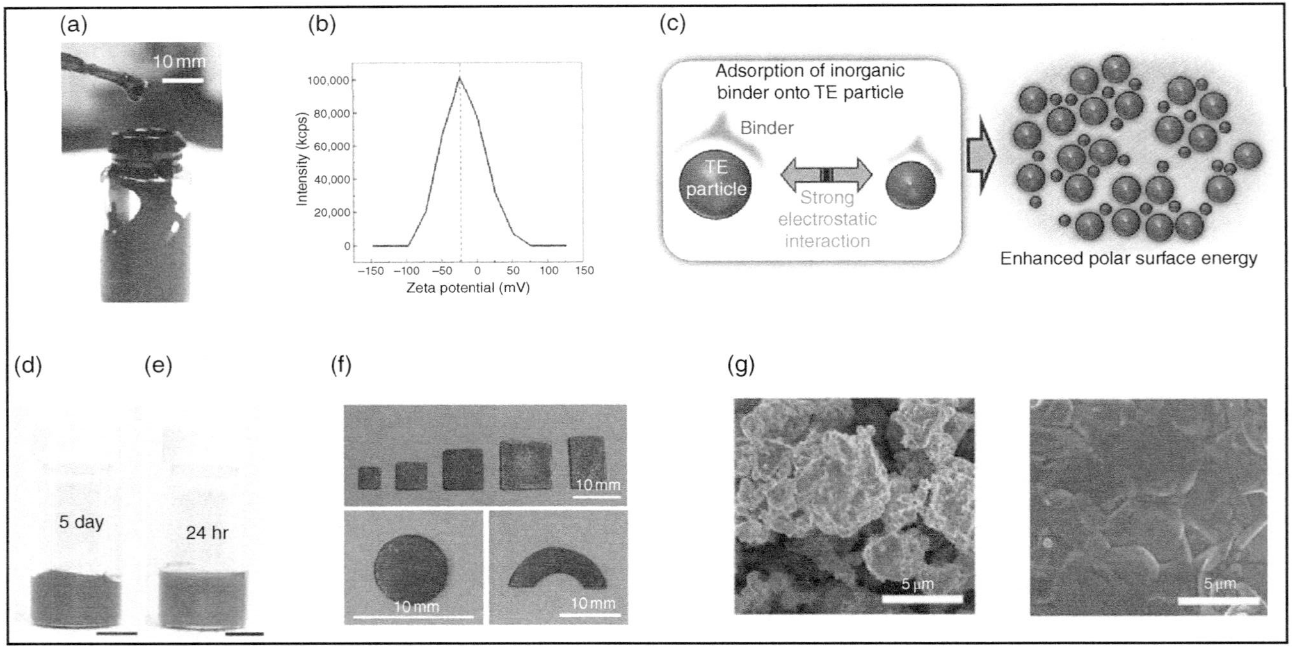


**Figure 13.2** (a) Photographs of Bi2Te3-based TE ink with ChaM [21]. *Source:* From F. Kim, B. Kwon, Y. Eom, J. E. Lee, S. Park, S. Jo, S. H. Park, B.-S. Kim, H. J. Im, M. H. Lee, T. S. Min, K. T. Kim, H. G. Chae, W. P. King, J. S. Son, 3D printing of shape-conformable thermoelectric materials using all-inorganic $Bi_2Te_3$-based inks. *Nature Energy* **2018**, *3*, 301. (b) Zeta potential curve of ChaM ions. (c) Schematic illustration of enhanced electrostatic interaction and polar energy of TE particles [27]. *Source:* (b) and (c) Y. Eom, F. Kim, S. E. Yang, J. S. Son, H. G. Chae, Rheological design of 3D printable all-inorganic inks using BiSbTe-based thermoelectric materials *Journal of Rheology* **2019**, *63*, 291. ©2019, AIP Publishing. Comparison of photographs showing the $Bi_2Te_3$-based TE ink (d) with ChaM (e) without ChaM. (f) Photographs of various shapes of printed materials. *Source:* From F. Kim, B. Kwon, Y. Eom, J. E. Lee, S. Park, S. Jo, S. H. Park, B.-S. Kim, H. J. Im, M. H. Lee, T. S. Min, K. T. Kim, H. G. Chae, W. P. King, J. S. Son, 3D printing of shape-conformable thermoelectric materials using all-inorganic $Bi_2Te_3$-based inks. *Nature Energy* **2018**, *3*, 301. (g) SEM images of the sintered materials prepared from $Bi_2Te_3$-based TE inks without and with ChaM, respectively.

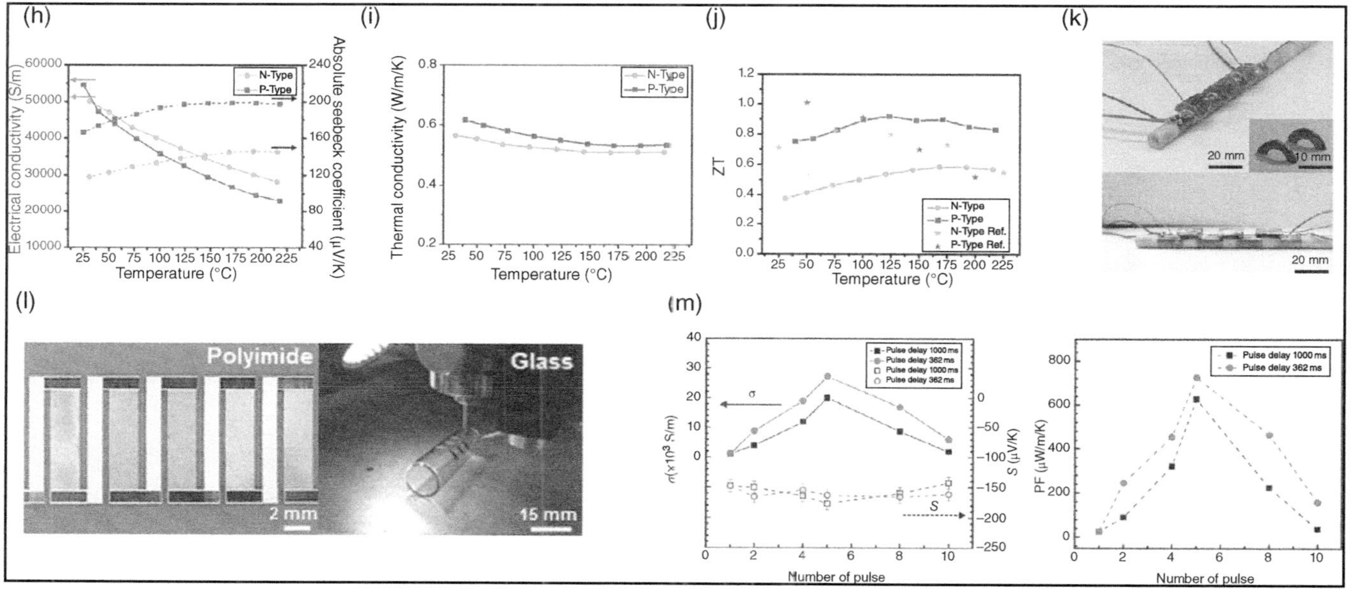


**Figure 13.2** (Continued)

Temperature-dependent TE properties of the n-type and p-type 3D-printed materials: electrical conductivity and absolute Seebeck coefficient (h), thermal conductivity (i), and ZT (j). (k) Photographs of the cylindrical TEG that conformed to the shape of the heat source [21]. (l) Photographs of 3D conformal aerosol jet printing on various 2D and 3D substrates. *Source:* (g), (h), (i), (j) and (k) From F. Kim, B. Kwon, Y. Eom, J. E. Lee, S. Park, S. Jo, S. H. Park, B.-S. Kim, H. J. Im, M. H. Lee, T. S. Min, K. T. Kim, H. G. Chae, W. P. King, J. S. Son, 3D printing of shape conformable thermoelectric materials using all-inorganic $Bi_2Te_3$-based inks. *Nature Energy* **2018**, *3*, 301. *Source:* (l) From M. Saeidi-Javash, W. Kuang, C. Dun, Y. Zhang, 3D conformal printing and photonic sintering of high-performance flexible thermoelectric films using 2D nanoplates *Advanced Functional Materials* **2019**, *29*, 1901930. © 2019, John Wiley & Sons. (m) Electrical conductivity (σ), Seebeck coefficient (S), and power factor (PF) as a function of number of pulses [28].

addition of binders due to ligands bonded to the nanoplates. The ligands were eliminated in the rapid photonic sintering process, which greatly shortened sintering time and posed no overheating issues. The corresponding electrical conductivity and power factors reached 270 S/m and 730 μW/mK$^2$, respectively (Figure 13.2m). Moreover, the $Bi_2Te_{2.7}Se_{0.3}$ nanoplate ink was printed on polyimide substrate to create a flexible device. The durability of the device was investigated through a bending cycles test. The electrical resistance of the device increased by about 4% after 500 bending cycles and had a 1.5 mm bending radius.

The applicability of an extrusion-based 3D printing of TE inks for designing TE materials and generators conformable to various heat sources has been previously introduced. The organic-binder-free TE ink exhibited excellent viscoelastic properties that enabled the fabrication of highly efficient TE materials. This concept of shape-conformable TEGs has enormous potential for efficiently converting waste heat into electricity.

### 13.2.2 Fused Deposition Modeling (FDM) Technique

Fused deposition modeling (FDM) or fused filament fabrication (FFF) is a well-established and rapid manufacturing technique used to produce high-resolution polymer-based 3D structures by melting polymers or polymer-based composites and depositing them in successive layers that are consolidated into one component [29]. Oztan, Ballikaya, Ozgun, Karkkainen, and Celik [30] utilized the FFF printing method to print efficient TE materials with an unlimited degree of geometrical freedom. $Bi_2Te_3$, one of the best commercial TE materials at room temperature, was mixed with an acrylonitrile butadiene styrene (ABS) polymer matrix that was extruded into composite TE filaments. ABS is widely used for 3D printing and prototype development due to its relatively inexpensive price and low melting point of 105 °C [31]. Composite TE filaments were loaded into a FFF printer and a 10 mm long cubic structure was printed. The printed material was then sintered at a temperature above 450 °C to remove the excess polymer matrix, which would otherwise deteriorate the electrical property of the resulting material, and enhance the mechanical properties of the material by consolidating the $Bi_2Te_3$ particles. The sintering process was performed in a brass enclosure in an Ar atmosphere to prevent the oxidation of the $Bi_2Te_3$ particles. Most of the remaining polymer matrix was removed; the exception was the sample sintered at 450 °C, where ABS particles were retained. The authors found that the post-processing temperature played a significant role in the microstructure and TE properties of the resulting material. They also systematically studied the correlation between the sintering temperature and the microstructure, composition, and TE properties. A higher sintering temperature was required to fully decompose any remaining ABS particles; however, a higher temperature led to a decrease in density due to the voids created by the removal of polymer particles and the disintegration of $Bi_2Te_3$. They observed an increase in the oxidation level and a decrease in carbon elements due to polymer burnout at the higher sintering temperature, although the $Bi_2Te_3$ elements remained constant. The $Bi_2Te_3$ elements are sensitive to oxidation levels, which can be detrimental to electrical properties. They sought to determine an adequate post-sintering temperature that could simultaneously remove the remaining polymer particles and prevent the oxidation of the thermoelectric material. Thermoelectric properties were measured as a function of the sintering temperature. Sintering above 500 °C resulted in a decreased Seebeck coefficient and electrical conductivity due to an increase in porosity, leading to a poor electrical interconnection. However, the porous nature of the printed material showed a significantly lower thermal conductivity than that of the bulk $Bi_2Te_3$, which ultimately led to a ZT value of 0.54 at room temperature. The ZT values of the FDM printed materials were lower than those of the bulk $Bi_2Te_3$ by almost half. A further optimization of the post-sintering process for binder removal and a reinforcement of mechanical properties are required to enhance the TE properties.

### 13.2.3 Stereolithography Apparatus (SLA) Process

Stereolithography, also known as the photo-solidification process, is a rapid 3D printing method that produces high-resolution 3D structures by photopolymerization; in this process liquid photoresins are selectively consolidated into solid parts, layer-by-layer, using an ultraviolet (UV) laser beam [32, 33]. The monomer carbon chains inside the photoresin are linked to produce polymers upon exposure to UV light. He et al. [34] utilized the stereolithography process to fabricate amorphous $Bi_2Te_3$ materials. They sought to revitalize a stagnant field of TE material fabrication by printing free-standing bulk thermoelectric materials that would replace the traditional energy-intensive process. Composite resin was prepared by fully dispersing Bismuth Antimony Telluride (BST) particles in a photoresist solution with a BST content of 40–60 wt%. SLA processes were carried out with a UV laser beam at 405 nm and 1.2 W, and a layer thickness of 100 μm. Figure 13.3a demonstrates schematic routes of the synthesis of the composite resin and SLA printing of the TE material, with a photographic image of the final product. Since the electrically insulating property of the photoresin would degrade the electrical properties of the printed material, like in the FDM process discussed above, the post-sintering process was necessary to remove resin residues inside the TE material. The authors performed thermogravimetric analysis (TGA) to determine the optimum temperature that would decompose the remaining resin without deteriorating the quality of the sample. TGA analysis demonstrated a weight loss at 425 °C; however, annealing at 425 °C resulted in the sample becoming cracked. Therefore, an annealing temperature of 350 °C was viable for the final product.

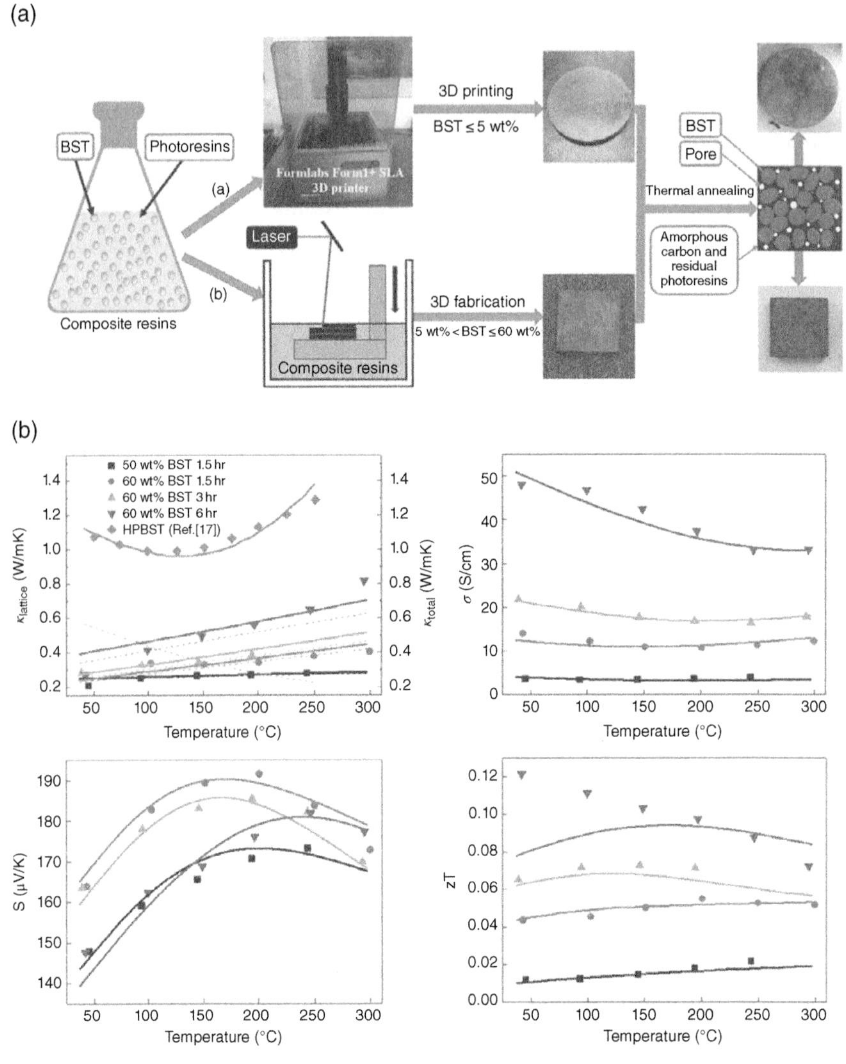


**Figure 13.3** (a) Schematic illustration of fabrication of SLA 3D printed samples. (b) Temperature dependence of TE properties [34]. *Source:* From M. He, Y. Zhao, B. Wang, Q. Xi, J. Zhou, Z. Liang, 3D Printing fabrication of amorphous thermoelectric materials with ultralow thermal conductivity. *Small* **2015,** *11,* 5889. © 2015, John Wiley & Sons.

Since the composition of the photoresist was almost half of the weight percentage of the composite resin, pores were formed upon the thermal annealing of photoresist, resulting in a considerably hollow structure. The density of the printed material with 60 wt% of BST was 2.631 g/cm$^3$, which is much smaller than the hot-pressed conventional BST sample (6.89 g/cm$^3$). Furthermore, the decomposition of the photoresist produced amorphous carbon. The energy-dispersive spectra (EDS) analysis revealed notable carbon and oxygen contents. The oxygen content was most likely due to the incomplete decomposition of photoresists. The presence of carbon and oxygen and the porous nature of the final product led a subpar electrical conductivity level as seen in Figure 13.3b. The resulting material had a low thermal conductivity of 0.2 W/m·K; however, poor electrical properties led to a ZT value of only 0.12 at 43 °C, which is one-tenth of the commercial bulk $Bi_2Te_3$ material.

### 13.2.4 Selective Laser Sintering (SLS) Process

Additive manufacturing techniques such as FDM and SLA require additional constituents to form a 3D structure; however, this compromises electronic properties of a material [35–37]. The selective laser sintering (SLS) process is an emerging manufacturing technique where successive layers of powder materials such, as metallic materials and alloys, are melted by a high-intensity laser that consolidates them into a 3D structure [38–41]. A highly complex design can be obtained simply by fusing particle granules directly into a 3D structure without adding any binding agents [42]. This laser manufacturing technique could potentially replace the typical fabrication route for commercial bulk TE materials. The technique involves zone melting (ZM) ingots followed by post-processing including multiple steps such as dicing, metallizing the TE material, and soldering; this results in considerable material loss (50–70%) [21]. The SLS process allows for the direct assembly of TE materials onto the electrode with an ultimate degree of geometrical freedom [43, 44]. However, a high operative temperature limits the adoption of TE materials to a SLS platform [45, 46]. In contrast to metallic materials, TE materials are semiconductors with low melting points, ductility, thermal conductivity, and weak mechanical strengths and thermal shock resistance [47]. These factors may generate defects at the micro- and macro-levels and are likely to produce a highly porous and brittle material. Moreover, laser processing of TE material may lead to variations in the chemical composition of the resulting materials [48]. When the laser volume energy density $E_V$ is excessively high, extremely high levels of evaporation leads to the formation of pores in the material and changes in its chemical composition [49]. TE materials are highly sensitive to their chemical composition since the slightest variation could lead to an undesirable change of conduction type and a lower TE performance. Thus, it is imperative to obtain an optimum laser energy density $E_V$ and the appropriate materials, i.e., those with a highly controlled chemical composition and excellent TE performance, after the high temperature laser-induced fusion and solidification process is complete.

Skutterudites and half-Heusler alloys, which are two of the best TE materials at intermediate temperatures, have high melting points, superior mechanical properties, and thermal conductivities similar to those of metals, which make them ideal for the SLS platform [50–52]. Yan, Ke, Yang, Uher, and Tang [53] selected n-type $CoSb_{2.85}Te_{0.15}$ skutterudites to prepare bulk materials that were deposited directly on a Ti substrate via SLS processing (Figure 13.4a). A $CoSb_{2.85}Te_{0.15}$ ingot produced by self-propagating high-temperature synthesis (SHS) was crushed to form a powder bed. Bulk samples of $CoSb_{2.85}Te_{0.15}$ with a thickness of 1.7 mm were fabricated on a Ti substrate. Ti was selected as the substrate due to the similarity of its thermal expansion coefficients and melting points with those of the bulk samples. The resulting material was deposited on the Ti substrate at a submicrometer grain size and attained a maximum ZT value of 0.56 at 823 K.

During laser processing, severe evaporation of materials may lead to uneven surfaces, pores and micro cracks, and variations in chemical composition. The optimal SLS processing window was determined by systematically analyzing the effect of processing parameters on the chemical composition and the overall quality of the printed layers. The quality of the layer sintered by the laser process is mostly dependent on laser volume energy density $E_v$. The formula is expressed as:

$$E_v = \frac{P}{vdh}$$

where $P$ is laser power, $v$ is scanning speed, $d$ is hatch spacing, and $h$ is layer thickness. For successful fabrication of the resulting material, the optimal processing window must be controlled. The forming quality of the first layer corresponds to laser power $P$ and scanning speed $v$ (mm/s). Figure 13.4b–e shows four morphological features as a function of scanning speed $v$ (mm/s) and laser power $P$. Figure 13.4b depicts the ideal morphology where the energy density is powerful enough to completely melt the powder in the scanned region without severe evaporation, which ultimately produces a flat and dense surface. Deviating from this energy density ($E_v$) region would produce a porous, unstable, and balling surface as seen in Figure 13.4c–e. In Figure 13.4c, an overheated single layer is formed when the laser power is too high and the scanning speed is too low, which develops a distinct concave profile since the high molten pool temperature leads to partial melting of the unscanned region. This forms a ridge-like surface feature and more importantly causes voids due to accelerated evaporation of the material (Figure 13.4d). Figure 13.4e depicts the balling effect on the surface due to the low power of the laser, which was not high enough to melt the powder completely.

As a follow up study, Yan et al. [56] prepared n-type ZrNiSn via the SLS process. ZrNiSn is an alloy with a different melting point for each element, which may potentially lead to accelerated evaporation of elements with a low melting point.

The optimal SLS processing window was determined by calculating the vaporization rates of each element as a function of the molten pool temperature to reduce the variation in chemical composition. This study also demonstrated direct bonding of bulk ZrNiSn samples on a Ti substrate without a transition layer and attained a maximum ZT value of 0.39 at 873 K.

Mao et al. [57] prepared n-type $Bi_2Te_{2.7}Se_{0.3}$ thermoelectric material using the SLS process and systematically analyzed the dependence of laser volumetric energy on the forming quality, phase structure, and TE properties. The authors found an appropriate range for the laser volume energy density required to fabricate a high quality $Bi_2Te_{2.7}Se_{0.3}$ layer at 10.0–33.3 J mm$^3$. Moreover, the composition of the starting raw material was modified in consideration of an inevitable Te and Se evaporation. The vaporization of Te and Se occurred even at the low laser energy density. The starting content of the raw material was adjusted to gain precise control over the chemical composition of the resulting material by using empirical formulas that calculated the vaporization rate of each element. The vaporization rate $J_i$ (g/cm$^2$·s) of the elements was calculated using the following formula:

$$J_i = 4.375\times10^{-4}\gamma_i X_i P_i^0\left(\frac{M_i}{T}\right)^{\frac{1}{2}}$$

where $g_i$ is the active coefficient of element $i$, $X_i$ is its molar fraction, $P_i^0$ is the vapor pressure of pure element $i$ (Pa), $M_i$ is its atomic weight (g/mol), and $T$ is the molten pool temperature (K). Using this formula, the vaporization rates of Se and Te are obviously much higher than that of Bi. The single phase $Bi_2Te_3$ structure was maintained without the introduction of secondary phases. The resulting material achieved a maximum ZT value of 0.84 at 400 K, which is comparable to the ZT values obtained by the hot-pressed materials.

Shi, Chen, Jia, and Wang [54] also reported a preparation of p-type BST using SLS processing. The authors believed that the porous nature of the final product was inevitable; therefore, they devised a new processing parameter that utilized the existence of pores as an advantage. Although pores reduced the mechanical and electrical properties of the resulting material, the micro-pores in the grain boundary induced phonon scattering, further reducing the thermal conductivity. The authors selected the ideal processing parameters that could produce micropores to a level that would not degrade the mechanical and electrical properties of the material. Figure 13.4f is the photo image of the SLS printed material and it reveals the existence of multiple random pore sizes varying from nanometers to micrometers. Moreover, the laser processing parameter was adjusted to a level of obtaining porous structures by only melting the boundaries of BST powders. By

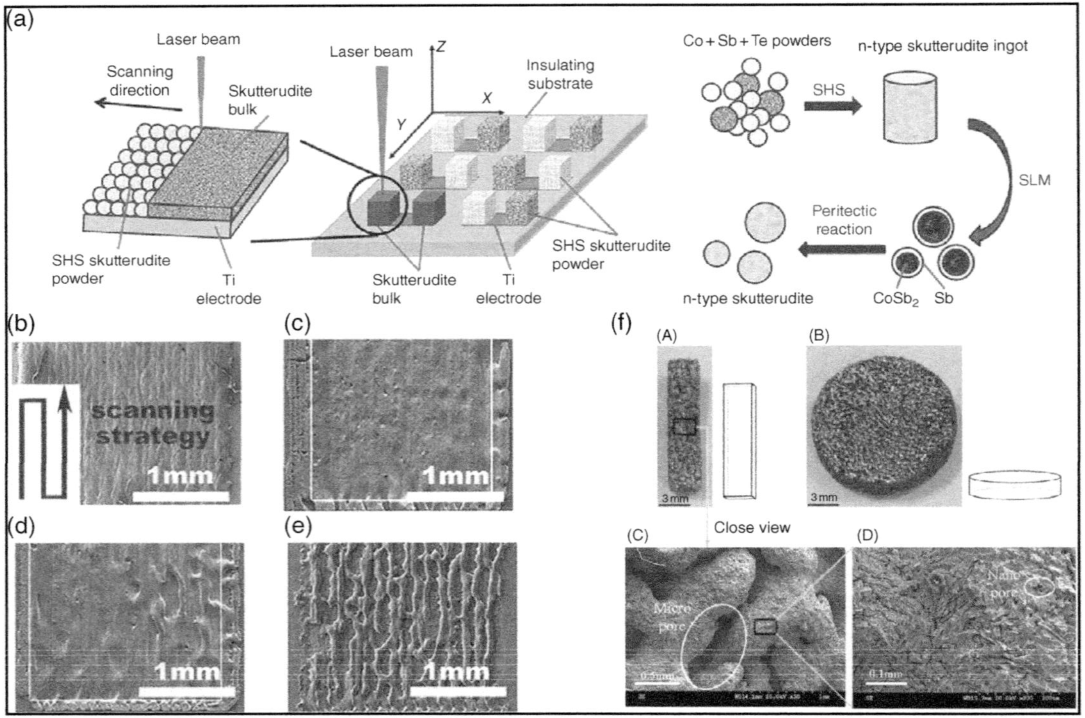


**Figure 13.4** (a) Schematic illustration of fabrication of 3D printed samples. SEM images of four typical surface morphologies of SLM-prepared layers: (b) flat, (c) vaporization, (d) unstable, and (e) balling [53]. *Source:* From Yan, Y., Ke, H., Yang, J., Uher, C., & Tang, X. Fabrication and thermoelectric properties of n-Type $CoSb_{2.85}Te_{0.15}$ using selective laser melting. *ACS Applied Materials & Interfaces*, **10**, 13669.
(f) Photographs and SEM image of the long square block and cylinder samples prepared using SLS 3D printing technology. Temperature dependence of (g) electrical conductivity, (h) Seebeck coefficient, (i) thermal conductivity, and (j) zT of the sample compared with that of commercial bulk (BST bulk) [54]. *Source:* From Shi, J., Chen, H., Jia, S., & Wang, W. 3D printing fabrication of porous bismuth antimony

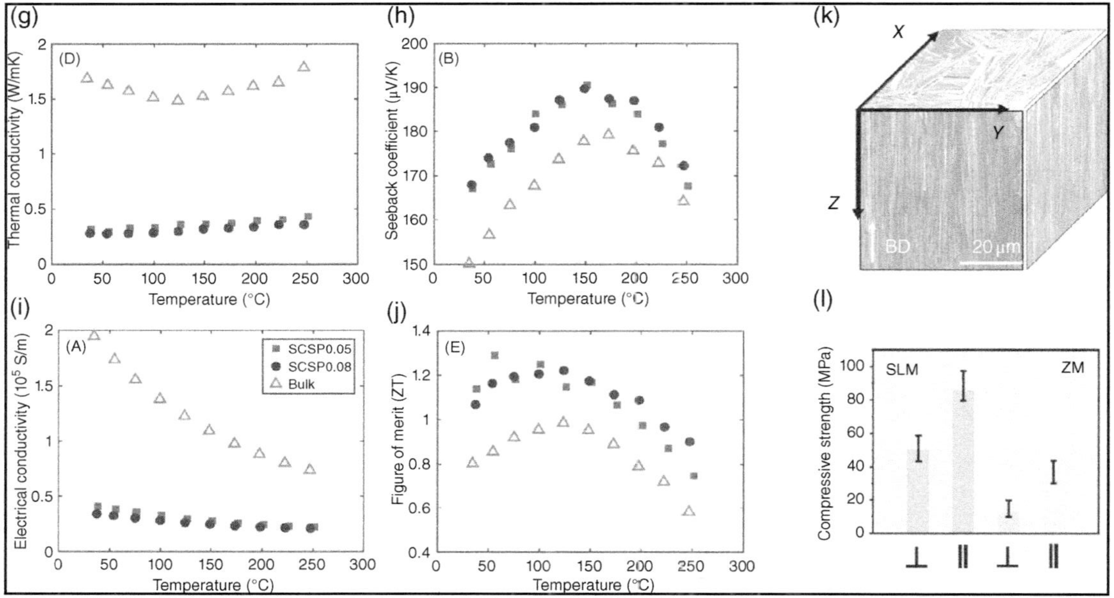


**Figure 13.4** (Continued)

telluride and study of the thermoelectric properties. *Journal of Manufacturing Processes*, **37**, 370. (k) 3D grain morphologies integrated by FESEM images of cross sections of the XY, XZ, YZ planes. (l) Comparison of compressive strengths of the printed and annealed sample and the ZM sample [55]. *Source:* From Qiu, J., Yan, Y., Luo, T., Tang, K., Yao, L., Zhang, J., ... Tang, X. 3D Printing of highly textured bulk thermoelectric materials: Mechanically robust BiSbTe alloys with superior performance. *Energy & Environmental Science*, **12**, 3106.

selectively melting the boundaries of BST powders, the crystalline structures in the particles remained intact, which helped intensify phonon scattering without altering the electrical transport behavior. As a result of intense phonon scattering at ground boundaries, the resulting material achieved an ultra-low thermal conductivity of 0.27 W/mK at 54 °C, which is lower than that of commercial bulk TE material (Figure 13.4g). The Seebeck coefficient and electrical conductivity reached their maximum values at 190.51 uV/K and $4.03 \times 10^4$ S/m, respectively (Figure 13.4h and i). The processing parameter produced a porous TE material that had an ultra-low thermal conductivity without severe degradation of its electrical property; a maximum ZT value of 1.29 was reached, which is higher than that of the commercial bulk BST material (Figure 13.4j). Although the modified processing parameter recorded a groundbreaking ZT, its inferior mechanical properties make it less suitable for adaptation to an actual TEG.

To resolve the inferior mechanical properties of SLS processing, Qiu et al. [55] suggested using selective laser melting (SLM) to fabricate textured $Bi_{0.4}Sb_{1.6}Te_3$ and enhance compressive strength along the building direction (BD) and TE properties. SLM, which is performed via metal and ceramic 3D printing technology, is very similar to SLS since it utilizes a high-energy laser beam. However, the irradiated region does not require an additional sintering process since the melted region immediately solidifies to form a bulk layer. The fast cooling rates can configurate fine and complex microstructures. In this report, bulk $Bi_{0.4}Sb_{1.6}Te_3$ fabricated by SLM had columnar-like grains that were highly oriented along the BD as shown in Figure 13.4k. Although the microstructures were strongly aligned, the Seebeck coefficient and ZT value of the printed sample were only about 112 μV/K and 0.72, respectively. This deterioration of thermoelectric properties resulted from the temperature difference caused by the depth of the molten pool. Additionally, previously printed layers were continuously damaged by the thermal stress of the subsequently formed layers. Accordingly, the printed sample was sintered at 673 K for 24 hr to decrease the concentration of anti-site defects, which improved the Seebeck coefficient to 195 μV/K. As a result, ZT reached 1.1 at 316 K. Moreover, highly oriented microstructures led to a strong enhancement in mechanical properties. The SLM-printed sample had a maximum compressive strength of 91 MPa, which is higher than that of the zone melted samples (Figure 13.4l).

SLS processing demonstrates the feasibility of one integrated step production of p- and n-type TE legs attached to electrodes, drastically reducing the production cost and material loss. Recent studies have reported disadvantages such as micro-level pores and cracks in the resulting material; however, with further optimizations of the processing window, this manufacturing technique has the possibility of developing highly efficient TE legs with a complex geometry that does not require post-processing for binder-removal. Furthermore, SLS opened a new approach that would enable a consolidation of TE material directly onto the electrode, thereby improving mechanical adhesion and minimizing contact resistance.

### 13.2.5 Summary and Outlook

In summary, we have reviewed the recent advancement in 3D printing technologies of TE materials and devices. The advantages of 3D printing process are the cost-effectiveness of manufacturing TE materials and devices with designed geometries, though the efficiencies of materials in the current status are not as high as those of materials synthesized by traditional energy-intensive processes yet. However, as the 3D printing of TE materials has explored very recently, there can be huge room for enhancing the efficiency of TE materials by well-established strategies in bulk materials. For example, nanostructuring has been widely studied in various TE materials to reduce the thermal conductivity by interface scattering of phonons and consequently to increase the ZT values of TE materials [58]. This strategy can be easily applied to TE inks by the formulation of inks with nanoparticles.

For the commercialization of 3D printing technologies of TE materials, further efforts are needed in the aspects of elemental technologies and materials durability. TE module systems consist of several components such as TE legs, electrodes, substrates, and several interlayer materials. However, compared with 3D printing of TE legs, other components have been less studied despite their importance for the commercialization of this technology [2]. Moreover, owing to lower density, the inferior mechanical durability of 3D-printed TE legs to traditionally prepared TE legs will be a critical issue in the industrialization. Recently, several researchers reported the improved mechanical properties of geometrically controlled inorganic materials, even higher than their intrinsic properties [59, 60]. Applying such a strategy to TE materials by 3D printing, one may enhance not only the mechanical properties but also the power generation performance of the 3D-printed TE system. We hope that 3D printing technologies of TE materials and devices will lead TE technologies to be a real solution to providing sustainable electricity supply.

## Acknowledgements

We acknowledge Samsung Research Funding Center of Samsung Electronics under Project No. SRFC-MA1801-05.

## References

**1** Panwar, N. L., Kaushik, S. C., & Kothari, S. (2011). Role of renewable energy sources in environmental protection: A review. *Renewable and Sustainable Energy Reviews*, 15, 1513.

**2** Jo, S., Choo, S., Kim, F., Heo, S. H., & Son, J. S. (2019). Ink processing for thermoelectric materials and power-generating devices. *Advanced Materials*, 31, e1804930.

3 Bell, L. E. (2008). Cooling, heating, generating power, and recovering waste heat with thermoelectric systems. *Science*, 321, 1457.

4 Sundarraj, P., Maity, D., Roy, S. S., & Taylor, R. A. (2014). Recent advances in thermoelectric materials and solar thermoelectric generators – A critical review. *RSC Advances*, 4, 46860.

5 Bahk, J.-H., Fang, H., Yazawa, K., & Shakouri, A. (2015). Flexible thermoelectric materials and device optimization for wearable energy harvesting. *Journal of Materials Chemistry C*, 3, 10362.

6 Champier, D. (2017). Thermoelectric generators: A review of applications. *Energy Conversion and Management*, 140, 167.

7 Sootsman, J. R., Chung, D. Y., & Kanatzidis, M. G. (2009). New and old concepts in thermoelectric materials. *Angewandte Chemie (International Ed. in English)*, 48, 8616.

8 Rull-Bravo, M., Moure, A., Fernández, J. F., & Martín-González, M. (2015). Skutterudites as thermoelectric materials: Revisited. *RSC Advances*, 5, 41653.

9 Wang, S., Li, H., Lu, R., Zheng, G., & Tang, X. (2013). Metal nanoparticle decorated n-type Bi2Te3-based materials with enhanced thermoelectric performances. *Nanotechnology*, 24, 285702.

10 Poudel, B., Hao, Q., Ma, Y., Lan, Y., Minnich, A., Yu, B., ... Ren, Z. (2008). High-thermoelectric performance of nanostructured bismuth antimony telluride bulk alloys. *Science*, 320, 634.

11 Miura, S., Sato, Y., Fukuda, K., Nishimura, K., & Ikeda, K. (2000). Texture and thermoelectric properties of hot-extruded Bi2Te3 compound. *Materials Science and Engineering A*, 277, 244.

12 M'Barki, A., Bocquet, L., & Stevenson, A. (2017). Linking rheology and printability for dense and strong ceramics by direct ink writing. *Scientific Reports*, 7, 6017.

13 Lower, G. W., Walker, W. C., & Zettlemoyer, A. C. (1953). The rheology of printing inks. II. Temperature control studies in the rotational viscometer. *Journal of Colloid Science*, 8, 116.

14 Wang, Z., Wang, X., & Fang, T. (1998). The rheology of the offset inks. *Surface Coatings International*, 81, 219.

15 Zhu, C., & Smay, J. E. (2011). Thixotropic rheology of concentrated alumina colloidal gels for solid freeform fabrication. *Journal of Rheology*, 55, 655.

16 Nadkarni, S. S., & Smay, J. E. (2006). Concentrated barium titanate colloidal gels prepared by bridging flocculation for use in solid freeform fabrication. *Journal of the American Ceramic Society*, 89, 96.

17 Lewis, J. A. (2006). Direct ink writing of 3D functional materials. *Advanced Functional Materials*, 16, 2193.

18 We, J. H., Kim, S. J., & Cho, B. J. (2014). Hybrid composite of screen-printed inorganic thermoelectric film and organic conducting polymer for flexible thermoelectric power generator. *Energy*, 73, 506.

**19** Du, Y., Shen, S. Z., Cai, K., & Casey, P. S. (2012). Research progress on polymer–inorganic thermoelectric nanocomposite materials. *Progress in Polymer Science*, 37, 820.

**20** Madan, D., Wang, Z., Chen, A., Wright, P. K., & Evans, J. W. (2013). High-performance dispenser printed MA p-type $Bi_{0.5}Sb_{1.5}Te_3$ flexible thermoelectric generators for powering wireless sensor networks. *ACS Applied Materials & Interfaces*, 5, 11872.

**21** Kim, F., Kwon, B., Eom, Y., Lee, J. E., Park, S., Jo, S., ... Son, J. S. (2018). 3D printing of shape-conformable thermoelectric materials using all-inorganic $Bi_2Te_3$-based inks. *Nature Energy*, 3, 301.

**22** Park, S. H., Jo, S., Kwon, B., Kim, F., Ban, H. W., Lee, J. E., ... Son, J. S. (2016). High-performance shape-engineerable thermoelectric painting. *Nature Communications*, 7, 13403.

**23** Jo, S., Park, S. H., Ban, H. W., Gu, D. H., Kim, B.-S., Son, J. H., ... Son, J. S. (2016). Simultaneous improvement in electrical and thermal properties of interface-engineered BiSbTe nanostructured thermoelectric materials. *Journal of Alloys and Compounds*, 689, 899.

**24** Jo, S., Park, S. H., Shin, H., Oh, I., Heo, S. H., Ban, H. W., ... Son, J. S. (2019). Soluble telluride-based molecular precursor for solution-processed high-performance thermoelectrics. *ACS Applied Energy Materials*, 2, 4582.

**25** Webber, D. H., & Brutchey, R. L. (2013). Alkahest for $V_2VI_3$ chalcogenides: Dissolution of nine bulk semiconductors in a diamine-dithiol solvent mixture. *Journal of the American Chemical Society*, 135, 15722.

**26** Kovalenko, M. V., Spokoyny, B., Lee, J. S., Scheele, M., Weber, A., Perera, S., ... Talapin, D. V. (2010). Semiconductor nanocrystals functionalized with antimony telluride zintl ions for nanostructured thermoelectrics. *Journal of the American Chemical Society*, 132, 6686.

**27** Eom, Y., Kim, F., Yang, S. E., Son, J. S., & Chae, H. G. (2019). Rheological design of 3D printable all-inorganic inks using BiSbTe-based thermoelectric materials. *Journal of Rheology*, 63, 291.

**28** Saeidi-Javash, M., Kuang, W., Dun, C., & Zhang, Y. (2019). 3D conformal printing and photonic sintering of high-performance flexible thermoelectric films using 2D nanoplates. *Advanced Functional Materials*, 29, 1901930.

**29** Dudek, P. (2013). FDM 3D Printing thecnology in manufacturing composite elements. *Archives of Metallurgy and Materials*, 58, 1415.

**30** Oztan, C., Ballikaya, S., Ozgun, U., Karkkainen, R., & Celik, E. (2019). Additive manufacturing of thermoelectric materials via fused filament fabrication. *Applied Materials Today*, 15, 77.

**31** Weng, Z., Wang, J., Senthil, T., & Wu, L. (2016). Mechanical and thermal properties of ABS/montmorillonite nanocomposites for fused deposition modeling 3D printing. *Materials and Design*, 102, 276.

**32** Mannoor, M. S., Jiang, Z., James, T., Kong, Y. L., Malatesta, K. A., Soboyejo, W. O., ... McAlpine, M. C. (2013). 3D printed bionic ears. *Nano Letters*, 13, 2634.

**33** Kolesky, D. B., Truby, R. L., Gladman, A. S., Busbee, T. A., Homan, K. A., & Lewis, J. A. (2014). Lewis, 3D bioprinting of vascularized, heterogeneous cell-laden tissue constructs. *Advanced Materials*, 26, 3124.

**34** M. He, Y. Zhao, B. Wang, Q. Xi, J. Zhou, Z. Liang, 3D Printing fabrication of amorphous thermoelectric materials with ultralow thermal conductivity. *Small* 2015, 11, 5889.

**35** Kumar, S., Hofmann, M., Steinmann, B., Foster, E. J., & Weder, C. (2012). Reinforcement of stereolithographic resins for rapid prototyping with cellulose nanocrystals. *ACS Applied Materials & Interfaces*, 4, 5399.

**36** Wang, X., Jiang, M., Zhou, Z., Gou, J., & Hui, D. (2017). 3D printing of polymer matrix composites: A review and prospective. *Composites Part B: Engineering*, 110, 442.

**37** Vandenbroucke, B., & Kruth, J. P. (2007). Selective laser melting of biocompatible metals for rapid manufacturing of medical parts. *Rapid Prototyping Journal*, 13, 196.

**38** Rafi, H. K., Pal, D., Patil, N., Starr, T. L., & Stucker, B. E. (2014). Microstructure and mechanical behavior of 17-4 precipitation hardenable steel processed by selective laser melting. *Journal of Materials Engineering and Performance*, 23, 4421.

**39** Riemer, A., Leuders, S., Thöne, M., Richard, H. A., Tröster, T., & Niendorf, T. (2014). On the fatigue crack growth behavior in 316L stainless steel manufactured by selective laser melting. *Engineering Fracture Mechanics*, 120, 15.

**40** Wang, Z., Guan, K., Gao, M., Li, X., Chen, X., & Zeng, X. (2012). The microstructure and mechanical properties of deposited-IN718 by selective laser melting. *Journal of Alloys and Compounds*, 513, 518.

**41** Song, B., Dong, S., Zhang, B., Liao, H., & Coddet, C. (2012). Microstructure and tensile properties of iron parts fabricated by selective laser melting. *Materials and Design*, 35, 120.

**42** Murr, L. E., Gaytan, S. M., Ramirez, D. A., Martinez, E., Hernandez, J., Amato, K. N., ... Wicker, R. B. (2012). Microstructural architecture, microstructures, and mechanical properties for a nickel-base superalloy fabricated by electron beam melting. *Journal of Materials Research and Technology*, 28, 1.

**43** Alhnan, M. A., Okwuosa, T. C., Sadia, M., Wan, K. W., Ahmed, W., & Arafat, B. (2016). Emergence of 3D printed dosage forms: Opportunities and challenges. *Pharmaceutical Research*, 33, 1817.

**44** Lin, D., Nian, Q., Deng, B., Jin, S., Hu, Y., Wang, W., & Cheng, G. J. (2014). Three-dimensional printing of complex structures: Man made or toward nature? *ACS Nano*, 8, 9710.

**45** Vrancken, B., Thijs, L., Kruth, J.-P., & Van Humbeeck, J. (2012). Heat treatment of Ti6Al4V produced by selective laser melting: Microstructure and mechanical properties. *Journal of Alloys and Compounds*, 541, 177.

**46** Kanagarajah, P., Brenne, F., Niendorf, T., & Maier, H. J. (2013). Inconel 939 processed by selective laser melting: Effect of microstructure and temperature on the mechanical properties under static and cyclic loading. *Materials Science and Engineering A*, 588, 188.

**47** El-Desouky, A. R. A., Bardet, P., Andre, M., & LeBlanc, S. (2015). Selective laser melting of a bismuth telluride thermoelectric materials. *Proceedings of Solid Freeform Fabrication Symposium*, 1043.

**48** El-Desouky, A., Carter, M., Mahmoudi, M., Elwany, A., & LeBlanc, S. (2017). Influences of energy density on microstructure and consolidation of selective laser melted bismuth telluride thermoelectric powder. *Journal of Manufacturing Processes*, 25, 411.

**49** El-Desouky, A., Carter, M., Andre, M. A., Bardet, P. M., & LeBlanc, S. (2016). Pulsed laser melting of bismuth telluride thermoelectric materials. *Materials Letters*, 185, 598.

**50** Su, X., Li, H., Yan, Y., Wang, G., Chi, H., Zhou, X., ... Uher, C. (2012). Microstructure and thermoelectric properties of $CoSb_{2.75}Ge_{0.25-x}Te_x$ prepared by rapid solidification. *Acta Materialia*, 60, 3536.

**51** Li, H., Tang, X., Su, X., & Zhang, Q. (2008). Preparation and thermoelectric properties of high-performance Sb additional $Yb_{0.2}Co_4Sb_{12+y}$ bulk materials with nanostructure. *Applied Physics Letters*, 92, 202114.

**52** Li, G., Bajaj, S., Aydemir, U., Hao, S., Xiao, H., Goddard, W. A., ... Snyder, G. J. (2016). P-type Co interstitial defects in thermoelectric skutterudite $CoSb_3$ due to the breakage of $Sb_4$-rings. *Journal of Materials Chemistry*, 28, 2172.

**53** Yan, Y., Ke, H., Yang, J., Uher, C., & Tang, X. (2018). Fabrication and thermoelectric properties of n-Type $CoSb_{2.85}Te_{0.15}$ using selective laser melting. *ACS Applied Materials & Interfaces*, 10, 13669.

**54** Shi, J., Chen, H., Jia, S., & Wang, W. (2019). 3D printing fabrication of porous bismuth antimony telluride and study of the thermoelectric properties. *Journal of Manufacturing Processes*, 37, 370.

**55** Qiu, J., Yan, Y., Luo, T., Tang, K., Yao, L., Zhang, J., ... Tang, X. (2019). 3D Printing of highly textured bulk thermoelectric materials: Mechanically robust BiSbTe alloys with superior performance. *Energy & Environmental Science*, 12, 3106.

**56** Yan, Y., Geng, W., Qiu, J., Ke, H., Luo, C., Yang, J., ... Tang, X. (2018). Thermoelectric properties of n-type ZrNiSn prepared by rapid non-equilibrium laser processing. *RSC Advances*, 8, 15796.

**57** Mao, Y., Yan, Y., Wu, K., Xie, H., Xiu, Z., Yang, J., ... Tang, X. (2017). Non-equilibrium synthesis and characterization of n-type $Bi_2Te_{2.7}Se_{0.3}$ thermoelectric material prepared by rapid laser melting and solidification. *RSC Advances*, 7, 21439.

**58** Dresselhaus, M. S., Chen, G., Tang, M. Y., Yang, R. G., Lee, H., Wang, D. Z., ... Gogna, P. (2007). New directions for low-dimensional thermoelectric materials. *Advanced Materials*, 19, 1043.

**59** Lin, C. Y., Wirtz, T., LaMarca, F., & Hollister, S. J. (2007). Structural and mechanical evaluations of a topology optimized titanium interbody fusion cage fabricated by selective laser melting process. *Journal of Biomedical Materials Research. Part A*, 83, 272.

**60** Xiao, Z., Yang, Y., Xiao, R., Bai, Y., Song, C., & Wang, D. (2018). Evaluation of topology-optimized lattice structures manufactured via selective laser melting. *Materials and Design*, 143, 27.

# 14

# Carbon Capture, Usage, and Storage

*Jason E. Bara*

*University of Alabama, Tuscaloosa, AL, USA*

Carbon capture, utilization, and storage (CCUS) can offer immediate and serious reductions in $CO_2$ emissions to the atmosphere from sources such as fossil fuel burning power plants. However, CCUS only achieves this goal of limiting $CO_2$ emissions and minimizing greenhouse effects if it is deployed broadly and at large scales (i.e., capturing at least 90% of emissions). Although there have been numerous calls to action for the installation of CCUS at coal-fired power plants and other industrial sources, meaningful progress has been slow to materialize. Inevitably, these large point sources of $CO_2$ emissions require very large infrastructures with which to handle the post-combustion flue gas or other stream containing $CO_2$. The most readily available process technology for $CO_2$ capture is "absorption-stripping" which has a long history of use for the removal of $CO_2$ from natural gas. The focus of this chapter is to analyze whether 3D printing technologies are available and capable of fabricating absorption and stripping columns on the scales at which $CO_2$ capture and CCUS would achieve the goals of greenhouse gas reductions.

## 14.1 Introduction

Carbon capture, usage, and storage (CCUS) is defined as the separation, compression, and storage/use of carbon dioxide ($CO_2$) emissions from industrial gas streams at point sources including the flue gas of fossil fuels burned for energy generation ($CO_2/N_2$), natural gas processing ($CO_2/CH_4$), syngas cleanup ($CO_2/H_2$) or in other industries such as steel, cement, or ethanol production. Furthermore, $CO_2$ capture concepts can also be applied to direct removal of $CO_2$ already in the atmosphere,

*3D Printing for Energy Applications*, First Edition. Edited by Albert Tarancón and Vincenzo Esposito.

which is known as "air capture." Several different types of processes can be utilized to perform the capture step. The most well-known and established processes for $CO_2$ removal are: absorption-stripping driven by chemical and/or physical reactions with liquid solvents, membrane-based separations driven by pressure/concentration differences across thin films, and adsorption/desorption on solid media driven by temperature, pressure, and interactions of $CO_2$ with surfaces. In order to prevent emission of the captured $CO_2$ to the atmosphere, the $CO_2$ can be compressed and stored (sequestered) in subterranean reservoirs and/or put to some beneficial use such as enhanced oil recovery (EOR) or as a feedstock for chemical or biological processes. However, if CCUS technologies are to make meaningful reductions to global greenhouse gas (GHG) emissions, the scales at which they must be deployed are massive and the pace of the deployment must be swift.

Widespread use of CCUS is thus potentially one of the grandest challenges of the twenty-first century. While a number of reports discuss the importance of CCUS in reducing GHG emissions and thus limiting future global temperature increases, deployment of CCUS technologies at any meaningful scale has been slow to materialize. Coal-fired power plants are the largest point sources of $CO_2$ emissions, and as such should be the logical starting place for deployment. However, there are significant infrastructure investments that would need to be made in order to capture $CO_2$ emissions from these plants. Given the urgency at which a reliable and proven technology for CCUS should be deployed, then a conventional absorption-stripping process is the only viable option that could be deployed immediately [1] as it is the most mature and developed process with nearly 90 years' worth of know-how [2]. This of course, does not taken into account whether the demand for the chemicals with which $CO_2$ is captured such as monoethanolamine (MEA) could be met [3], but rather the readiness of such systems to be constructed and operated.

In order to understand the scale of just one $CO_2$ capture process applied to the flue gas of a coal-fired power plant (and there will be hundreds to thousands of such processes needed worldwide), consider an absorption-stripping system (Figure 14.1) modeled for a 750 MW supercritical coal-fired power plant as reported by Nittaya, Douglas, Croiset, and Ricardez-Sandoval [4].

The flue gas is distributed equally to three absorbers (A101, A201, A301) each of which are 11.8 m in diameter and 34 m tall ($V = 3{,}718\,m^3$). It has been reported that the maximum diameter for a cylindrical column absorber is 12–15 m due to limitations associated with fabrication techniques, weight, and support structures. In these absorbers, the flue gas is contacted with an aqueous solution of MEA resulting in a chemical reaction which binds the $CO_2$ to the amine while the other components of flue gas (mostly $N_2$) are vented to the atmosphere. This $CO_2$-rich amine solution is then passed through a heat exchanger (HX101) before being equally split and fed to two strippers (D101, D201) are 10.4 m in diameter and 16 m tall ($V = 1{,}359\,m^3$). In the strippers, most of the $CO_2$ is released from solution to be compressed. The $CO_2$-lean solution is then returned to the absorbers to repeat the process.

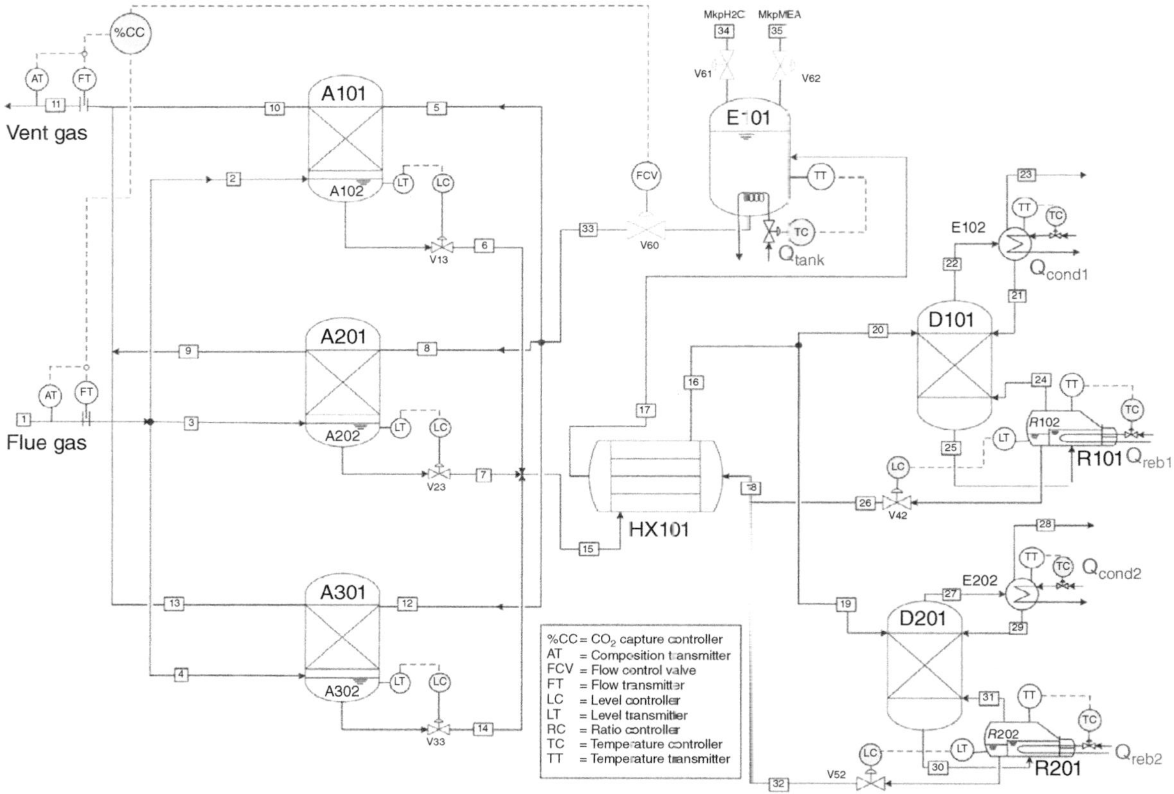


**Figure 14.1** Flowsheet of a commercial $CO_2$ capture plant [4]. *Source:* From Nittaya, T.; Douglas, P. L.; Croiset, E.; Ricardez-Sandoval, L. A., Dynamic modeling and evaluation of an industrial-scale CO2 capture plant using monoethanolamine absorption processes. *Industrial & Engineering Chemistry Research* **2014,** 53, (28), 11411–11426. © 2014 American Chemical Society.

It is quite clear that the volume of the vessels required for a full-scale $CO_2$ capture process is gargantuan, with the absorbers each being ~1.5× an Olympic-sized swimming pool ($V = 2{,}500\,m^3$) and the strippers ~0.5× of the same reference volume. However, the sheer volume is not the sole factor to be considered in construction of these vessels. In order to provide gas–liquid contact which drives the absorption and stripping processes, the absorber and stripper are filled with "packing" (or trays) which creates surface area on which these processes occur. These packings have a specific surface area of ~$100\,m^2/m^3$.

Two other major vessels in the flowsheet (Figure 14.1) are the heat-exchanger (HX101) and the buffer tank (E101). HX101 is a single-pass shell and tube heat exchanger of ~3 m diameter and 3 m long ($V = 21.2\,m^3$) containing 14,459 tubes. The buffer tank was reported to have a diameter of 12 m but no details are given on the height or total volume. Nonetheless, the volume of the buffer tank is $>100\,m^3$ for every 1 m of height. Figure 14.1 also depicts two reboilers (R101, R201) which are reported to have diameters equal to those of the strippers (10.4 m), again indicating volumes $>100\,m^3$ for each 1 m of height.

Although there is no absorption-stripping process that has been built to such a scale, conventional manufacturing and fabrication techniques are applied to large vessels such distillation columns used in oil refineries and chemical production. The world's largest distillation column was fabricated in China and shipped to Nigeria where it is being installed at the Dangote Oil Refinery in Nigeria. This distillation column is reported to have a diameter of 12 m and a height of 112.56 m ($V = 12{,}730\,m^3$) [5]. Given the sheer size of the vessels in a full-scale $CO_2$ capture process at a coal-fired power plant and the relatively complex internals of some vessels, there is an obvious question to ask...

## 14.2 Can 3D Printing Be Used to Fabricate a $CO_2$ Capture Process at Scale?

Because of the caustic/corrosive nature of aqueous amine solvents, stainless steel must be the first material to be considered for the fabrication of the major equipment of a $CO_2$ capture process. There is no doubt that current commercially available metal-based laser sintering 3D printing techniques could be applied at relatively small scales (10–100 L volumes) to produce the requisite vessels and internal components such that a small, benchtop $CO_2$ capture process could be fabricated primarily on additive manufacturing. For example, the *X Line 2000R* printer from Concept Laser is touted as the "largest metal fusion machine in the world" with a build volume of $0.8\,m \times 0.4\,m \times 0.5\,mm$ ($0.160\,m^3 = 160\,L$) [6]. However, the use of 3D printers at small "benchtop" scales may not be advantageous, especially if used to fabricate relatively simple, hollow vessels (e.g., absorber, stripper, tanks) as conventional manufacturing techniques will almost certainly be much faster.

However, the case for 3D printing for $CO_2$ capture technologies becomes much more compelling if it is applied at much larger scales, as it could potentially accelerate deployment and allow for on-site fabrication. 3D printing may also reduce $CO_2$ emissions compared to conventional manufacturing, however, of course these emissions are relatively small compared to the magnitude of global $CO_2$ emissions from coal-fired electric power generation, which were 11.5 million metric tons in 2018 [7].

Recent examples of large 3D printed objects/structures include boats, buildings, and even bridges [8]. The current efforts in the aerospace industry to 3D print rockets are most encouraging for the future of chemical process technology as rockets, much like absorbers, strippers, and distillation columns, essentially start as hollow, cylindrical vessels.

Relativity Space is a company based in Los Angeles, USA that is working toward fully automated 3D printing of rockets at launch sites. Relativity Space's first rocket, dubbed "Terran 1" will be about 100 feet (30.5 m) tall and can be completed within 60 days [9]. 95% of the process is automated, with only 5% requiring human interaction, primarily in the final testing and assembly tasks. The height of this rocket is already comparable to that needed for absorbers and strippers described earlier meaning that this 3D printing process would be able to build these vessels the requisite heights. However, rocket diameters are not as large as those needed for the full-scale absorbers and strippers, thus this large-format 3D printing technology would need to be expanded to fabricate wider objects. For reference, SpaceX's Falcon 9 FT has a height of 70 m but with a diameter of only 3.7 m [10]. The wall thickness is reported to be about 5 mm, which would also be consistent with the requirements for absorption and stripping columns.

Perhaps the scaling of Relativity Space's technology to make absorbers and strippers with the larger diameters needed for $CO_2$ capture processes would not be particularly difficult, but it would certainly add time to the fabrication process and more than 60 days would be needed for the fabrication of these vessels. The wider diameter of absorbers and strippers in chemical process equipment (as compared to the aforementioned rockets) is critical to ensure uniform distribution and flow of the liquid and gas phases.

Clearly, a $CO_2$ capture process requires more than just absorbers and strippers. However, as Relativity Space plans to 3D print all of the rocket parts, including the engines, this may also be a positive indicator that the heat exchangers and pumps required for a $CO_2$ capture process could also be fabricated by this technology.

3D printing with concrete is gaining momentum as a technology for large-scale construction projects, allowing for low cost and rapid construction of buildings, homes, and other structures. But can concrete-based 3D printing be a viable option for $CO_2$ capture processes? The answer appears to be yes, at least with respect to absorption columns.

As the interior of the absorption column for removing $CO_2$ from flue gas will experience only moderate temperatures (~60 °C), a thermally insulating material such as concrete is a viable option, provided that the walls of the absorber are lined with a chemically resistant material (e.g., high-performance polymers). Although fabricated by "slip forming" and not by 3D printing techniques, the $CO_2$ capture process at Technology Centre Mongstad (TCM) in Norway is the world's first example of the use of concrete for this purpose. The absorption column at TCM is rectangular in shape with dimensions of 3.5 m × 2 m with a height of 62 m, requiring only 20 days to erect [11]. An image of the concrete absorber at TCM is shown in Figure 14.2.

Developments in large-scale 3D printing of rockets and buildings hold promise for bringing large-scale $CO_2$ capture processes and CCUS as a whole closer to reality. If the world is serious about using CCUS as a means to drastically curb $CO_2$ emissions, meaningful changes will only be achieved through processes which capture, sequester, and/or utilize nearly all of the $CO_2$ from large point sources. Of course, the capture step is only the beginning and there is still a need for compression, pipelines, and underground storage (or sources for utilization). While it would not seem worthwhile to 3D print pipes, it is possible that 3D printing could also be used for compressors or other essential equipment.

## 14.3 A Brief Note on 3D Printing and $CO_2$ at Smaller Scales & Research Efforts

3D printing is undoubtedly an exciting technique and tool for research that captivates the imagination, greatly reduces fabrication challenges, and expedites the time between conceptualization and functioning product. Some forms of 3D printing also allow for new functional materials (e.g., adsorbents, catalysts) to be added to existing resins/powders/feedstocks such that the desired functionality can directly incorporated during the printing without a need for post-processing.

Although $CO_2$ capture from large point sources is clearly the most pressing need with respect to global environmental concerns, $CO_2$ levels must also be controlled/managed in enclosed environments such as submarines and space travel. Thus, 3D printing can also offer some new approaches for the design and production of small-scale equipment. Examples of such objects include novel designs for gas–liquid contactors and packings [12, 13] as well as for solid adsorbents containing zeolites [14–16], metal-organic frameworks (MOFs) [17–20], aminosilica [21], and

**Figure 14.2** Photograph of concrete absorber (left) at TCM [11]. *Source:* Image reproduced with permission from de Koeijer, G.; Enge, Y.; Sanden, K.; Graff, O. F.; Falk-Pedersen, O.; Amundsen, T.; Overå, S., CO2 technology centre mongstad–design, functionality and emissions of the amine plant. *Energy Procedia* **2011,** 4, 1207–1213. © 2011 Elsevier.

carbonates [22]. Figure 14.3 shows illustrations of a gas–liquid contactor formed using 3D printing and molding techniques.

As the types, scale and access to 3D printing continues to grow, there will certainly be continued interest in using 3D printing at the small scale to manage $CO_2$ in enclosed environments.

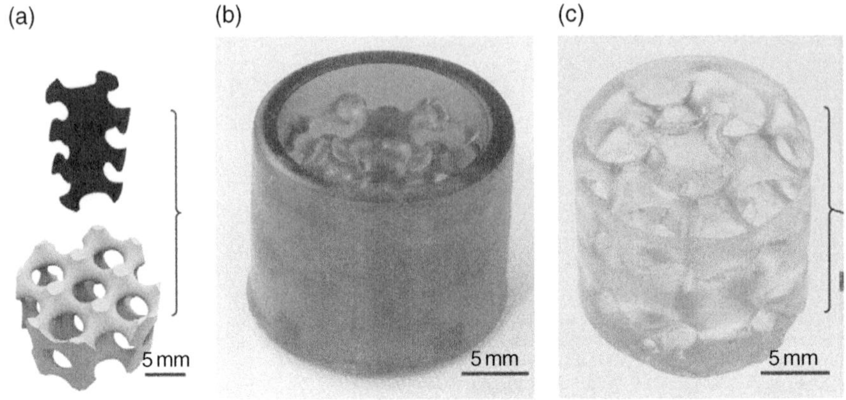


**Figure 14.3** (a) Design of contactor surface [12]. *Source:* Reproduced with permission from Femmer, T.; Kuehne, A. J. C.; Torres-Rendon, J.; Walther, A.; Wessling, M., Print your membrane: Rapid prototyping of complex 3D-PDMS membranes via a sacrificial resist. *Journal of Membrane Science* **2015,** 478, 12–18. © 2015 Elsevier; (b) 3D printed sacrificial mold; (c) gas–liquid contactor cast from PDMS in mold.

## 14.4 Conclusions

3D printing may have a key role to play in CCUS and global greenhouse gas reduction if the innovations that are taking place in the aerospace and construction industries could also be directed toward the design of $CO_2$ capture processes from large point sources. 3D printing of absorber and stripper columns along with other large process vessels now appears to be highly feasible, and the potential to use concrete as the material of construction for the absorber is particularly intriguing. However, it should be noted that there are a number of other hurdles that have prevented the widespread deployment of CCUS to date and that 3D printing will not necessarily solve the technical, economic, and societal questions that remain. Nonetheless, if CCUS does move forward with the purpose of curbing nearly all $CO_2$ emissions from electric power generation and other industries, it appears very likely that 3D printing will factor into the construction of these processes.

## References


1 Rochelle, G. T. (2009). Amine Scrubbing for CO2 Capture. *Science*, 325(5948), 1652–1654.

2 Bottoms, R. R. (1930). Separating acidic gases. *US1783901*.

3 Bara, J. E. (2012). What chemicals will we need to capture CO2? *Greenhouse Gases: Science and Technology*, 2(3), 162–171.

4 Nittaya, T., Douglas, P. L., Croiset, E., & Ricardez-Sandoval, L. A. (2014). Dynamic modeling and evaluation of an industrial-scale CO2 capture plant using monoethanolamine absorption processes. *Industrial & Engineering Chemistry Research*, 53(28), 11411–11426.

5 Osae-Brown, A., Alake, T., & Nightingale, A. (2019, December 4). World's biggest oil-refining tower completes 11,000-mile voyage. Retrieved from https://www.bloomberg.com/news/articles/2019-12-04/world-s-biggest-oil-refining-tower-completes-11-000-mile-voyage

6 3DNatives. (2020, August 25). A comprehensive list of all the metal 3D printer manufacturers. Retrieved from https://www.3dnatives.com/en/metal-3d-printer-manufacturers/

7 Administration, U. S. E. I. (2020, March27). How much of U.S. carbon dioxide emissions are associated with electricity generation? Retrieved from https://www.eia.gov/tools/faqs/faq.php?id=77&t=11

8 Anonymous. (2019). Giant 3D printers for making boats, bridges, buildings and rockets. *The Economist*.

9 Grush, L. (2019). Aerospace startup making 3D-printed rockets now has a launch site at America's busiest spaceport. *The Verge*.

10 SpaceX. (2020, August 25). Falcon 9. Retrieved from https://www.spacex.com/vehicles/falcon-9/

11 de Koeijer, G., Enge, Y., Sanden, K., Graff, O. F., Falk-Pedersen, O., Amundsen, T., & Overå, S. (2011). CO2 technology centre mongstad–design, functionality and emissions of the amine plant. *Energy Procedia*, 4, 1207–1213.

12 Femmer, T., Kuehne, A. J. C., Torres-Rendon, J., Walther, A., & Wessling, M. (2015). Print your membrane: Rapid prototyping of complex 3D-PDMS membranes via a sacrificial resist. *Journal of Membrane Science*, 478, 12–18.

13 Bara, J. E., Hawkins, C. I., Neuberger, D. T., & Poppell, S. W. (2013). 3-D printing for CO2 capture and chemical engineering design. *Nanomaterials and Energy*, 2, 235–243.

14 Couck, S., Cousin-Saint-Remi, J., Van der Perre, S., Baron, G. V., Minas, C., Ruch, P., & Denayer, J. F. M. (2018). 3D-printed SAPO-34 monoliths for gas separation. *Microporous and Mesoporous Materials*, 255, 185–191.

15 Thakkar, H., Lawson, S., Rownaghi, A. A., & Rezaei, F. (2018). Development of 3D-printed polymer-zeolite composite monoliths for gas separation. *Chemical Engineering Journal (Amsterdam, Netherlands)*, 348, 109–116.

16 Thakkar, H., Eastman, S., Hajari, A., Rownaghi, A. A., Knox, J. C., & Rezaei, F. (2016). 3D-printed zeolite monoliths for CO2 removal from enclosed environments. *ACS Applied Materials & Interfaces*, 8(41), 27753–27761.

**17** Thakkar, H., Eastman, S., Al-Naddaf, Q., Rownaghi, A. A., & Rezaei, F. (2017). 3D-printed metal-organic framework monoliths for gas adsorption processes. *ACS Applied Materials & Interfaces*, 9(41), 35908–35916.

**18** Lahtinen, E., Precker, R. L. M., Lahtinen, M., Hey-Hawkins, E., & Haukka, M. (2019). Selective laser sintering of metal-organic frameworks: Production of highly porous filters by 3D printing onto a polymeric matrix. *ChemPlusChem*, 84(2), 222–225.

**19** Evans, K. A., Kennedy, Z. C., Arey, B. W., Christ, J. F., Schaef, H. T., Nune, S. K., & Erikson, R. L. (2018). Chemically active, porous 3D-printed thermoplastic composites. *ACS Applied Materials & Interfaces*, 10(17), 15112–15121.

**20** Lawson, S., Al-Naddaf, Q., Krishnamurthy, A., Amour, M. S., Griffin, C., Rownaghi, A. A., ... Rezaei, F. (2018). UTSA-16 growth within 3D-printed Co-Kaolin monoliths with high selectivity for CO2/CH4, CO2/N2, and CO2/H2 separation. *ACS Applied Materials & Interfaces*, 10(22), 19076–19086.

**21** Thakkar, H., Eastman, S., Al-Mamoori, A., Hajari, A., Rownaghi, A. A., & Rezaei, F. (2017). Formulation of aminosilica adsorbents into 3D-printed monoliths and evaluation of their CO2 capture performance. *ACS Applied Materials & Interfaces*, 9(8), 7489–7498.

**22** Nguyen, D., Murialdo, M., Hornbostel, K., Pang, S., Ye, C., Smith, W., ... Stolaroff, J. (2019). 3D printed polymer composites for CO2 capture. *Industrial and Engineering Chemistry Research*, 58(48), 22015–22020.

# Index


## a

AconityMINI system 138
acrylonitrile butadiene styrene (ABS) 293
additive manufacturing (AM) 104–105
  capacitors
    advantages 215–216
    binder jetting 228–229
    complete capacitor device fabricate 230–232
    disadvantages 216
    existing reviews 216
    general selection criteria 216–217
    material extrusion 217–223
    material jetting 227–228
    multi-technology 229–230
    powder bed fusion 225–227
    process categories 217–229
    vat polymerization 223–225
  ceramics 3D printing technologies
    extrusion 38–44
    jetting 49–52
    lamination object modeling 37–38
    laser-based technologies 47–49
    photopolymerization 44–47
  challenges 153
  components 5
  direct material deposition 12–15
  extrusion based 7
  geometrical gradients in 6
  hybrid 19–22 (*see also* hybrid additive manufacturing)
  industrial application 5
  LPBF
    characterization techniques 139–140
    geometrical design 137
    grain size via adaptive processing strategies 143–145
    lattice structures 148–150
    machines/materials 138–139
    material chemistry 137
    material properties by using powder blends 146–148
    material qualification and process development 140–143
    parameters 137
    relative density 141, 142
    sample preparation 138–140
  material gradients in 6–7
  metal 5–6
  physical and digital links of process chain 3–4
  powder bed fusion 7–12
  solid-state 16–19
  thermoelectrics
    extrusion-based additive manufacturing process 328–336


*3D Printing for Energy Applications*, First Edition. Edited by Albert Tarancón and Vincenzo Esposito.

additive manufacturing (AM) (*cont'd*)
fused deposition modeling 336–337
selective laser sintering 339–345
stereolithography apparatus 337–339
air capture 352
alkaline cells (ACs) 273
Al-oxide phosphate thin layers 51
alumina housing 117, 118
AM. *See* additive manufacturing (AM)

## b

barium titanate, Sr-substituted 51
$BaTiO_3$ 46
batteries
3D printed electrode structures 186–195
3D printed full batteries 197–200
hybrid supercapacitors 183
large scale energy storage 183
lithium ion batteries 183–186
solid state 195–197
supercapacitors 183
transportation sector batteries 183
binder jetting
benefits 229
challenges 229
operating principle 228
process considerations and feedstock requirements 228–229
biofuel cells (BFCs) 294–296
biomedicine 36
bioprinting, of realistic and vascularized tissue 162–163

## c

$CaCu_3Ti_4O_{12}$ (CCTO) 49
CAL. *See* computed axial lithography (CAL)
calcium carbonate ($CaCO_3$) 110
capacitors
additive manufacturing technologies
advantages 215–216
binder jetting 228–229
complete capacitor device fabricate 230–232
disadvantages 216
existing reviews 216
general selection criteria 216–217
material extrusion 217–223
material jetting 227–228
multi-technology 229–230
powder bed fusion 225–227
process categories 217–229
vat polymerization 223–225
classifications 206
electrochemical 210–211
electrolytic 209–210
electrostatic 206–209
components 211–213
demand for electricity 205
manufacturing challenge 214–215
performance 213–214
carbon capture, utilization, and storage (CCUS)
$CO_2$ capture process 354–356
definition 351
3D printing 356–358
flue gas 352
carbon nanotube (CNT) 73, 74
electrical actuation of 82–85
honeycomb structure 85, 86
L-shaped 84–85
printing of 73–75
reinforced polymer nanocomposites
resistive heating of 71
strain sensing of 70–71
steady-state temperature 82
thermal gradient 81
CCUS. *See* carbon capture, utilization, and storage (CCUS)
cellular solids 8

ceramics 33
  CAGR 35
  fabrication 46
  high quality monoliths 48
  inkjet 49
  microstereolithography 46
  morphology 41, 45
  particle size distribution 41, 45
  3D printing technologies
    extrusion 38–44
    jetting 49–52
    lamination object modeling (LOM) 37–38
    laser-based technologies 47–49
    photopolymerization 44–47
chemical passivation 158
CNC. *See* computer numeric controls (CNC)
Coaxial extrusion of multilayered hydrogel lumens 162
complexity
  geometrical 154–155
  material 155–156
compound parabolic concentrator (CPC) 257
compression-compression test 121, 123
compressive force sensing
  multicomponent printed device 122–124
  piezoceramic sensors in metallic pipes 119–120
computational design
  design scenarios 98–99
  overhang angle
    definition 95
    local angle control 96
    physics-based constraints 97
    simplified printing process 97–98
  3D-printing constraint 99–104
    fluid flow TO 100–104
    front propagation 102–103
    pipe bend, topology optimization 101, 103, 104
computed axial lithography (CAL) 164
  capability 170–171
  computation for 166–167
  current challenges in 169–170
  material-process interactions in 167–169
  volumetric fabrication 165
computer-aided design (CAD) 20, 69
  smart coupling fabrication 116
computer aided manufacturing of laminated engineering materials (CAM-LEM) 37
computer numeric controls (CNC) 12–13, 16, 72
condensation 117
continuous liquid interface production (CLIP) 159
conventional fabrication techniques 69
crystalline structure analysis 126–128
Curie temperature 118

## d

Darcy's law 100
dead zone 161
debinding process 21–22, 42
decoupling 113
DED. *See* direct energy deposition (DED)
deformation mechanisms 138, 139, 146
Design for Additive Manufacturing (DfAM) 95
digital light printing (DLP) 159
digital light processing (DLP) technique 45, 46
digital printing 69
*N,N*-dimethylformamide (DMF) 40
direct energy deposition (DED) 9, 12, 13, 110, 307
  characteristics 308
  exemplary process chains 308, 309

direct energy deposition (DED) (*cont'd*)
  hybrid manufacturing 314–315
    additive manufacturing 315–317
    nozzle rings 320–323
    production cell 319–320
    turbocharger nozzle ring 317–319
  materials gradients in 15
  of turbomachinery components repair 309–310
    process 310
    repair of turbine blade 310–314
    work environment 310
direct ink writing (DIW) 39–42, 187
  benefits 222
  challenges 223
  operating principle 221–222
  process parameters 222
direct material deposition (DMD) 12–15
  functional material gradients in 13–15
  powder and wire feedstock 12–13
Direct Metal Laser Sintering (DMLS) process 291
direct print (DP) 109
direct write (DW) process 110
DIW. *See* direct ink writing (DIW)
DMD. *See* direct material deposition (DMD)

## e

Eikonal equation 102
electrical resistance 78, 79
electric heating performance analysis 79–82
electrochemical capacitors 210–211
electrolytic capacitors 209–210
electron backscatter diffraction (EBSD) 138, 139
electron beam melting (EBM) 114, 115
electron beam powder bed fusion (EB-PBF) 111
electrophoretic deposition 158
electrostatic capacitors 206, 209
energy devices. *See* batteries
energy dispersive spectroscopy (EDS) 137–139
external quantum efficiency (EQE) 262
extrusion
  ceramics 38–39
    direct ink writing 39–42
    fused deposition modeling 42–44
    robocasting 39–42
  coaxial, multilayered hydrogel lumens 162

## f

facile hybridization 220
FDC. *See* fused deposition of ceramics (FDC)
field-effect transistors 51
filament extrusion 20
finite element analysis (FEA) 14, 95
4D printing 156
freeform reversible embedding of suspended hydrogels (FRESH) 162
frozen turbulence approach 100
fuel cells and electrolyzers
  alkaline cells 273
  biofuel cells 294–296
  polymer exchange membrane cells 273
    PEMEC 293–294
    PEMFC 287–293
    stack operation and components 286–287
  solid oxide cells 273
    SOEC 274, 283–284
    SOFC 274–283
    stacks and components 284–286
fully printed PEMEC 294
functionally gradient additive manufacturing (FGAM) 8

fused deposition modelling (FDM) 38, 39, 72, 153, 291, 336–337
fused deposition of ceramics (FDC) 42–44
fuse deposition modelling (FDM) 109–110
fused filament fabrication (FFF) 153
- benefits 220
- challenges 221
- operating principle 217
- process parameters 217, 220

*g*

GE Leap fuel nozzle 5
geometrical complexity 154–155
geometric light management, in solar cell technology
- characterization 261–266
- compound parabolic concentrator 257
- design and 3D-printing 260–261
- external light trap 256
- internal light trap 255–256
- optical model 257–260

Gibson-Ashby models 8
glass transition temperature 85
graphene oxide (GO) 188
green body sintering
- additive manufacture 20–21
- AM technologies for 19
- CAD design 20
- debinding process 21–22
- functionally graded components 21–22
- shrinkage compensation 20

*h*

high-manganese steel (HMnS) 138
holographic methods 164
human umbilical vein endothelial cells 162
hybrid additive manufacturing (HAM) 12, 19–22, 314–315
- additive manufacturing 315–317
- nozzle rings
  - choice of DED strategy 320–321
  - DED process 321–322
  - milling 322
  - optical metrology 322
  - top ring 322–323
- production cell 319–320
- turbocharger nozzle ring 317–319

hybrid printing 229–230
hybrid supercapacitors 183
hydrodynamic stress 160, 161

*i*

interdigitated electrode structures 192–195
internal quantum efficiency (IQE) 262
ionic diffusion 188

*j*

jetting 49–52
jetting-powder bed system 49–50
Joule printing 158

*k*

knowledge based decision support systems (kb-DSSs) 216

*l*

lamination object modeling (LOM) 37–38
laser-based technologies 47–49
- selective laser melting 47–48
- selective laser sintering 47–48

laser powder bed fusion (LPBF)
- tailoring AM component properties via
  - characterization techniques 139–140
  - geometrical design 137

laser powder bed fusion (LPBF) (*cont'd*)
grain size via adaptive processing strategies 143–145
lattice structures 148–150
machines 138–139
material chemistry 137
material properties by using powder blends 146–148
material qualification and process development 140–143
materials 138–139
parameters 137
relative density 141, 142
sample preparation 138–140
laser-powder bed system 47
laser volume energy density 340
lattice
Al-Si10-Mg 8
functionally graded 8
TiNi 8–9
layer-by-layer deposition 7
layer-by-layer fabrication process 113
lead magnesium niobate (PMN) 40
lead nickel niobate (PNN) 40
lead zirconate titanate (PZT) 114
light optical microscopy (LOM) 137
liquid gas diffusion layer (LGDL) 293
lithium ion batteries 183–186
lithium molybdate 41
LOM. *See* lamination object modeling (LOM)
Low Temperature Direct Writing (LTDW) 187
LPBF. *See* laser powder bed fusion (LPBF)

***m***

macroscopic CPCs 257
material complexity 155–156
material exchange 159
material jetting 227–228
material-process interactions, in CAL 167–169
Mbot printer 72
mechanical stiffness 163
metal
additive manufacturing 5–6
deposition 157–159
metallic pipes
piezoceramic sensors in 114–122
compressive force sensing 119–120
impact in sensing properties 119
materials 116–117
re-poling of PZT 118–119
sensor housing 117–118
smart coupling component 119, 120
smart coupling fabrication process using EPBF technology 114–116
temperature sensing 120–122
methacrylated gelatin (GelMA) 162
microelectromechanical system (MEMS) 51
$MoS_2$-graphene aerogels 155
motion blurring 169
MTS Landmark servohydraulic test system 119
multicomponent printed device
AM impacts 122–128
compressive force sensing 122–124
crystalline structure analysis 126–128
temperature sensing 124–126
multicomponent printing 109–111
piezoceramic sensors in smart pipes 111–128
multilayered hydrogel lumens, coaxial extrusion of 162
multiplexing, temporal 159–160
multi-technology 229–230
multi-walled carbon nanotubes 73

***n***

Navier-Stokes equations 100

## o

Oakridge National Lab (ORNL) 5
Ohnesorge number 50
optical model 257–260
overhang angle
 definition 95
 local angle control 96
 physics-based constraints 97
 simplified printing process 97–98
overhang filter 102–104

## p

particle dispersion 50
particle-laden resin 161
PBF. *See* powder bed fusion (PBF)
photopolymerization 44–47, 170
piezoceramic sensors
 in metallic pipes 114–122
  compressive force sensing 119–120
  impact in sensing properties 119
  materials 116–117
  re-poling of PZT 118–119
  sensor housing 117–118
  smart coupling component 119, 120
  smart coupling fabrication process using EPBF technology 114–116
  temperature sensing 120–122
 for smart metering 111–114
piezoelectrics 40
 coefficient 116
 effect 112
piezo-resistive effect 70
pipe bend, topology optimization (TO) of 101, 103, 104
point-by-point electrodeposition 158
polydimethylsilicone (PDMS) strain sensor 73, 74
polylactic acid (PLA) 293
polymer exchange membrane cells (PEMCs) 273, 286–287
 PEMEC 293–294
 PEMFC 287–288
  bipolar plates and flow fields 290–293
  catalysts layer 288–289
  electrolyte 288
  gas diffusion layer 289–290
 stack operation and components 286–287
polymer exchange membrane electrolysis cells (PEMEC) 293–294
polymer exchange membrane fuel cells (PEMFC) 287–288
 bipolar plates and flow fields 290–293
 catalysts layer 288–289
 electrolyte 288
 gas diffusion layer 289–290
polyvinylidene fluoride (PVDF) 40
powder bed fusion
 benefits 225 226
 challenges 226–227
 operating principle 225
 process parameters 225
powder bed fusion (PBF) 7, 8, 110
 directed energy deposition processes 12
 electron-beam 7
 geometric gradients in 8–9
 laser 7, 11
 layer-by-layer deposition 7
 material gradients in 9–12
powder blends 146–148
printed nanocomposites, strain sensing of 73–79
pyroelectric coefficient 121
pyroelectric current 124
pyroelectric effect 120
PZT
 based ultrasound transducer array 46
 re-poling 118–119

*r*

Radon transform 166–167
Ram-Lak filter 166
relativity space 355
re-poling, of PZT 118–119, 129
resin formulations 160
robocasting. *See* direct ink writing (DIW)

*s*

scanning electron microscopy (SEM) 137, 139
selective laser melting (SLM) 47–48, 96, 292
selective laser sintering (SLS) 47–48, 339–345
Selective Thermoplastic Electrophotographic Process (STEP) 162
sensor housing 117–118
shape memory alloys (SMA) 8
shape memory polymers (SMPs) 71, 73
shape recovery ratio (SR) 82
shrinkage, compensation 20
single-walled carbon nanotube (SWCNT) 70
sintering 21–22
  laser 48
SLA. *see* stereolithography (SLA)
smart coupling fabrication 114–117
sodium-ion batteries 188
solar cell technology
  efficiency 249
  geometric light management
    characterization 261–266
    compound parabolic concentrator 257
    design and 3D-printing 260–261
    external light trap 256
    internal light trap 255–256
    optical model 257–260
  3D-printing for PV 250–255
solid electrolytes
  conventional electrodes 195
  3D-print solid electrolyte structures 197
  interdigitated full cell battery 196
  ionic conductivity 195
  lithium-ion batteries 196
solid oxide cells (SOCs) 273–275
  SOEC 274, 283–284
  SOFC
    electrodes 278–283
    electrolyte 276–278
    yttria stabilized zirconia 275–276
  stacks and components 284–286
solid oxide electrolysis cells (SOEC) 274, 283–284
solid oxide fuel cells (SOFC) 41
  electrodes 278–283
  electrolyte 276–278
  yttria stabilized zirconia 275–276
solid-state additive manufacturing (AM) 16–19
solid state batteries 195–197
spatial resolution 169
spray deposition modeling (SDM) 72
stacking fault energy (SFE) 138, 146
"stair-step" effect 161
stereolithography (SLA) 40, 44, 159
stereolithography apparatus (SLA) process 337–339
strain sensing
  of carbon nanotube reinforced polymer nanocomposites 70–71
  of printed nanocomposites 73–79
strain sensors 70
  carbon nanotube 70–71
  CNT/PDMS 73, 78
  multidirectional 86
  printed nanocomposite 78
  rosette-shaped 79, 80
  single-walled carbon nanotube 70
  strain-dependent responses 78

supercapacitors 183
surface mounting, of sensor 113

## t

Technology Centre Mongstad (TCM) 356
temperature sensing
 multicomponent printed device 124–126
 piezoceramic sensors in metallic pipes 120–122
temporal multiplexing 159–160
tensile tests 139
thermal jet nozzle 72
thermal resistance 126
thermocouples, platinum-based 112
thermoelectrics (TE)
 additive manufacturing techniques
  extrusion-based additive manufacturing process 328–336
  fused deposition modeling 336–337
  selective laser sintering 339–345
  stereolithography apparatus 337–339
 energy conversion 327
 module manufacturing 328
 output power 327
thin-film capacitors 51
3D printed full batteries 197–200
topology optimization (TO) 93
 bracket evolution 94–95
 flowchart 94, 95
 fluid flow 100–102
  with 3D printing constraint 103–104
 front propagation-based 3D printing constraint 102–103
 of pipe bend 101, 103, 104
transformation induced plasticity (TRIP) 138, 146
transportation sector batteries 183
Triply Periodic Minimal Surfaces (TPMS) 9
turbine blades repair 310
 adaptive machining 314
 DED process 313
 generation of tool paths 313
 machining & preparation 310–311
 reverse engineering 311–313
turbocharger nozzle ring 317–319
twinning induced plasticity (TWIP) 138, 147
two-photon polymerization (2PP) 46–47

## u

UAM. *See* ultrasonic additive manufacturing (UAM)
ultrasonic additive manufacturing (UAM) 16–18, 110
ultrasonic metal welding 113
ultrasound acoustic fields 163

## v

vat polymerization
 benefits 224
 challenges 224–225
 operating principle 223
 process parameters 224
viscosity 45, 161
volumetric additive process 163–166

## x

X-ray diffraction (XRD) 113, 119

## y

$YBa_2Cu_3O_{7-\delta}$ (YBCO) 49
Young's modulus 160
yttria stabilized zirconia (YSZ) 275–276

## z

ZnO 40

Printed and bound by CPI Group (UK) Ltd, Croydon, CR0 4YY

17/09/2026

14964244-0002